Benchmark Papers
in Analytical Chemistry

PUBLISHED VOLUMES AND VOLUMES IN PREPARATION

ION-EXCHANGE CHROMATOGRAPHY
Harold F. Walton
THERMAL ANALYSIS
W. W. Wendlandt and L. W. Collins
DIGITAL COMPUTERS, PARTS I AND II
Thomas L. Isenhour and Joseph B. Justice, Jr.

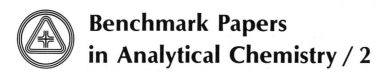

Benchmark Papers
in Analytical Chemistry / 2

A BENCHMARK ® Books Series

THERMAL ANALYSIS

Edited by

W. W. WENDLANDT
University of Houston

and L. W. COLLINS
**Monsanto Research Corporation
Miamisburg, Ohio**

Dowden, Hutchinson
& Ross, Inc.

STROUDSBURG, PENNSYLVANIA

Distributed by

HALSTED
PRESS

A Division of
John Wiley & Sons, Inc.

78 77 76   1 2 3 4 5
Manufactured in the United States of America.

LIBRARY OF CONGRESS CATALOGING IN PUBLICATION DATA
Main entry under title:
Thermal analysis
   (Benchmark papers in analytical chemistry / 2)
   1. Thermal analysis—Addresses, essays, lectures. I. Wendlandt,
Wesley William. II. Collins, L. W.
QD79.T38T44   543'.086   76-10768
ISBN   0-87933-238-7

Exclusive Distributor: **Halsted Press**
A Division of John Wiley & Sons, Inc.
ISBN: 0-470-98949-1

PREFACE

The papers collected in this volume are concerned with the development of only two of the many thermal analysis techniques, *differential thermal analysis* (DTA) and *thermogravimetry* (TG). There will no doubt be differences of opinion among workers in the field as to why these papers, from among the many thousands available, were selected. It should be kept in mind that this series of Benchmark volumes is on the subject of analytical chemistry and that the papers included here must reflect this viewpoint. Both differential thermal analysis and thermogravimetry contained, at one time, the word *analysis*; the latter was called *thermogravimetric analysis* (TGA) before it was shortened to thermogravimetry. As pointed out by the early workers in thermogravimetry, the first applications of this technique were analytical in nature. This was also true, to a certain extent, of DTA, although the physicochemical applications are also evident—for example, the determination of heats of transition, reaction kinetics, and so on.

The Editors are, of course, solely responsible for the selection of papers. They hope that their choices represent the most important papers in the development of DTA and TG in analytical chemistry.

W. W. WENDLANDT
L. W. COLLINS

CONTENTS

Contents

PART II: COMMENTARY ON THERMOGRAVIMETRY

CONTENTS BY AUTHOR

THERMAL ANALYSIS

INTRODUCTION

GENERAL

Interest in the effect of heat on various materials predates recorded history. Early man, since the discovery of fire, observed the effect of heat on fish and game, rocks and minerals, organic materials, and so on. From these observations, he developed the art of extracting copper, silver, tin, lead, and iron from their respective ores, the cooking of food, and other processes. Recorded history describes man's attempt to distinguish between various rocks and minerals by the effect of heat on them. Theophrastus, ca. 300 B.C., recorded many of these effects, and from that time onward, there was a general increase in the technological applications of heat to numerous processes. In the Middle Ages, an increasingly larger number of metals were extracted from various ores by the application of heat. Various analytical procedures, such as the assay methods for gold and silver in alloys, the blowpipe, and other procedures, were based on the controlled application of heat. Few analytical chemists today are aware of the importance of the blowpipe in the analysis of rocks and minerals (1). With it, the qualitative composition of many minerals were ascertained as well as the discovery of new elements.

In the eighteenth century, accurate pyrometric measurements of high temperatures were developed, thus placing the application of heat to a chemical process on a more quantitative basis. An early air thermometer was demonstrated by Amontons

1

in Paris in 1700 (2), but its development into a temperature measuring device had to wait until the nineteenth century. The most widely used temperature-detection technique during this period, according to Mackenzie (2), was the Wedgwood pyrometer, developed in 1782. It consisted of a piece of china clay cut to a certain geometrical shape after firing at a dull red heat. The clay fragment was then placed into a furnace whose temperature was to be determined, and after removal and cooling, it was measured for shrinkage by a calibrated V-notched indicator. From the change in dimensions of the clay fragment, the temperature of the furnace could be calculated in degrees Wedgwood. Wedgwood attempted to correlate his temperature scale with the Fahrenheit scale and found that 1 degree Wedgwood was equal to 130°F at temperatures above 1077°F, the beginning of his scale. However, this gave highly erroneous results, because the shrinkage of the clay fragment was not linear with temperature; the melting point of cast iron was calculated to be about 10,000°C.

According to Mackenzie (2), Wedgwood performed experiments that were later to lead to the thermal-analysis techniques of dilatometry and evolved gas analysis (EGA). For example, he noted that, on heating a piece of clay, the clay first decreased in size owing to dehydration, then expanded, and finally decreased in volume continuously. He also heated china clay fragments in a sealed container and collected the evolved gases in an attached bladder. The evolved gases were then analyzed, which is probably one of the first experiments in EGA.

Most of the methods of temperature measurement were developed in the nineteenth century. The thermoelectric effect was observed by Seebeck in 1821 and applied to high-temperature measurements by Becquerel in 1826. Many difficulties were experienced by this new method of measurement because of the uncertainty of the composition of the alloys employed. Owing largely to the efforts of Le Châtelier, the thermocouple was eventually developed into an accurate temperature-measuring device by the 1880s. The resistance thermometer was developed by Siemens in 1871 and the optical pyrometer by Le Châtelier in 1892.

As a result of the development of accurate temperature-measurement devices, it was perhaps inevitable that they would soon be applied to chemical systems at elevated temperatures. This indeed was the case. In 1877, Hannay applied isothermal mass-change determinations to "amorphous" materials, although he did not heat them above 100°C. Le Châtelier, who was interested in both clay minerals and pyrometry, introduced the identification of clays by the use of heating curves in 1887. This application was to give rise eventually to the thermal analysis technique of differential thermal analysis (DTA). One year earlier, Lovejoy had published mass-loss data of clays at the melting points of different metals and found, for example, that kaolin decomposed near the melting point of antimony, 624°C (3). This work, as well as that of Nerst and Riesenfeld in 1903, probably foreshadowed the advent of thermogravimetry (TG). A more detailed description of the development of DTA and TG will be given later.

THERMAL ANALYSIS

Thermal analysis is a general term covering a group of related techniques whereby the dependence of the parameters of any physical property of a substance on temperature is measured (4). The physical parameter is determined as a dynamic function of temperature; measurements made at fixed or isothermal temperatures are normally not included. Using this definition, the principal thermal-analysis techniques are thermogravimetry and differential thermal analysis. Thermogravimetry is the technique in which the change in mass of the sample is recorded as a function of temperature. In order for a mass change to be detected, a volatile component must be evolved or absorbed by the sample. The former is the usual mode of measurement, but many examples are also known for the latter. Since elevated temperatures are normally required for the evolution of volatile materials, mass-change measurements are made at increasing rather than decreasing temperatures. Routine measurements can now be made at temperatures from −190° to 2400°C or higher.

Differential thermal analysis is a thermal technique in which the temperature of a sample, compared to the temperature of a thermally inert material, is recorded as the sample is heated or cooled at a uniform rate (5). It is a differential method in that the temperature of the sample, T_s, compared to the temperature of an inert reference material, T_r, is recorded as a function of the sample, reference, or furnace temperatures. The name of the technique has been a source of confusion for many years (6) and attempts to change it have been made. Since not all applications of DTA have an implicit analytical goal, perhaps the term *differen-*

3

tial thermometry would be a more accurate description of the technique.

There are many other thermal analysis techniques, each based on a characteristic physical property of the system. In order to keep this volume to a reasonable length, only TG and DTA (along with differential scanning calorimetry) will be discussed; interested readers should refer to the book by Wendlandt (5) for a description of the other thermal analysis techniques.

REFERENCES

1. F. Szabadvary, *History of Analytical Chemistry,* Pergamon Press, Oxford, 1966, p. 50.
2. R. C. Mackenzie, *Differential Thermal Analysis,* Vol. I, Academic Press, London, 1970, p. 17.
3. C. Keattch, *An Introduction to Thermogravimetry,* Heyden/Sadtler, London, 1969, p. 1.
4. R. C. Mackenzie, *Talanta,* **16,** 1227 (1969).
5. W. W. Wendlandt, *Thermal Methods of Analysis,* 2nd ed., Wiley–Interscience, New York, 1974, p. 134.
6. W. W. Wendlandt, "Thermal Methods of Analysis," in *Analytical Chemistry—Part 2,* T. S. West, ed., MTP International Review of Science, Butterworth, London, 1973, p. 177.

Part I

COMMENTARY ON DIFFERENTIAL THERMAL ANALYSIS AND DIFFERENTIAL SCANNING CALORIMETRY

Editor's Comments
on Papers 1 Through 25

24   WENDLANDT and BRADLEY
The Automation of Thermal Analysis Instrumentation: Differential Thermal Analysis

25   MURPHY
Differential Thermal Analysis

Because of his interest in both pyrometry and clay minerals, it was to be expected, perhaps, that Le Châtelier would combine the two into an identification procedure for clays and minerals. This he did in 1887 (Paper 1), when he described a heating curve method for identifying various clays, using the photographic recording of the reflection from a galvanometer mirror (activated by a thermocouple) of a spark generated by an induction coil. The recorded heating curve consisted of a series of lines whose spacing on the photographic paper indicated the absorption or evolution of heat by the sample.

The absorption and evolution of heat (endothermic and exothermic reactions) occurred at different temperatures and hence were characteristic for each of the clay samples. Although Le Châtelier is generally acknowledged as the founder of differential thermal analysis (1), he used a single thermocouple immersed in the sample and recorded the derivative, dT/dt, rather than the differential temperature, $T_s - T_r$, currently employed.

The differential method, in which the temperature of the sample is compared to that of an inert reference material, was conceived by Roberts-Austen in 1899 (Paper 2). Since his field of interest was metallurgy, this method was first applied to iron and steel rather than to clay samples.

Following the introduction of the technique by Roberts-Austen, further developments had to wait until more accurate and reliable recording devices became available. In 1904, Kurnakov (2) developed a photographic recording pyrometer, which is still in use at the present time. This instrument led to the development of the Soviet school of thermal analysis. Their prolific efforts in this field led to the first book ever published on thermal analysis, *Termografica*, by Berg, Nikolaev, and Rode in 1944.

In a classic paper published in 1908 (Paper 3), Burgess reviewed all the methods of thermal analysis to date. He summarized the recording systems then in use as well as the various methods of recording the data such as T vs. t, $T - T'$ vs. T, T vs. $T - T'/T$, T vs. dT/dt, and so on. This paper was long neglected and was not fully appreciated until the late 1950s.

In 1913, Wallach (3) in France and Fenner (4) in the United States independently applied DTA to the study of clays and silica minerals. The technique was applied with great vigor to this area in the 1920s and 1930s and soon became a standard method for the identification and determination of clays of all types. The paper by Norton (Paper 5) is typical of this type of investigation.

The modern development of DTA started in the late 1940s when more suitable recording techniques, such as strip-chart potentiometric recorders, were introduced. These techniques were probably first employed by Kerr and Kulp (5) and Kauffman and Dilling (6).

As early as 1915, Hollings and Cobb (7) introduced control of the atmosphere around the sample in their studies on the pyrolysis of coal. Saunders and Giedroyc (8) and Rowland and Lewis (9) made more extensive use of this technique some 35 years later. In 1952 (Paper 9), Stone described his dynamic gas atmosphere DTA apparatus, which was the beginning of precision, controlled-atmosphere DTA. Provision was made to obtain DTA curves under flowing gaseous atmospheres containing known partial pressures of water, carbon dioxide, and other gases. He stated that "The equipment presented in this paper is far from perfect, but it is hoped that the presentation of the fundamental principles will serve as a guide to those interested in this field." This equipment soon became commercially available from the Robert L. Stone Company, and it started the trend toward commercial rather than home-built equipment. Commercial DTA equipment enabled investigators to obtain reproducible results and agreement with other workers around the world, if the pyrolytic conditions were identical. Stone's dynamic gas atmosphere equipment was well accepted and he continually modified and improved it (see Paper 15) until his untimely death in 1969.

Prior to the work of Boersma in 1955 (Paper 10), most of the DTA instruments employed block-type sample holders, which were the source of many complex heat-transfer interactions, as described by Smyth in 1951 (Paper 7). Boersma was the first to place the sample and reference materials in small nickel cups, which were positioned on a ceramic plate. When an exothermic reaction occurred, heat transfer occurred through the ceramic plate to the nickel heat shield. Thus the temperature of the sample chamber was solely dependent on the heat of reaction and the heat-transfer properties of the ceramic plate. With this arrangement, the calibration coefficient of the system was independent of the sample size, contrary to block-type sample holders. He then showed that the peak area was related to other system parameters by the expression

$$\text{peak area} = \int_{t_1}^{t_2} (T_s - T_r) \, dt = \frac{mq}{G}$$

where $(T_s - T_r)$ is the differential temperature, m the sample mass, q the heat of reaction per unit mass, and G a heat-transfer coefficient between the nickel cup and the nickel heat shield.

The first quantitative application of DTA to a chemical system appears to be the study of the polymorphism of Na_2SO_4 by Kracek in 1929 (Paper 4). Five different phases and their transition temperatures were detected. Vold called attention to the use of DTA to measure the heats of transformation of various long-chained fatty acids in 1949 (Paper 6) and also developed a new theory to describe the DTA curve. It was Wittels, however, who first described a DTA apparatus as a micro-calorimeter while applying the technique to the decomposition of $CaCO_3$ (Paper 8). His apparatus could be used from 300 to 1100°C at a sensitivity from 30 to 100 times greater than the instruments described at that time.

Using organic samples, Barrall and Rogers in 1962 (Paper 18) investigated the effects of instrument variables in the quantitative estimation of these materials. David, using a Boersma cup-type system, also studied the effects of the many operating variables on the determination of specific heat and heats of fusion of various materials (Paper 20). One of the first quantitative applications of DTA to liquid crystals was by Barrall et al. in 1964 (Paper 21), when the heats of transition were measured for the *solid* → *nematic* and *nematic* → *isotropic* states.

Perhaps the most influential work in modern DTA was that by Borchardt. His investigation of inorganic hydrate systems and homogeneous kinetics by this technique, based on his Ph.D. thesis at the University of Wisconsin in 1956, did much to encourage further work in DTA (Papers 12 and 13). The novelty of determining nonisothermal reaction kinetics from a single DTA curve was of great interest. Although his kinetics method was developed for a stirred, homogeneous system, it was soon

applied rather haphazardly to heterogeneous samples, which resulted in erroneous results. Borchardt and Daniels also developed an expression for the rate constant based on heat
changes rather than temperature data; this expression became
very useful when differential scanning calorimetry (DSC) was introduced some seven years later.

The applications of DTA to analytical chemistry were first
pointed out by Garn and Flaschen in 1957 for inorganic systems
(Paper 11). In 1960 Schwenker and Beck (Paper 14) applied DTA to
textile fibers and polymers while in 1962 Chiu (Paper 16) and
Vassallo and Harden (Paper 17) studied the analysis of organic
systems. The equipment used by the latter were the basis for the
furnace and sample holder of the well-known DuPont thermal
analysis system.

The technique of differential scanning calorimetry was introduced by Watson et al. of the Perkin-Elmer Corporation in 1963
(Paper 22). The DSC instrument, in contrast with DTA-type instruments, provided thermochemical data on chemical systems in
millicalories of heat per second of transition heat changes over
the temperature range $-100°$ to $500°C$. These data were shown to
be independent of the specific heat of the sample, sample geometry, and heating rate. Qualitative information was said to be
equivalent or superior to conventional DTA in terms of speed,
sensitivity, resolving power, and operational convenience. The
principal innovation of the instrument was the capability of direct, convenient, and precise quantitative measurements.

The usual sample size in present-day DTA instruments is 1 to
20 mg. To use still smaller samples, say 1 to 100 μg, Mazières in
1964 (Paper 23) developed the technique of micro DTA (μDTA), or
if larger, 0.1- to 10-mg, samples are employed, semimicro DTA. In
the case of the former, the sample is placed inside the junction of
the detecting thermocouple. Using such small samples, a high-
sensitivity ΔT system must be employed which requires isolated
sample and reference holders.

The importance of timely reviews on DTA should not be overlooked in the development of the technique. These reviews enabled the prospective thermal analyst to assess the development
of the field and its potential applications. One of the more important reviews on DTA instrumentation was that by Gordon in 1963
(Paper 19). All the commercially available instruments at that time
were described including their applications to chemical and
technological problems. Also, Murphy has written a biennial re-

view on DTA (later on thermal analysis) starting in 1958 (Paper 25). The first review is given here because of its historical content and the stimulation it provided in the field.

A look into the future of differential thermal analysis instrumentation is provided by the automated DTA apparatus described by Wendlandt and Bradley in 1970 (Paper 24). It was the first instrument capable of changing samples automatically which permitted operator-free data collection.

REFERENCES

1. R. C. Mackenzie, *Differential Thermal Analysis,* Vol. I, Academic Press, London, 1970, p. 19.
2. N. S. Kurnakov, *Z. Anorg. Chem.,* **42,** 184 (1904).
3. R. Wallach, *Compt. Rend.,* **137,** 48 (1913).
4. C. N. Fenner, *Am. J. Sci.,* **36,** 331 (1913).
5. P. F. Kerr and J. L. Kulp, *Am. Mineralogist,* **33,** 387 (1948).
6. A. J. Kauffman and E. D. Dilling, *Econ. Geol.,* **45,** 222 (1950).
7. H. Hollings and J. W. Cobb, *J. Chem. Soc.,* **107,** 1106 (1915).
8. H. L. Saunders and V. Giedroyc, *Trans. Brit. Ceram. Soc.,* **49,** 365 (1950).
9. R. A. Rowland and D. R. Lewis, *Am. Mineralogist,* **36,** 180 (1951).

1

CONCERNING THE ACTION OF HEAT ON CLAYS

H. Le Châtelier

Translated expressly for this Benchmark volume by Dr. L. E. Nesbitt, Southern Colorado State College, Pueblo, Colorado, from "De l'Action de la chaleur sur des argiles," Bull. Soc. Fr. Minéral. Cristallogr., 10, 204–211 (1887)

The hydrated aluminum silicates (clays, kaolins, etc.), in spite of their importance due to their abundance in nature as well as their numerous industrial uses, are still little understood from the point of view of their composition. They generally form mixtures too complex for chemical analysis to furnish any precise data on their structure. I have thought that in studying the temperature of dehydration of these substances, one could perhaps begin to characterize a small number of chemical species and to distinguish the presence of each of these in various mixtures.

If one rapidly heats a small quantity of clay, there is produced, at the moment of dehydration, a deceleration in the temperature rise (elevation); this "point of arrest" can be used to establish a distinction among the various hydrated aluminum silicates. Experiment shows, in effect, that the position of this point of arrest in the temperature scale is appreciably independent of the experimental conditions, notably of the heating rate. This is a result similar to that which I have already noted for the calcination of limestone. This fact holds in that the speed of chemical reactions, as soon as they have an appreciable value, undergo enormous increases for extremely small elevations in temperature.

In these experiments, I have used the thermoelectric couple, pure Pt and 10% Rh in Pt, for temperature measurement, which had already been used in our previous research. The observations have been recorded by the photographic method. A brilliant induction spark, at regular intervals of 2 seconds, gives, after reflection on the mirror of a galvanometer, images whose distance (apart) precisely measures the speed of heating. The weld of the couple was placed in the middle of a small mass of clay enclosed in a platinum cone of 5 millimeters opening, which itself is contained in a crucible, completely filled with calcined magnesia, and heated in a de Forquignon furnace. Under the conditions where I have worked, the temperature takes 10 minutes to ascend to

1000°, which corresponds to a mean heating rate of 4° for 2 seconds. The calibration of the couple has been made by taking as fixed points the fusion or boiling of the following substances:

H₂O	S	Se	Au
100	448	665	1045

The chief result of these experiments has been to show that during the heating of the clays, one observes not only the decelerations corresponding to the dehydrations, but occasionally also abrupt accelerations indicating the production of phenomena accompanied by a release of heat.

The comparison of observations carried out on a very large number of clays shows that the complexity of these substances is much less than one would have originally been led to believe. One can refer them to five distinct types, not always present, but at least in the samples that were in our hands, with gradual transition from one to another. These samples were from the Mineralogy Collection of the School of Mines or from the Adam Collection, which also belongs to the School of Mines.

The sketch below [Fig. 9], traced from my negatives, gives an example of the heating law observed for each of the five types. The top line is the reproduction of experiments made for the calibration of the thermocouple. At the moment of fusion or vaporization, the temperature remains stationary, which leads to the superposition of several consecutive images of the spark. The line thus reinforced has been lengthened a bit on the drawing to render it more apparent. The following lines [Fig. 9], numbered from 1 to 5, refer to the various aluminum silicate hydrates.

No. 1: Halloysite from Miglos (Ariege). One observes a first deceleration a bit marked between 150 and 200°; a second very important deceleration ending at 700° and finally a sharp acceleration beginning at 1000°.

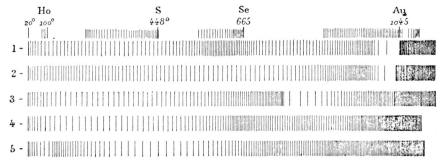

Figure 9 Reproduced from *Bull. Soc. Fr. Minéral. Cristallog.*, **10**, 209 (1887). Copyright © 1887 by the Société Française de Minéralogie et de Cristallographie.

Some identical plates have been obtained with some sedimentary clays or chemical clays from the following sources: refractory clay of Forges (Lower Seine) and of Bolene (Vaucluse); plastic clay of Gentilly (Seine); halloysite of Angleur (Belgium), of Russia, of Miglos (Ariege), of Laumede (Dordogne), of Huelgoat (Finistere), of Breteuil (Eure); white silicious bauxite of Brignoles (Var); lenzinite of La Vilate (Upper Vienne), of Eiffel (Germany); white soap of Plombiers (Vosges); severite of St. Sever (Landes).

No. 2: Allophane of Saint Antoine (Oise). The only distinct deceleration is situated between 150 and 220°; there is produced, as in the preceding, a sharp acceleration at 1000°.

The same results have been obtained with the allophanes of Saint Antoine (Oise), of Vize, of Utah (America), and the collyrite of the Pyrenees.

No. 3: Crystalline Kaolin from Red Mountain, Colorado, U.S.A. One observes a single deceleration, very marked, which ends at 770°, and a slower acceleration around 1000°, which varies in importance from one sample to another. This last ought, without doubt, be attributed to the presence of a small amount of colloidal clay whose existence in the kaolins has been recorded by Mr. Schloesing. Similar results have been obtained with various kaolins from France and China.

No. 4: Pyrophyllite of Beresow (Ural Mtns.). One observes a first deceleration ending at 700° and a second indefinitely at 850°.

A pagodite from China behaves in the same fashion.

No. 5: Montmorillonite of St. Jean de Cole (Dordogne). A very important first deceleration ends around 200°, a second less marked at 770°, and a third dubious one at 950°. One does not see any heat release as with the halloysites of the first group, which by their physical character and the nature of their strata seem to approach the montmorillonites.

Some similar plates have been obtained with the montmorillonites of St. Jean de Cole (Dordogne), of Confolens (Charente), the "confolensite" of Confolens (Charente), the steargilite of Poitiers (Vienne), the cymolites of the Isle of Cymolis (Greece), and the smectic clays of Reigate (England), and of Styrie. With these last two clays, nevertheless, the second time of arrest is not quite so marked; perhaps they should be placed in a separate category.

The clays could now be classified, according to their pyrogeneous decomposition, in five distinct categories, which are not generally represented by the mixtures among them. But they could be extracted from the silica or free alumina in whose presence they are originally found.

Under the influence of a progressive heating, the hydrated silica gives a deceleration between 100 and 200°. The hydrated alumina behaves in a

15

very variable fashion, according to its source. Precipitated from sodium aluminate, it shows a first time of arrest before 200° and a second ending at 360°. Precipitated from aluminum salts or obtained by the gentle calcination of the nitrate, it gives appreciably the same decelerations, and, in addition, a sharp acceleration at 850°. It is following this evolution of heat that the alumina becomes insoluble in acids. Finally, the hydrated alumina from bauxite gives a deceleration terminating at 700°, that is, at the same temperature as that of the halloysites. This means that the presence of hydrated silica cannot be shown in any of the silicates of hydrated alumina which nearly all give a deceleration between 100 and 200°; the first two hydrates of alumina could not, on the contrary, be encountered in any case; that of bauxite can only exist in the halloysite group.

One also sees that the evolution of heat observed in the calcination of clays of the first and second groups ought to be attributed to the molecular transformation of the alumina. It is, in fact, after this phenomenon that the alumina of the clays becomes insoluble in acids. This free alumina, which does not exist originally in the clays, results from the decomposition of this latter at the moment of dehydration.

To complete this study, it remains to determine the chemical composition of aluminum silicates which communicates to each group its distinctive characteristics. Two among these are already completely known. The fourth group is that of pyrophyllite, $4SiO_2 \cdot Al_2O_3 \cdot OH$. The third group is that of kaolin, $2SiO_2 \cdot Al_2O_3 \cdot 2OH$. The second group, that of theallophanes, contains only a small number of clays, whose formula appears to be, according to known analyses, $SiO_2 \cdot Al_2O_3 \cdot aq$. The fifth group, that of montmorillonite, contains a very large number of clays; but they are generally very impure, containing some alkalis, some lime, some iron, some manganese—whose presence is manifested by the fusibility of the material. Among these, the steargilite of Poitiers appears to me to furnish the purest product. Washed in slightly acidulated water, it leaves behind, besides its lime content, an extremely fine red clay which remains in suspension, and it leaves as a residue a very homogeneous white clay. Its analysis gives:

SiO_2	49
Al_2O_3	23.1
Fe_2O_3	2.4
CaO	0.5
OH at 250°	16.7
OH at red heat	7
	98.7

The analysis previously made on confolensite and on montmorillonite come near to the preceding; that of the cymolite gives a bit more silica.

Their composition could be represented by the formula

$$4SiO_2 \cdot Al_2O_3 \cdot HO \cdot aq$$

The first group, that of the halloysites, is much the most important; it includes the totality of sedimentary clays and the major part of the chemical clays. These first are formed, according to the work of M. Schloesing, by a mixture of quartz, of crystalline aluminum silicate, and of colloidal clay. They are too complex for one to deduce any consequence from their raw chemical analyses. The chemical clays are met, on the contrary, often in an extremely pure state, recognizable by their whiteness, by their infusibility, by their homogeneity, and by the fineness of their grain. They thus present a very regular composition, represented by the formula $2SiO_2 \cdot Al_2O_3 \cdot 2OH \cdot aq$, as is shown in the table of analyses, made on the samples sorted with care which we have used for the calcination experiments. The material had been previously heated to 250° to drive out the hygroscopic water. The sources were the following: (1) Angleur, (2) Huelgoat, (3) Miglos, (4) Breteuil, (5) Laumede, (6) Eifel, (7) Russia.

	1	2	3	4	5	6	7	Calc.
SiO_2	46.3	47.9	46.3	48.3	48.7	46.6	47.7	46.4
Al_2O_3	39.5	38	38.7	35.6	36.5	39.3	38.8	39.7
HO	14.3	14.3	14	14.3	13.6	13	14	13.9
	100.1	100.2	99.0	98.2	98.8	98.9	100.2	100.0
Hygroscopic water	8.5	5.4	6.5	12.5	4	3.5	7	

The water is very nicely divided into two parts: one part to 150° after 24 hours or to 250° in ¼ hour; the other part does not commence to go until almost above 400°. The proportion of this is always very exactly 2 equivalents for 1 equivalent of alumina. The relation of the silica and of the alumina, which in the majority of cases is that which is given in the analysis above, diverges in some cases from the normal value on account of the mixture of free silica or alumina. I shall cite in the first place the silicious bauxites; the white bauxite of Brignoles, on which my experiments were done, contains a bit less than 1 equivalent of SiO_2 for 1 equivalent of Al_2O_3. The presence of free alumina in this product should be concluded without doubt. On the contrary, the soap of Plombiers gives a quantity of silica varying between 3 equivalents and 4 equivalents; but again in this case, I have concluded that the proportion of water of combination assigned to the alumina maintains its normal value. This renders very probable the presence of free silica; in addition, this product, in spite of its appearance, is sufficiently impure and very irregular; it contains some sulfates of calcium, magnesium, etc.

But these divergences of composition are rare, and one can accept as the formula for this group of aluminum silicates,

$$2SiO_2 \cdot Al_2O_3 \cdot 2OH \cdot aq$$

That is the same formula as that of the kaolin group, but one should not reunite these two compounds, which possess some completely different properties. The action of acids after dehydration, and the release of heat at 1000°, establish a perfectly distinct cut.

2

Reprinted from *Proc. Inst. Mech. Eng.*, 35–41 (Feb. 1899)

FIFTH REPORT * TO THE ALLOYS RESEARCH COMMITTEE: STEEL.

By Sir William C. ROBERTS-AUSTEN, K.C.B., D.C.L., F.R.S.,
HONORARY LIFE MEMBER.

Before dealing with the subject to which this Report is devoted, the author would point out that the Research Committees of this Institution exerted a noteworthy influence in connection with the preliminary enquiry, the results of which have led to the recommendation that a National Physical Laboratory should be established. The report of the Council of the Royal Society recently published states that: "The deliberations of the Committee, appointed by Her Majesty's Treasury to consider the desirability of establishing a National Physical Laboratory, have resulted in an important addition to the responsibilities of the Royal Society. The Committee reported in favour of the establishment of such a Laboratory, and recommended that its control should be vested in the Royal Society, who should also nominate its governing body. Her Majesty's Government, having accepted the recommendations of the Committee, has invited the Royal Society to undertake the charge proposed in their report, including the administration of the funds which the Treasury has offered to furnish for the equipment and maintenance of the institution. The Council has decided in general terms to accept the trust offered to the Society." The author was a member of the Treasury Committee to which reference is made; and this Institution was also represented by Professor Kennedy and Mr. W. H. Maw, both of whom, as witnesses before the Committee,

---

* For the First, Second, Third, and Fourth Reports, *see* Proceedings 1891, page 543; 1893, page 102; 1895, page 238; and 1897, page 31.

gave evidence of a highly important character. It may be added that, in the evidence which was taken by the National Physical Laboratory Committee, frequent reference was made to the work of the Alloys Research Committee. Lord Kelvin, who was also a witness, pointed out that continuing the investigation of alloys in a National Laboratory would be of great importance. The same view was taken in the evidence given by the President of the Institution of Civil Engineers, by the late Sir William Anderson, Professor Oliver Lodge, and several other witnesses. The research work therefore, on which the Committee of this Institution has been so long engaged, is fully and officially recognised, not only as being of national importance, but also as having rendered important service to industry.

Carburized Iron considered as a solution.—In the course of the last Report to the Committee, an attempt was made to 'connect the freezing-point curves of carburized iron with those of ordinary alloys. The series of Reports as a whole had abundantly shown that alloys behave like saline solutions. It will be evident therefore that, if carburized iron, that is, steel and cast-iron, can be brought into line with ordinary solutions, a point of the utmost metallurgical importance will be gained. The series of results plotted in Plate 11 of the last Report* made it quite clear that the analogy of steel to an ordinary solution of salt is of the closest possible character, even if the evidence rests only on cooling curves taken in the usual way.

Recording Pyrometer.—It was certain however that the question is one of great complexity, and that a far more delicate method than that hitherto in use must be adopted, if conclusive and trustworthy results are to be obtained. The recording pyrometer, which has rendered such good service throughout this enquiry, had therefore to be rendered more sensitive than had hitherto been the case. The improvements described in the last Report consisted, it will be

---

* Proceedings 1897, pages 70 and 90.

remembered, in employing a galvanometer so adjusted that the current produced by heating a thermo-junction attached to it would deflect the mirror through an angle of 50°. The current from the thermo-junction however was not allowed to pass unchecked through the galvanometer. It was opposed by a current from a large Clark cell, and the amount of the latter current could be accurately adjusted and measured by a potentiometer, so that only a small portion of the current from the thermo-junction really passed through the galvanometer. The result is shown in delicate curves, one of which was given in Plate 9 of the last Report (1897). The angular deflection of the galvanometer mirror is small, even at the highest temperatures: the sensitiveness of the instrument is not diminished, its resistance is not increased, but a portion of the current from the thermo-junction is balanced. It was found better to replace the astronomical clock of the older recorder by a water-clock consisting of a float moving upward between guides, and bearing a photographic plate. Since this description was published in 1897, enquiries have from time to time been made as to the details of the appliance. It is therefore shown in Plate 1 of the present Report, which represents that portion of the Research Laboratory where the new recorder is placed. The galvanometers are shown at G and G¹ with sources of light for illuminating their mirrors at L and L¹. The potentiometer is shown at B, and consists of four sets of resistance coils. The current flowing through the potentiometer is maintained by means of a Clark cell g with large electrodes. The water-clock H and its float F, the photographic plate P, and the hood K, are shown at the right of the drawing. S is the slit through which the rays of light from the mirrors of the galvanometers G G¹ are admitted to the sensitized plate. The arrangements of the galvanometers are shown diagrammatically in Figs. 1 and 2, page 38 ; and the system has been described in detail elsewhere * by Dr. Alfred Stansfield, who has assisted in conducting this series of investigations almost from their outset. As the result of many experiments, I adopted the following new method, of which a

---

* Philosophical Magazine, July 1898, page 59.

brief description will be sufficient. In the ordinary method, the twisted
end of the thermo-junction A, Fig. 1, is placed in the heated mass of

Fig. 1.

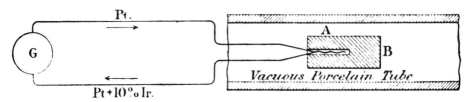

metal B under examination, and its free ends are connected with
the galvanometer G. In the new method, Fig. 2, two thermo-
junctions A and A_1 are employed. One of these is placed in the piece

Fig. 2.

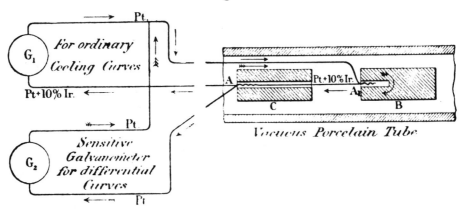

of metal B, and the other in a compensating piece of copper platinum
or fire-clay C. A sensitive galvanometer G_2, connected to both
thermo-junctions, measures on a large scale the difference between
the temperatures of B and C; and magnified records of the evolutions
of heat in B can thus be obtained, which are not affected by the
general fall of temperature of the system. The actual temperature
of the piece of metal B is simultaneously registered by the less
sensitive galvanometer G_1 in the usual way. In the new method
therefore the heat lost by the cooling mass of metal B, Fig. 2,
is compensated or balanced by the heat lost by a mass of
platinum C. The result is that the effect on the galvanometer G_2

of any evolution of heat by the cooling mass B is greatly augmented. As has already been indicated, the heat suddenly evolved by the mass B of iron or steel, which is liable to molecular change, is not masked by the fact that the mass is itself rapidly losing heat; because the temperature of the entire system does not affect the sensitive galvanometer G_2, and the heat which is evolved by the mass B is free to make itself felt. Hence the curves recorded by the mirror of G_2 possess extraordinary sensitiveness. In Figs. 1 and 2 the arrows show the directions of the currents. Those with feathers indicate the direction of the current which is due to the difference of temperature; this difference is caused by the excess of heat in the iron B, as compared with the platinum C. The featherless arrows show the direction of the current through the unsensitive galvanometer G_1, which records ordinary cooling curves.

Reference to a special case, furnished by the cooling of electro-iron from a white heat, will serve to make this clear. A bead of electro-iron was deposited on a thermo-junction protruding from a glass tube into which the wires were fused. The iron was deposited from a solution of ferrous chloride which had been purified with scrupulous care. The anode was a plate of electro-iron; but the method of preparing the solution and depositing the iron need not be given, as it would break the continuity of this description. The deposited iron weighed five grammes = 0·18 ounce, and its appearance magnified 4 diameters is shown in Fig. 3 (page 40). A transverse section showing one of the wires of the thermo-junction, magnified 50 diameters, is shown in Fig. 4 (page 41). Its hardness was about that of fluor-spar, and when placed in water heated to 70° C. or 158° F. it freely evolved hydrogen, which ceased to come off after some hours. The bead of electro-iron was then arranged as shown in Fig. 2, and was placed in a porcelain tube glazed inside and out, and rendered vacuous by the aid of a mercurial pump, which also enabled the gas evolved from the iron to be collected. More hydrogen was freely evolved as the portion of the tube containing the iron was gradually heated; but, although the evolution of gas never absolutely ceased, the amount of hydrogen delivered by the mercurial pump was very small when the iron attained a temperature of some 1,300° C. or 2,370° F.

A cooling curve of this iron after four successive heatings is shown in Plate 2, on the actual scale on which it was recorded; and it is at once evident that at least three hitherto unobserved points are here revealed. The co-ordinates are time and temperature, as usual; but the temperature represents on a large scale molecular evolutions of heat, and not the temperature of the mass under examination. There is at A the point at 1,132° C. or 2,069° F., which was first observed by my friend and former student, Dr. E. J. Ball. Then at B there is the ordinary Ar 3 of Osmond, which in this case occurs, not as in mild steel at the normal temperature of 850° C. or 1,562° F., but at 895° C. or 1,643° F. When the mass continues to cool down, there is, as was anticipated, the point Ar 2, which in this case occurs at 766° C. or 1,411° F. The carbon point Ar 1 could not be expected to occur in iron of so high a degree of purity, and it does not exist; but there is evidence of evolution of heat at a point which I believe to be between 550° and 600° C. or 1,020° and 1,110° F. It is difficult to fix this point accurately; it seems to vary somewhat in successive curves. The next point, at which heat evolved, is a new one of extraordinary interest. It occurs between 450° and 500° C. or 840° and 930° F.; and evidence will subsequently be adduced to show that it is connected with the retention of hydrogen by the mass of iron, even though it had been heated to 1,300° C. or 2,370° F. Finally there is a small evolution of heat at about 261° C. or 502° F., that is, at a temperature of no less than 400° C. or 720° F. below redness. The significance of these new points will now be considered.

[*Editors' Note:* Figures 3 and 4 have been omitted because the original halftones were unavailable.]

3

Reprinted from *Bull. Bur. Stand.*, 5(2), 199–225 (1908)

ON METHODS OF OBTAINING COOLING CURVES.

By G. K. Burgess, Associate Physicist.

CONTENTS.

The rôle of thermal analysis in many metallurgical and chemical problems, involving in many instances the constitution and behavior of very complex substances, is of increasing importance; and great advances are being made in our knowledge of the properties of many alloys, minerals, and chemical compounds at high temperatures by means of the pyrometer.

Any internal change in the physical or chemical nature of a substance usually alters many of its physical properties, as, for example, its magnetic and thermoelectric behavior, electrical resistance, specific heat, dimensions, density, and microscopic structure. A large internal change at a definite temperature or within a small range of temperature—in other words, "a transformation"—will cause sudden or very rapid changes in some or all of these physical properties, and several of them may be used with advantage in the detection and study of such transformations. In this paper, however, we shall confine ourselves to the consideration of such changes as may be detected, measured, and recorded by thermometric means.

METHODS OF THERMAL ANALYSIS.

All methods of thermal analysis are based upon the principle that chemical and physical transformations within a substance are, in general, accompanied by an evolution or absorption of heat. The detection of the temperature and the measurement of the extent of these transformations, and in many cases their interpretation, may be carried out by taking the cooling curve of the substance, which in its simplest form consists in plotting the temperature of the cooling substance against the time. It is evident that the heating curve of a substance may also be taken to find its characteristics, and this is sometimes done, but, in general uniform results are obtained more easily by taking the cooling curve, mainly on account of the greater experimental difficulties in maintaining a uniform rate of heating in the containing furnace. In some problems it is desirable to have both curves, while occasionally, as in transformations involving loss of water of crystallization or of constitution, the heating curves alone are of significance. The same apparatus will evidently serve for both.

The cooling curve, however exactly taken, will of course give no indication of those transformations for which there is no evolution or absorption of heat. If the cooling is at constant pressure, as we shall assume in all of what follows, the absence of a transformation—physical or chemical—may be assumed only when both the energy and the volume changes are zero. The detection of changes in volume, unaccompanied by changes in internal energy, may be effected by the use of an apparatus measuring linear expansion, such as the recording differential dilatometer of Sahmen and Tammann.[1] These cases are exceptional, however, and we shall not consider them further.

---

[1] *Sahmen* und *Tammann:* Über das Auffinden von Umwandlungspunkten mit einem selbst-registrierenden Dilatographen. Ann. d. Phys. **10**, p. 879–896; 1903.

There have been developed a considerable number of methods for obtaining cooling curves which are adapted to the study of recalescence points in steels as well as to the investigation of the composition and properties of alloys and chemical compounds. There has as yet been no general discussion[2] of the different methods nor of their availability for special problems and it may therefore be of some interest to have at hand an outline of the principles of the methods that may be used in obtaining cooling curves as well as a brief mention of the various types of apparatus available, with a discussion of their advantages and limitations.

The methods may be classified in various ways; thus we have to distinguish between those adapted for slow cooling, which is the case most commonly met with, and for very rapid cooling as in quenching steels; those methods which require an auxiliary body possessing no thermal transformations on cooling as compared with those requiring only the substance studied; and finally we may have on the one hand methods involving the time, and on the other hand those in which the time may be eliminated. In this paper we have considered in detail only such methods as are adapted for slow cooling, and have classified them in terms of the forms of the curves representing the experimental data.

Many operations which can be carried out in the laboratory can not be applied conveniently in industrial works, so that it will be necessary also to distinguish the various types of apparatus as regards their adaptability either for purely scientific researches or for industrial needs. For the latter, especially, it is very desirable to make all operations as automatic as possible, so that the different methods of autographic and photographic recording should be considered. Again, we shall have occasion to point out those methods which are the most advantageous to use when very minute quantities of heat or differences in temperature are to be detected, as, for example, the secondary breaks in the cooling

---

[2] A paper by *W. Rosenhain*, on "Observations on Recalescence Curves" was read before the London Physical Society, January 24, 1908, an abstract of which indicates he has compared the merits of the "Inverse Rate" and "Differential Methods."

The Differential Method has also been studied in detail by *Portevin*, Notes sur l'Emploi du Galvanomètre Différentiel, Rev. de Métallurgie, **5**, p. 295; 1908.

curves of many alloys and of numerous compounds and mixtures. In such cases it becomes necessary to use methods of the highest possible sensitiveness, which usually necessitates the discarding of autographic and photographic recording.

We shall mention those methods which are suitable for taking cooling curves in the range of temperatures up to 1500° C, but much of what is said will apply also to higher temperatures if proper precautions be taken to eliminate the effects upon the measuring apparatus of the electrical conductivity of the materials and contents of the furnace. Although several methods of measuring temperature may be used over most of the range indicated, such as the change of electrical resistance of platinum with temperature and the various optical and radiation pyrometers, we shall confine ourselves to the thermoelectric pyrometer made of the platinum metals as being on the whole the most generally suitable over this range for this kind of work, although undoubtedly particular problems may occur in which the use of some other type of pyrometer is preferable.

Use of the Thermocouple.—It may be recalled that the thermocouple possesses most of the desirable attributes of a temperature indicator. With its insignificant volume it may be introduced into a very small space, and so be used with small samples, and it takes up the temperature of the sample with great promptness. When made of the platinum metals, it is very durable and retains the constancy of its indications in a most satisfactory way, even when subjected to contaminating atmospheres, and after deterioration it may usually be restored to its former condition by glowing. Temperatures may be obtained by means of a simple form of galvanometer without any accessories. Such a galvanometer, it is true, indicates electromotive forces, while in general the temperature of a thermojunction is not strictly proportional to the electromotive force generated by it, although such a linear relation, which it is desirable to realize in order to simplify the interpretation of the indications of some types of recording instruments, holds very nearly in the case of the platinum-iridium couple of the composition Pt, 90 Pt—10 Ir.

The following table shows the E.M.F.—Temperature relation, and the E.M.F. in microvolts per degree centigrade (dE/dT) for couples composed of a wire of pure platinum joined to one of the approximate composition: *90 Pt—10 Rh*, *90 Pt—10 Ir*, and pure *Ni*, respectively.

Temp. Cent.	Pt, 90 Pt—10 Rh		Pt, 90 Pt—10 Ir		Pt, Ni	
	E.M.F. Microvolts	dE/dT	E.M.F. Microvolts	dE/dT	E.M.F. Microvolts	dE/dT
300	2290	9.0	4080	15.9	7940	11.8
500	4160	9.7	7300	16.7	10510	14.4
700	6170	10.4_5	10720	17.5	13670	17.1
900	8340	11.2	14300	18.3	17400	19.7
1100	10630	11.9	18030	19.1	21640	22.4
1300	13070	12.6	21940	19.9	26300	25.0
1500	15600	13.3_5	26010	20.7	....	....

It will be seen that the platinum-iridium couple is nearly twice as sensitive as the platinum-rhodium couple beside having a more nearly linear E.M.F.—Temperature curve. These advantages are in part offset, however, by the fact that the iridium couple deteriorates more rapidly and is less constant in its indications. The platinum-nickel couple, although very sensitive, is less reliable than the others and the *Ni* wire soon becomes brittle and breaks. Moreover, the effect of the nickel recalescence point (about 375° C) has some influence on readings taken in its neighborhood. There are other thermojunctions, made of alloys of the more refractory baser metals, which are much more sensitive than the above and which may be suitable over particular ranges. These couples are usually constructed of wire of considerable diameter and are, therefore, not adapted for work with small samples. For exact work one should make sure of their constancy of indication over the temperature range to be studied. A more robust and less sensitive indicating instrument may evidently be used with this type of thermocouple, although recently pivot instruments, suitable for use with the platinum couples, have been constructed by Paul of London, the Cambridge Scientific Instrument Company, and by Siemens and Halske.

29

Methods of Recording.—Before describing the various methods that have been suggested for the taking of cooling curves, it may be well to consider the ways in which the observations may be recorded. Either the observer may himself read and record the indications of the instruments and discuss the data so obtained, analytically or graphically; or he may use, if the method and desired precision permit, an autographic or a photographic self-registering instrument, when it may or may not be necessary to make further reductions, depending upon the method used and the interpretation sought.

It is evidently of great advantage to use self-recording apparatus when possible, and it then becomes necessary to choose between the photographic type and the autographic.[3] The latter possesses the advantage that the experimenter may watch any part of the record, and can therefore control the operation and at any moment vary the conditions affecting the experiment; whereas with a photographic recording apparatus, as usually constructed, the observer does not know whether or not the experiment is progressing properly until it is finished and he has developed the sensitive plate. The manipulation by the photographic method is usually also more delicate and time consuming and the adjustment less sure, and the record often requires further graphical interpretation. The autographic method is in general not adapted for interpreting phenomena taking place within an interval of a few seconds, so that for very rapid cooling it is necessary to employ the photographic method. It is possible to construct the photographic recorder so as to obtain a very considerable range of speeds with the same apparatus, while it is difficult and costly to construct an autographic recorder having more than two speeds.

TEMPERATURE—TIME CURVES.

I. θ *vs t.*—The simplest method of obtaining a cooling curve is to take simultaneous measurements of the temperature of the cooling substance and of the time, from which a plot may be made showing the variation of temperature with time. (See Plate I, Fig. I.)

---

[3] The term *autographic* is here used to designate an instrument which is self-recording by any other than photographic means.

The most obvious defect of this method is that it will not distinguish between phenomena proper to the substance studied and those due to outside conditions, such as accidental variations in the rate of cooling of the furnace due to air drafts and like causes. Again, where measurements over a considerable range of temperature are to be taken continuously, it becomes impracticable without elaborate experimental arrangements to combine great sensibility with this great range, so that the experimenter is in general forced to choose between great sensibility over a small temperature range or a relatively small sensibility over a large range; and this is especially true if it is desired to record the data automatically.

This method was naturally the first used[4] and it is to-day perhaps the most common one for taking cooling curves in both metallurgical and chemical laboratories. In its most elementary form it requires a minimum of apparatus—a thermocouple and an indicating galvanometer. Any desired sensibility and range may be had by substituting for the direct reading galvanometer a potentiometer and sensitive galvanometer. With this arrangement it is advantageous, when rapidity of observation is an object, to measure the last increment of temperature in terms of the galvanometer deflection rather than try to balance the potentiometer exactly while the temperature is changing. It is possible in this way to take readings as often as every 5 seconds with a properly devised set-up and quick period galvanometer.[5] A precision of $0°.1$ C at $1000°$ C may be attained. Independently of the method of measurement used, the certainty of the detection of slight transformations may usually be increased by increasing the size of the sample under observation, thus making available larger quantities of heat.

The constant attention of the observer is of course required for either of the above systems of measurement. There have been devised, however, many kinds of self-recording apparatus for using this method, the earlier forms being photographic, while many of the later ones are autographic.

---

[4] Frankenheim, Pogg. Ann. der Physik. **37, 38** (1836–1837).

[5] See W. P. White, Potentiometer Installation, Phys. Rev. **25**, 334; 1907.

Photographic Recorders.—Among the earliest photographic recorders we may mention the apparatus of Roberts-Austen [6] (Fig. 1) in which the photographic plate P was given a vertical motion, either by clockwork, or, in order to secure maximum sensibility for several rates of cooling, by means of buoying the plate on water whose rate of flow could be regulated. The vertical motion of the plate then gave the time while the deflection of the

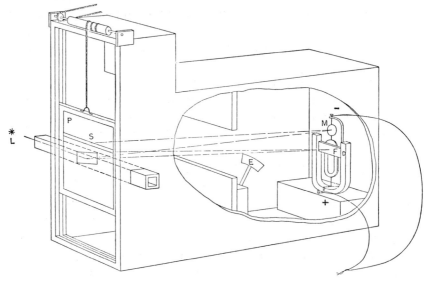

Fig. 1.

galvanometer gave the corresponding temperature, and a beam of light from the source L reflected from the galvanometer mirror M and incident on the plate, after passing through a fine slit S placed horizontally before the moving plate, gave directly on the latter the time-temperature curve. Light reflected from a fixed mirror F and interrupted at equal intervals by a pendulum E gave a fixed zero line as well as a measure of the regularity of the motion of the plate. It was the practice later,[7] when taking measure-

[6] Proc Royal Society, **49**, p. 347; 1891. Nature, **45**, p. 534; 1892. First Report of the Alloys Research Committee in Proc. Inst. Mech. Engs. 1891.

[7] Fourth Report of the Alloys Research Committee, Proc. Inst. Mech. Engs., 1897. A. Stansfield, Phil. Mag. **46**, p 59; 1898.

ments over short temperature ranges, to increase the sensibility by balancing the greater part of the E.M.F. of the thermo-couple with an auxiliary measured E.M.F., and giving the galvanometer the maximum sensibility that would keep its deflections on the plate.

In the apparatus used by Charpy,[8] or in its very elaborate form as constructed by Toepfer, of Potsdam, for Kurnakow,[9] the vertically moving plate is replaced by a rotating cylinder wound with the sensitized paper on which the deflections of the galvanometer are registered. This form of recorder had also been used and discarded by Roberts-Austen. Kurnakow's apparatus, which must be placed in a dark room, is furnished with an auxiliary telescope and scale system using red light, so that the experiment may be controlled during the taking of a record. As constructed, five speeds may be given to the cylinder; and there is provided an E.M.F. compensating system for maintaining the maximum sensibility over a series of temperature ranges.

There is another device, used by C. L. A. Schmidt,[10] by which the experiment may be watched while a photographic record of a cooling curve is being taken. It consists in shunting the sensitive photo-recording galvanometer G (Fig. 2), in series with a high resistance R, across a direct reading millivoltmeter V. If the resistance of R+G is great compared with that of V, the readings

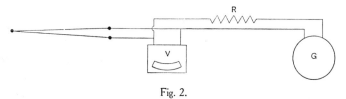

Fig. 2.

of the millivoltmeter will not be altered appreciably by this operation. Schmidt moves the photographic plate, mounted as in the apparatus of Roberts-Austen, by means of a screw driven by a small motor. In this way any desired speed may be given to the plate.

---

[8] G. Charpy, Bull. de la Soc. pour l'Encouragement, **10**, p. 666; 1895.
[9] N. S. Kurnakow, Zs. f. Anorg. Chemie, **42**, p. 184; 1904.
[10] C. L. A. Schmidt, Chem. Eng., **6**, p. 80; 1907.

In practice it has been found difficult to realize conveniently a sufficiently steady motion of the plate in the Roberts-Austen system of recording, and attempts have been made to devise methods in which the photographic plate remains fixed in position. This has been successfully accomplished by Saladin, whose apparatus (Fig. 8, p. 215) has been modified by Wologdine[11] to give the temperature-time curve by removing the prism M and substituting for the second galvanometer G_2 a plane mirror turning about an horizontal axis. This mirror may be controlled by an hydraulic system as in Roberts-Austen's apparatus, or by clockwork as in the model constructed by Pellin, of Paris. The deflection of the galvanometer G_1 gives to the beam of light an horizontal motion over the plate proportional to the temperature, while the vertical motion of the beam of light is given by the mirror turning at a uniform rate, and is therefore approximately proportional to the time as registered on a flat plate.

Autographic Recorders.—To obtain a satisfactory autographic or pen record without sacrifice of sensibility of the galvanometer it is necessary to eliminate the friction of the pen or stylus upon the paper. This has been accomplished by the use of mechanisms which cause the pen or stylus at the end of the galvanometer boom to make only momentary contact with the moving paper.[12]

In the Siemens and Halske[13] form of instrument, Fig. 3, the paper P is driven forward by the same clockwork that controls the pressing down, by means of the arm B, of the stylus N, which imprints dots periodically on the paper by means of a typewriter ribbon running across and beneath the record sheet. This system permits of taking a record continuously over very long periods of time. In most of the other recorders the paper is wound upon a

---

[11] S. Wologdine, Rev. de Métallurgie, **4**, p. 552; 1907.

[12] There are a considerable number of thermoelectric recorders. Among the manufacturers of these instruments are: Siemens and Halske, Berlin; Hartmann and Braun Frankfort a. M ; Pellin, Chauvin and Arnoux, Carpentier, and Richard, Paris; Leeds and Northrup, and Queen, of Philadelphia; The Scientific Instrument Company, of Cambridge, England and Rochester, N. Y.; The Bristol Company, Waterbury, Conn.

[13] Zs. f. Instrk., **24**, p. 350; 1904.

drum, and various devices are used to obtain the record; thus in the Hartmann and Braun type a silver stylus makes sulphide dots on a prepared paper, and in the Cambridge thread recorder rectangular coordinates are obtained by having the galvanometer boom strike an inked thread which runs parallel to the drum.

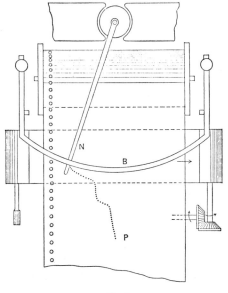

Fig. 3.

As previously stated, these autographic instruments all give intermittent records and are limited to one or two speeds, and although they may be made very sensitive they are not adapted for the detection of transformations which take place very rapidly, since the recording interval can not readily be shortened much below 10 seconds, and in most instruments this interval is greater than 15 seconds. In other words, they can be used advantageously only for slow cooling.

II. θ *vs* t, θ *vs* t' *(or* θ *vs* θ').—In order to eliminate the effect of irregularity of outside conditions which influence the rate of cooling, a method commonly used when endeavoring to detect small transformations consists in placing a second thermocouple in the furnace, but sufficiently removed from the substance studied to be uninfluenced by its behavior. Alternate readings on the temperature of the test piece (θ) and of the furnace (θ') are then taken, preferably at definite time intervals. The data are most readily discussed by plotting the two temperature-time curves side by side as shown in Fig. 4, or by plotting the difference in temperature $\theta-\theta'$ against the temperature θ of the test piece.

This method may be made recording either by using two instruments or by modifying one of the above mentioned autographic recorders so as to trace the curves of two thermocouples on the same sheet.[14] In practice, however, this method is usually resorted to only when great sensibility is desired, as in detecting minute internal energy changes, when the potentiometer combined with the deflection galvanometer is the most sensitive and quick-working arrangement for taking the measure-

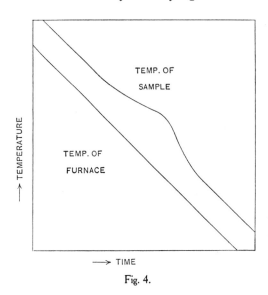

Fig. 4.

ments. (See p. 205.) It is convenient to use thermocouples of the same composition so as to have readings of both the temperature of the sample and of the furnace given by the same potentiometer setting, and so depend upon the galvanometer deflections for measuring the residual parts of θ and θ'.

Regarding the precision of this method, it is to be noted that the quantity it is really desired to measure is θ–θ' in terms of θ, and this is accomplished by measuring θ and θ', hence the sensibility of θ–θ' is no greater than that of θ or θ'. In other words the method requires the maximum refinement of measurement to obtain the quantity sought, as well as the maximum of computation or plotting to reduce the observations.

DIFFERENTIAL CURVES.

III. θ *vs t,* θ–θ' *vs t.*—The preceding method, II, may readily be modified so as to give θ–θ', the difference in temperature between the test piece and furnace, by direct measurement instead of by computation, with the added advantage that the

---

[14] The Siemens and Halske instrument has been so arranged. See Zs. f. Instrk., **24**, p. 350; 1904.

precision of $\theta-\theta'$ may be made very great as compared with that of θ, the temperature of the sample. This may be accomplished, for example, by placing a commutator in the thermocouple circuit at A, Fig. 5, so that alternate measurements on θ and $\theta-\theta'$ may be taken in terms of the time. Evidently the connections may be so made that either the galvanometer G_2 of the same direct reading or potentiometer system that measures θ, or a separate instrument G_1, as shown in the figure, may be used to measure $\theta-\theta'$.

Use of a Neutral Body.—Accidental variations in the indications of the auxiliary thermocouple giving θ', the furnace temperature may largely be eliminated by placing this couple within a blank or neutral substance. The material of the neutral body should be such that it undergoes no transformations involving an absorption or evolution of heat within the temperature range

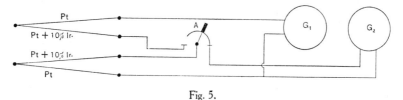

Fig. 5.

studied, such as a piece of platinum, porcelain, or even in some cases nickel or nickel steel. It is also desirable that the sample and neutral have as near as may be the same heat capacities and emisivities. The sample to be studied and the neutral piece are placed near together and arranged symmetrically with respect to the temperature distribution within the furnace.

To Roberts-Austen [15] again was due the credit of first devising a sensitive differential method using the neutral body. He also modified his photographic recorder (Fig. 1) so as to give, by means of a second galvanometer, the $\theta-\theta'$ *vs* t curve on the same plate with the θ *vs* t curve, from which a curve giving $\theta-\theta'$ in terms of θ could be constructed. His arrangement of the direct reading and differential thermocouple and galvanometer circuits

[15] Fifth Report of the Alloys Research Committee, Proc. Inst. Mech. Engs., p. 35; 1899. Metallographist, **2**, p. 186; 1899.

is shown in Fig. 6 in which S is the sample or test piece and N the neutral body possessing no transformations; the galvanometer G_2 measures the temperature θ of the sample, and G_1 measures the difference in temperature $\theta-\theta'$ between the sample and the neutral. Curves for steels and alloys were usually taken with the samples in vacuo.

It is evident that Roberts-Austen's final photographic apparatus, although very sensitive, was also complicated and very delicate of adjustment, and in practice it took great skill in its use, requiring for instance some three or four successive exposures adjusted to the proper adjacent temperature ranges, to take the cooling curve of a steel from 1100° C to 200° C.

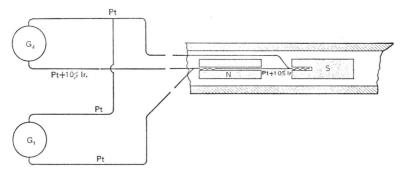

Fig. 6.

Most of the recent exact work [16] employing the principle of this method has been done by taking the observations of θ directly on a potentiometer and $\theta-\theta'$ on the same or an auxiliary galvanometer. In this case of direct reading, the simpler arrangement of thermocouples indicated in Fig. 5 may advantageously replace Roberts-Austen's (Fig. 6), or the modification shown in Fig. 7, such as used by Carpenter and others, thus dispensing with one thermocouple and the drilling of a second hole in the sample.

This method is evidently capable of attaining maximum sensitiveness since the galvanometer connected to the differential thermocouple, giving $\theta-\theta'$ *vs t*, may be made as sensitive as desired independently of the θ *vs t* system. There is the further

---

[16] See for example: Carpenter and Keeling, Collected Researches, Natl. Phys. Lab., **2**, 1907.

advantage that no limits are set to the range of temperatures over which a given precision in $\theta-\theta'$ may be had. There is, however, a limitation on the certainty of interpretation of results by this method, especially when the rate of cooling is rapid, due to the fact that it is practically impossible to realize the ideal condition of having $\theta-\theta' = a$ constant, or keeping the cooling curves of the test piece and neutral parallel for temperature intervals within which there are no transformations of the test piece. The rate of cooling, and hence the value of $\theta-\theta'$, is influenced by several factors, among the most important of which are the mass of each substance—the unknown and the neutral—its specific heat, conductivity, and emissivity, as well as the relative heat capacities of the furnace and inclosed samples. The $\theta-\theta'$ vs t line is, how-

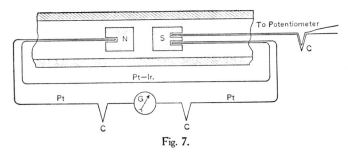

Fig. 7.

ever, always a smooth curve, except for the regions in which there are transformations in the substance under study.

The autographic system of recording may also be used, and it is possible to construct an apparatus by means of which both the θ vs t and $\theta-\theta'$ vs t curves shall be recorded simultaneously on the same sheet by the same galvanometer boom. In order to accomplish this we have made use of a Siemens and Halske recording millivoltmeter having a total range of 1.5 millivolts and a resistance of 10.6 ohms. The E.M.F. generated by the differential thermocouple, proportional to $\theta-\theta'$, is recorded directly by this instrument. 1° C corresponds to from 16 to 19 microvolts between 300° and 1100° C for a platinum-iridum couple, or to about 1.8 mm on the record paper. In series with the Pt-Ir thermocouple giving temperatures is a suitable resistance, about 200 ohms in this case, so that the galvanometer boom may be kept within the limits of the paper when recording values of θ. The circuit is made alternately through the direct and the differential thermocouple circuits in series with the recorder by means of a polarized relay actuated by the same battery that depresses the galvanometer boom when the mark is made on the paper. The thermocouple circuits may be those of either Figs. 5, 6, or 7, but with the galvanometer G_2 indicating temperatures suppressed. It is evident that by recording the two curves, $\theta-\theta'$ *vs* t and θ *vs* t, on the same sheet there is some sacrifice in the ability to detect small and rapid transformations, since the spacing is doubled. Usually also, with such an arrangement, the galvanometer will not be completely aperiodic for one or the other system. On the other hand, it is of great advantage to have the curves together and obtained independently of inequalities in clock rates, which are a serious source of error in locating transformation points exactly when two separate instruments are used.

When it is desired merely to detect the existence of a transformation without measuring its temperature exactly, the sensitive form of recording millivoltmeter may be connected directly to the differential thermocouple without other accessories.[17]

IV. θ *vs* $\theta-\theta'$.—It is sometimes of advantage to be able to record and discuss the data independently of the time, and so express $\theta-\theta'$, the difference in temperature between sample and neutral, directly in terms of θ, the temperature of the sample. This may evidently be accomplished by replotting the results obtained from the curves of the previous differential methods which involve the time. It was reserved, however, to Saladin, engineer of the Creusot Works, to invent, in 1903, a method [18] that would record photographically the θ *vs* $\theta-\theta'$ curve directly, thus obviating any replotting. His method possesses also the advantage of having the photographic plate fixed in place. The forms of curve obtained in this way are illustrated in Plate I, Fig. IV.

---

[17] Hoffmann und Rothe, Zs. Instrk., **25**, p. 273; 1905.

[18] Saladin: New Autographic Method to Ascertain the Critical Points of Steel and Steel Alloys, Iron and Steel Metallurgy and Metallography, **7**, p. 237; 1904. First presented at Reunion des Membres français et belges de l'association Internationale des Methodes d'Essais, 28 Fev., 1903.

The arrangement of the apparatus in its simplest form, due to Le Chatelier,[19] is shown in Fig. 8. Light from the source S strikes the mirror of the sensitive galvanometer G_1 whose deflections measure the differences in temperature $(\theta-\theta')$ between the sample under study and the neutral body. The horizontal deflections of the beam of light are now turned into a vertical plane by passing through the totally reflecting prism M placed at an angle of 45°. A second galvanometer G_2, whose deflections are a measure of the temperature of the sample and whose mirror in its zero position is at right angles to that of G_1, reflects the beam horizontally upon the plate at P. The spot of light has thus

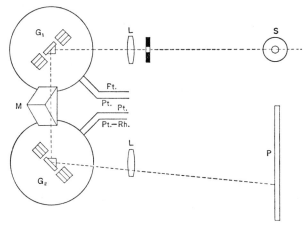

Fig. 8.

impressed upon it two motions at right angles to each other, giving, therefore, on the plate a curve whose abscissæ are approximately proportional to the temperature θ of the sample and whose ordinates are proportional to $\theta-\theta'$. The sensitiveness of the method depends upon that of the galvanometer G_1, which may readily be made to give five or six millimeters for each degree centigrade. The arrangement of the thermocouple circuits is the same as in Figs. 6 or 7. If so desired, the time may also be recorded by means of a toothed wheel driven by a clock and placed in the path of the beam of light. Compact forms of

---

[19] H. Le Chatelier, Rev. de Métallurgie, **1**, p. 134; 1904.

this apparatus, which is used considerably in metallurgical laboratories, are made by Pellin, Paris, and by Siemens and Halske, Berlin.

When steels and metallic alloys in the solid state are being investigated, advantage may be taken of the thermoelectric behavior of the sample itself to record the critical regions with Saladin's apparatus. Thus Boudouard [20] measures $\theta-\theta'$ by means of platinum wires set into crevices at each end of the sample, taking advantage of the fact that the transformation will usually be progressive along the sample. This modification eliminates the neutral piece and one platinum or alloy wire, but, as Le Chatelier has shown,[21] is less accurate than the usual form of Saladin's apparatus; and its indications may even be indeterminate or ambiguous, as the reaction may start midway between the embedded wires or at either end.

Saladin's method, it should be noted, is a perfectly general one for recording the relations between any two phenomena which may be measured in terms of E.M.F. or as the deflections of two galvanometers. This method can not readily be made autographic, as this would require the simultaneous action of two galvanometer systems on a single pen. When used as a direct reading method it reduces to III.

IVa. θ *vs* $\theta-\theta'/\Delta\theta$.—For even moderately rapid cooling the differential method gives distorted curves which are often difficult of interpretation. This distortion is due largely to the different heat capacities and emissivities of the sample and neutral piece, thus causing differences in their rates of cooling in the furnace. Rosenhain has suggested (l. c.) that these irregularities may be eliminated by taking what he calls the "Derived Differential Curve," or expressing the temperature θ of the sample in terms of the difference in temperature $\theta-\theta'$ between the sample and neutral for equal temperature decrements $\Delta\theta$ in the sample. The experimental method is the same as in IV, but in addition it is necessary to replot the data obtained from the θ vs $\theta-\theta'$ measurements so as to give the value of $\theta-\theta'$ per degree change

---

[20] O. Boudouard, Rev. d. Metallurgie, **1**, p. 80; 1904.
[21] Rev. de Metallurgie, **1**, p. 134; 1904

in temperature of the sample in terms of its temperature; but this appears to be worth while when the cooling curves of the sample and of the neutral differ considerably.

DIRECT AND INVERSE RATE CURVES.

V. θ *vs* $d\theta/dt$.—In many problems it is of interest to measure the speed of the transformations under observation. This may be done by determining directly the rate of change of temperature of the sample in terms of its temperature. Le Chatelier [22] used this method in 1887 in his study of the properties of clays. He was also the first to employ a photographic apparatus for the recording of cooling or heating curve data, using an arrangement in which the plate remained stationary. The sparks from an induction coil were made to pass at intervals of two seconds before a slit and gave upon the plate, after reflection from the galvanometer mirror, images of the slit whose spacing was a measure of the speed of heating, which in his experiments was about 2° C per second.

Another method of recording the rate of heating or cooling in terms of the temperature has been devised by Dejean.[23] The new feature of this method is the use of an induction galvanometer or relay which may be inserted in the circuit of the more sensitive galvanometer G_1 of the Saladin system (Fig. 8). The principle of the apparatus is shown in Fig. 9. The induction

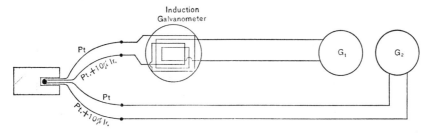

Fig. 9.

relay is a modified d'Arsonval galvanometer having an electro-magnet and a movable coil, the latter consisting of two distinct

---

[22] H. Le Chatelier, Comptes Rendus (Paris), **104**, p. 1443; 1887.
[23] P. Dejean, Rev. de Metallurgie, **2**, p. 701; 1905. **3**, p. 149; 1906.

insulated windings, one of which is connected to a thermocouple. Heating or cooling one junction of this couple causes the coil to be deflected and its motion in the field of the electromagnet induces an E. M. F. in the second winding of the coil which is proportional to its angular speed and hence to the rate of change of E. M. F. of the couple, or approximately to the rate of cooling or heating, i. e., to $d\theta/dt$. The induced E. M. F. is measured by joining this winding to the sensitive galvanometer G_1. The galvanometer deflection passes through a minimum when the heating or cooling passes through a minimum, that is for a region in which there is an absorption or evolution of heat. A second thermocouple in series with the other galvanometer G_2 of the Saladin system gives the temperature of the sample. We have, therefore, on the plate P (Fig. 8) when the record is taken photographically, the temperatures as abscissæ and the rate of cooling $d\theta/dt$ as ordinates as shown in Plate I, Fig V. Dejean has used this method in the study of steels and has also investigated with it the copper-cuprous oxide system. The transition temperatures are very sharply marked. If desired, direct reading may be substituted for the photographic recording, with an increase in precision, as discussed on p. 205. This method is evidently a perfectly general one for recording the rate of change of E.M.F. (dE/dt).

For neither Le Chatelier's nor Dejean's arrangement can differences in the rate of heating or cooling due to the substance itself be distinguished from those due to external causes, since no neutral piece is used.

VI. θ *vs* $dt/d\theta$.—Among the methods adapted for slow cooling, we shall mention last one of the earliest which was used to throw into prominence the abnormalities of a cooling curve, namely, the *inverse rate method*, which was employed as early as 1886 by Osmond[24] in his classic researches in metallurgy. It consists in noting the intervals of time required for a substance to cool by equal decrements of temperature and plotting this quantity ($dt/d\theta$) in terms of the temperature. (See Plate I, Fig. VI.) Osmond thus describes his method: "The time taken by the ther-

---

[24] F. Osmond, Comptes Rendus (Paris), **103**, p 743, p. 1122; 1886. **104**, p. 985; 1887. Annales des Mines, **14**, p. 1; July, 1888.

mometer during the heating or cooling of the sample to rise or fall one division of the scale (1 mm) was registered by means of a Morse telegraph ribbon or on a rotating cylinder turned by a small electric motor. * * * A halt of the thermometer is transcribed as a cusp and a slowing down by a swell of the curve, whose area is proportional to the quantity of heat set free "

But one thermocouple is needed and no neutral piece is used, so that the apparatus is the same as required for a θ *vs* t curve, I, although it is necessary, if work of precision is to be undertaken, to record very exactly by means of a chronograph and key the intervals of time ($\varDelta t$) required to pass over a given number of degrees ($\varDelta \theta$), say 5° C. or 10° C intervals.[25] The method, however, can not readily be made automatically recording for the variables θ and $dt/d\theta$ in terms of each other, and therefore requires the active presence of the observer. It has the same disadvantage as method II in that the precision of a difference in temperature ($\theta - \theta'$ or $\varDelta\theta$) can be made no greater than that of the temperature θ. The inverse rate method is perhaps best considered as one for interpreting and plotting the θ *vs* t data in such a way as to emphasize its irregularities and so the more readily permit the detection of any critical regions.

RAPID COOLING.

None of the experimental arrangements so far described is adapted for measuring the very rapid cooling, i. e., several hundred degrees in a few seconds, met with in quenching or chilling.

Le Chatelier,[26] in an investigation of the quenching of small samples of steel and the effect of various baths, made use of a galvanometer having a period of 0.2 second and a resistance of 7 ohms, whose deflections were recorded on a photographic plate moving vertically at a speed of 3 mm per second. A half second's pendulum vibrating across the path of the beam of light, from a Nernst glower as source, gave a measure of the time. He succeeded in recording satisfactorily temperature intervals of 700° C in 6 seconds, using as samples cylinders 18 mm on a side. It

---

[25] This method is well illustrated by F. Wüst, Metallurgie (German), **3**, p. 1; 1906.

[26] H. Le Chatelier, Rev. de Metallurgie, **1**, p. 473; 1904.

52839—08——5

would be desirable to increase the precision and sensitiveness of this method, which might be done, as Le Chatelier himself suggests, by using an oscillograph arrangement, or a string galvanometer such as Einthoven's, in which the displacements of a silvered quartz fiber of high resistance in an intense magnetic field are measured photographically.

CHARACTERISTICS OF COOLING CURVES.

In conclusion, let us consider briefly the forms that the different cooling curves may take and their approximate interpretation, for three typical kinds of transformation:

(a) The substance remains at a constant temperature throughout the transformation.

(b) The substance cools at a reduced rate, which may or may not be constant over a portion of the transformation.

(c) The substance undergoes an increase in temperature during the first part of the transformation.

An approximation to the first case, that of a strictly isothermal transformation, is met with in the freezing of chemically pure substances of sufficient thermal conductivity which do not undercool appreciably; in the formation of eutectics; and occasionally in other transformations. The second case is perhaps the most common; and the third is represented by the phenomena of recalescence and of undercooling preceding crystallization.

In the case of an isothermal transformation (a) heat is generated at the same rate that it is lost by radiation, convection, and conduction, or, considering the phenomena as confined to the sample alone:

$$\frac{dQ}{dt} = Ms \frac{d\theta}{dt} \quad \text{------------------------}(a)$$

where M is the mass, and s the specific heat of the body supposed constant during the transformation, $\frac{dQ}{dt}$ the rate of generation of heat, and $\frac{d\theta}{dt}$ the rate of cooling the body would have, when passing through the temperature of the transformation, if there were no transformation. The heat Q generated in such a transformation lasting a finite time Δt, is therefore:

$$Q = Ms \frac{d\theta}{dt} \varDelta t \dotfill (\beta)$$

in which $\frac{d\theta}{dt} \varDelta t$ is a measure of the fall of temperature, $\varDelta\theta$ the substance would undergo if there were no evolution of heat during the time $\varDelta t$, that is, during the transformation, whence:

$$Q = Ms . \varDelta\theta \dotfill (\gamma)$$

In the case of the transformation (*b*) we have

$$\frac{dQ}{dt} < Ms \frac{d\theta}{dt}$$

and for the transformation (*c*):

$$\frac{dQ}{dt} > Ms \frac{d\theta}{dt}$$

In Plate I are illustrated these three types of transformation as given by the following cooling curves:

I.	θ *vs t*	Temperature-Time.
IV.	θ *vs* $\theta-\theta'$	Differential.
V.	θ *vs* $d\theta/dt$	Temperature-Rate.
VI.	θ *vs* $dt/d\theta$	Inverse rate.

The first horizontal line of figures refers to case (*a*), the second line to (*b*), and the third line to (*c*). For all of the curves the ordinates are temperatures; and the corresponding parts of the several curves are indicated by the same letters. The vertical lines O, O, represent the zero of abscissæ for each group of curves.

In the case of the temperature-time curves, Figs. I, I', I'', they are drawn for an accompanying neutral piece (p. 211), A'B'C', as well as for the sample under study, ABCD. In I the sample and neutral are cooling at the same rate, while in I' and I'' they are cooling at different rates. In each figure the point C_1 indicates the temperature that would have been reached by the sample if there had been no transformation. C_1 may be considered as practically coinciding with C_2, the temperature that would have been

47

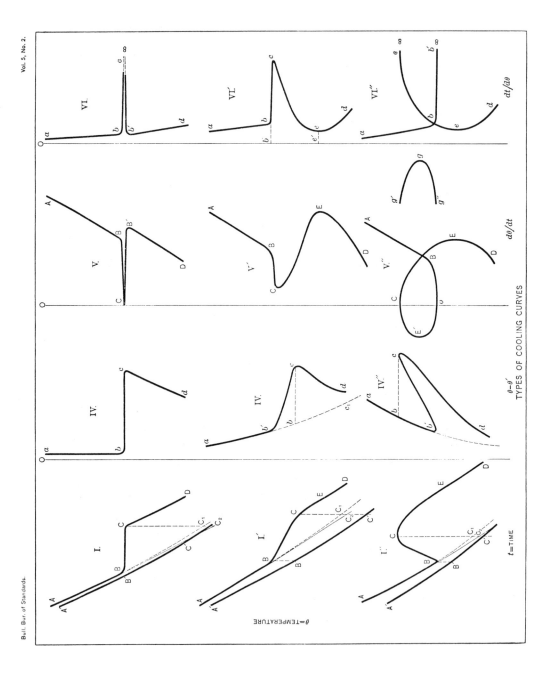

TYPES OF COOLING CURVES

reached if the sample had continued to cool at the same rate as at B, the beginning of the transformation.

For each of these transformations (*a*), (*b*), and (*c*), as represented by the curves I, I′, I″, the heat evolved is approximately proportional to $\varDelta\,\theta \doteq CC_1 \doteq CC_2$, or to the maximum difference in temperature produced by the phenomenon. It is to be noticed that in general the rate of cooling just to the right of C will be greater than that just to the left of B, on account of the presence of the furnace, for the furnace walls, to which the heat is being lost, have continued to cool during the time BC. Even if the furnace were at a constant temperature, the rates at B and C might still be different due to the difference in specific heats and in emissivities of the substance before and after the transformation. The end of a transformation is always marked by a point of inflexion as shown at E (Figs. I′, I″).

From the temperature-time curve, therefore, we may determine the temperatures of beginning and ending of a transformation, its duration, an approximate measure of the heat evolved by the transformation, as well as the rate of cooling and therefore of transformation, at any instant. The interpretation of the variations in these factors forms the basis of thermal analysis, which has already been so productive in the determination of the constitution of alloys and chemical compounds.[27]

Although the forms of the curves representing the various methods are those corresponding to what is actually obtained from a sample cooling inside a furnace, yet it should be noted that the equations representing the evolution of heat are written in a form that practically neglects the influence of the furnace.

A complete discussion of the cooling within a furnace of a substance possessing transformation points would be more complicated than contemplated by this paper; requiring a knowledge of the law of cooling of the furnace and that of the inter-

---

[27] The principles of thermal analysis as based on the θ *vs t* curve are quite fully described in the following papers by *Tammann:* Zs. Anorg. Chem., **37**, p. 303; 1903. **45**, p. 24; 1905. **47**, p. 289; 1905, a résumé of which is given by *Portevin* in Rev. de Metallurgie, **4**, p. 979; 1907. See also *Bancroft*, J. Phys. Chem., **6**, p. 178; 1902. *Shepherd*, ibid., **8**, p. 92; 1904, Iron and Steel Mag., **8**, p. 222; 1904, and *Bakhuis Roozeboom*, Die Heterogenen Gleichgewichte, 1901.

change of heat between the sample and the furnace, the latter depending on their relative heat capacities and conductivities, the emissivity of the sample, and the distribution of temperature within the furnace.

If small quantities of heat are involved in a transformation, the simplest assumptions that can be made are that at each instant the sample is losing heat to the furnace at a rate proportional to their difference in temperature, and that the parts of the furnace receiving this heat are all at the same temperature throughout at any instant. On this basis, the quantity of heat Q, lost by the sample during a transformation, as interpreted by the θ *vs* t curve, would be more nearly proportional to $\overline{CC_1^2}$ than to CC_1.

The practical conditions of cooling, however, are in general somewhere between the two extremes of the sample cooling independently of the furnace as required by our equations, and within the simplified furnace above described; so that CC_1 may be considered a minimum measure of the heat evolved, the actual measure of this quantity varying with every different experimental arrangement.

In what follows, the limitations just discussed, concerning the measurement of the heat evolved during a transformation, are assumed to apply to each of the types of cooling curve.

Figs. IV, IV', and IV" give the forms of the curves traced when the difference in temperature $(\theta-\theta')$ between the sample and the neutral piece (p. 211) of approximately the same thermal capacity is plotted in terms of the temperature (θ) of the sample. In Fig. IV, the line bc, since it is the approximate equivalent of $CC_1 \doteq \Delta\theta$ of Fig. I, may be taken as a measure of the heat evolved during the transformation. If the cooling curves of the sample and of the neutral are not parallel, as shown in Figs. I' and I", it is necessary, in estimating the heat evolved from the difference curve $ab'cd$ (Figs. IV' and IV") to take into account the variations in $\theta-\theta'$ during the transformation. This change in $\theta-\theta'$ is given by BB'—C_1C' (Figs. I', I"). Furthermore, successive differential curves for the same sample are comparable only when the sample and neutral are cooled in the different experiments so as to always maintain the same temperature-time relation over a given range. When this is the case, *bc* (IV′ and IV″) is still an approximate measure of the heat generated.

It can be shown that, except for horizontal tangents, the points of inflexion E of the θ *vs t* curves do not correspond to points of inflexion at the same temperatures on the differential curves, hence the end of a transformation can not be determined from the latter. If equal time intervals are marked on the differential curve, the rate of cooling or of transformation over an interval $\varDelta t$ may be obtained by finding the value of $\varDelta \theta / \varDelta t$.

Although, as we have seen (p. 210), the differential method may be made more sensitive and certain than the direct method, yet, from the above, it is evident that it furnishes a less complete basis for the interpretation of the physico-chemical phenomena involved in the indicated transformations.

Figs. V, V′, V″ represent the *temperature-rate* curves (θ *vs* $d\theta/dt$) and Figs. VI, VI′, VI″ the *inverse rate* curves (θ *vs* $dt/d\theta$). The two types may be considered as reciprocals of each other. Comparing them with the θ *vs t* curves (Figs. I, I′, II′) it is seen that a sharp break in the latter corresponds to a perpendicular to the temperature axis for both the former; and that for an isothermal transformation, $d\theta/dt$ becomes zero at C and $dt/d\theta$ infinite at *c*. When the θ *vs t* curve is convex to the θ axis, the $d\theta/dt$ curve is concave and the $dt/d\theta$ curve convex. Both curves give negative values (OE′C, Fig. V″, and $g'gg''$ Fig. VI″) for a region of recalescence.

The end of a transformation, corresponding to a point of inflexion E on the θ *vs t* curve, is sharply indicated (E, *e*) on both these rate curves by the tangent becoming parallel to the θ axis and the curve having a maximum or minimum abscissa. For any region of cooling at a constant rate, the tangent remains parallel to the θ axis during this interval.

From neither of the rate curves can the relative amounts of heat evolved during a transformation be readily computed for the different kinds of transformation. The area of the inverse rate curve, however (Fig. VI′), taken between the limits *b′*, *e′* is proportional to $\varDelta t$ or to $Q/d\theta/dt$ (see equation β); that is, to the quantity of heat generated divided by the rate of cooling.[28]   If the rates of cooling at the beginning of two transformations are equal, then the areas, taken as above for the inverse rate curves (Fig. VI′), become an approximate measure of the heat generated. This condition, however, is rarely realized in practice.   An examination of Fig. VI″ shows the practical impossibility of constructing any instrument other than one using an optical system, which would record the complete inverse rate curve.

We see, therefore, that both these rate curves mark the limits of a transformation more sharply than either the temperature-time or differential curves, and in general show greater changes in form for slight heat changes, but the rate curves do not in general give a simple measure of the heat evolved, nor is a neutral piece used to eliminate extraneous heat fluctuations.

A comparison of the properties of these various cooling curves indicates that the one from which the most comprehensive view of a transformation can be obtained is the simple temperature-time curve (I) when this method is made sufficiently sensitive; the one giving the least information is the temperature-rate curve (V); and those which cannot readily be recorded directly by any form of instrument yet devised are the inverse rate curve (VI) and the derived differential curve (IV*a*); while the one that can be made the most sensitive and certain is the differential curve (IV).   The analytical discussion, therefore, leads to the same result as the examination of the experimental methods, namely, that for great range combined with greatest sensitiveness, the most certain and complete data may be obtained by combining the temperature-time observations with those obtained by the differential method.

WASHINGTON, August 3, 1908.

---

[28] Area $b'bcee' = \int_{\theta_1}^{\theta_2} \frac{dt}{d\theta} d\theta = t_2 - t_1 = \varDelta t$, or time occupied in falling a temperature $\varDelta \theta$.

4

Reprinted from *J. Phys. Chem.*, **33**, 1281–1303 (1929)

THE POLYMORPHISM OF SODIUM SULFATE:

I. THERMAL ANALYSIS

BY F. C. KRACEK

Introduction

Altho the study of inversions or transitions in crystals has occupied the attention of chemists and crystallographers for at least a hundred years, particularly during the last twenty years of the preceding and the first decade of this century, it must be admitted that our knowledge of these phenomena is still largely incomplete. This is forcibly borne out by the extensive work of E. Cohen[1] and co-workers, among others.

Inversions proceed with a great variety of speeds, varying from the extremely sluggish transitions characteristic of the principal forms of silica (quartz-tridymite-cristobalite) to the very rapid ones such as high $\rightleftarrows$ low quartz or the rapid enantiomorphic transformations in many salts, e.g., KNO_3, $AgNO_3$, K_2SO_4. These are called high-low inversions by Sosman.[2] The sluggish and high-low inversions, however, are the extreme cases, and a very great number of polymorphic changes proceed at moderate speeds in both directions. Some enantiomorphic inversions are pseudo-monotropic, that is, while reversible under certain conditions, they ordinarily proceed with measurable speed in one direction only, usually at a temperature much above the point of thermodynamic equilibrium. The calcite-aragonite inversion appears to be of this type. Another example is found in the behavior of anhydrous sodium acetate. Some of the phenomena described in this paper fall into this class.

It has long been known that anhydrous sodium sulfate inverts in the neighborhood of 234°. While engaged in the recalculation of the vapor pressure and solubility data for the system H_2O—Na_2SO_4[3] I had my attention drawn to apparent discrepancies in the published statements with regard to this inversion.[4] The results of optical studies and of thermal analysis are not in agreement. The temperature 234° quoted above is based on the principal arrest obtained with cooling curves, and represents, as will be seen later, only one of several arrests occurring in heating and cooling curves made on preparations of known thermal history. Optical examination also reveals the existence of several forms. A brief recapitulation of published optical studies on this substance is given in Table I.

---

[1] E. Cohen: "Physico-chemical Metamorphosis and Problems in Piezochemistry," (1926).

[2] R. B. Sosman: "The Properties of Silica," pp. 41-179 (1927).

[3] F. C. Kracek: P-T-X relations for systems of two or more components and containing two or more phases (L-V, L_I-L_{II}-V and S-L-V systems). Int. Crit. Tables, **3**, 371 (1927).

[4] Most of the pertinent references are given by H. W. Foote: Int. Crit. Tables, **4**, 7 (1928), who estimates the temperature of the inversion to be 236 ± 3°. An error is found in Int. Crit. Tables, **1**, 150, where it is stated that thenardite changes at 100° to a monoclinic form which then changes at 500° to hexagonal. No basis has been found for these values.

TABLE I

Characteristic Observations on Polymorphism of Na₂SO₄

1. Mügge: Neues Jahrb. Mineral. Geol., **1884** (2), 1-14; microscopic examination of preparations after heating to various temperatures. Optical properties of thenardite are permanently altered, the birefringence being markedly decreased to a low value, by heating. Beginning of change 185°-205°, end of change 215°-225°, in some cases not till 240°. Always completely changed at 260°; no further permanent change by heating to 330° and higher. Temperature of beginning of alteration variable for different fragments of same crystal as well as for different crystals. Fragments do not alter completely at one temperature after change starts. Hot slide placed quickly under the microscope with nicols crossed shows on cooling a momentary intensification of birefringence accompanied by violent motion of the particles; the same phenomenon is exhibited by slides prepared from molten Na₂SO₄. In both cases the final form is that characteristic of altered thenardite.

2. Wyrouboff: Bull. Soc. min. France, **13**, 311-6 (1890); optical examination with a heating microscope. Na₂SO₄ exists in 4 forms:

   α, ordinary thenardite, orthorhombic, moderate birefringence, stable to 200°.

   β, probably monoclinic and analogous to ordinary Li₂SO₄. High birefringence. Stable above 200° and below 230°; can be cooled to room temperature. Always mixed with other forms.

   γ, orthorhombic, analogous to K₂SO₄. Low birefringence, optically negative. Very stable when cooled to ordinary temperature. Density 2.696.

   δ, hexagonal, isomorphous with high-temperature form of K₂SO₄, low birefringence. Stable only at high temperature.*

   Transformation is indirect and sluggish for the first three forms; rapid and easily reversible for the last two.

3. Nacken: Neues Jahrb. Mineral. Geol. Abt. A, Beil. **24**, 27-9, (1907). Optical examination with heating microscope. A gradual decrease in birefringence on heating toward 230°. Irreversible. Further heating to near melting produces no other sudden change in optical properties.

4. Müller: Neues Jahrb. Mineral. Geol. Abt. A, Beil., **30**, 33-5, (1910). "Na₂SO₄ is orthorhombic when crystallized from solution, and 'pseudo-hexagonal' at ordinary temperature when crystallized from melt. High temperature form is hexagonal, isomorphous with the hexagonal 4Na₂SO₄. CaSO₄." The irreversible transformations observed by others assumed to be incorrect.

---

* Wyrouboff estimated the corresponding inversion temperature as 500°. Jaenecke (Z. physik. Chem., **91**, 548 (1916)) recorded an indefinite heat effect at 400 ± 50°, which he later retracted (Z. physik. Chem., **91**, 676 (1916)).

Correlation of the statements in Table I indicates that the behavior of Na_2SO_4 is far from simple. The existence of three modifications is certain, namely: (a) thenardite, which separates above 32.5°C from aqueous solutions saturated at atmospheric pressure; (b) the low birefracting form which appears when Na_2SO_4 heated to high temperature is cooled down to room temperature; and (c) the hexagonal high-temperature form. The existence of two other forms is probable. Wyrouboff observed a modification of higher birefringence than thenardite, which he called the β modification, and which he would keep at room temperature for some time. This β form results from thenardite and changes on further heating into his γ modification considered above under (b). It is reported to be always admixed with thenardite (α) and the γ modification. Careful microscopic work is required to detect the β form, as it is never present in very large amount. Mügge, on the other hand, noticed that Na_2SO_4 cooled from high temperature under the microscope passes momentarily thru a phase which is more birefringent than either the high-temperature hexagonal δ form or the (metastable) γ form of Wyrouboff which is the inert modification resulting from cooling to room temperature. This Mügge modification, according to our experience, can be obtained and preserved for only a short time by quenching from above 260°.

Smits and Wuite's[1] work on the vapor pressures and solubilities of saturated aqueous solutions of Na_2SO_4 yields a definite break at 235° with an uncertainty of about 5°. The small number of points established by them and the low accuracy obtainable in this difficult region do not warrant conclusions about the course of the curve below this break.

In connection with the consideration of the complex behavior of Na_2SO_4 it is of interest to note that the anomalies were recognized, from studies on the heats of solution of the salt, several years before its polymorphism was established. Berthelot[2] and Thomsen[3] differed on the heat of solution by amounts considerably in excess of the probable error of their determinations; moreover, they found some difficulty in reproducing their own values, dependent upon the thermal treatment of the salt before dissolving. Pickering[4] took up the question in the same year that Mügge published the results of his optical study; he found that the heat of solution of 1 mol of Na_2SO_4 in 420 mols H_2O was about 60 cal. for salt dried at a temperature not exceeding 150°; if the salt was dried considerably above 200°, but below the melting point (884°) he obtained an average value of 760 cal., or an apparent discrepancy of 700 cal. for the two heat treatments. This is quite in order, in view of the optical results which clearly show that previously ignited Na_2SO_4 differs from thenardite.

No detailed thermal analysis of the inversions in Na_2SO_4 exists to date and no heating curves appear to have been published. The latter are greatly

[1] Smits and Wuite: Versl. kon. Akad. Amsterdam, Sept. (1909); Wuite: Z. physik. Chem., **86**, 349 (1914).

[2] Berthelot: Ann. Chim. Phys., **14**, 445 (1878); **29**, 295 (1883).

[3] J. Thomsen: J. prakt. Chem., **17**, 171 (1878); **18**, 5 (1878).

[4] Pickering: J. Chem. Soc., **45**, 686 (1884).

needed in order to confirm and extend the optical results, which indicated that the heating and cooling curves should show quite different characteristics. This expectation has been completely verified; indeed the complexity of behavior exceeds that of any single system hitherto investigated in detail if we exclude the behavior of silica and analogous substances, in which the complexity is of a totally different character. According to the present data, Na_2SO_4 exhibits pentamorphism within the comparatively narrow temperature range from *ca.* 190° to 250°, accompanied by pseudo-monotropic behavior of some of the inversions. Precedent for this type of behavior is not definitely established; it may, however, be of interest to direct attention to the work of Bowen and Greig[1] on carnegieite in which anomalous heating curves were also encountered.

This paper forms the initial communication of a series in which it is proposed to investigate the nature of the polymorphism exhibited by this substance as completely as is possible with the methods at our disposal.

Experimental Part

1. *Materials.* The sodium sulfate used in this work was prepared in the form of thenardite, by purification and recrystallization in various ways, from Baker's "Analyzed" or Kahlbaum's Anhydrous Na_2SO_4. The purchased salt was dissolved in water and filtered free of mechanical impurities and of the precipitate due to small amounts of Fe and Al present. No other impurities were found in detectable amounts. The clear solution of pure Na_2SO_4 was then crystallized by evaporation above 50° in open or covered vessels, or precipitated by pouring the warm solution into warm 95 per cent alcohol. In addition to these methods, samples of anhydrous Na_2SO_4 were prepared by slow dehydration of the deca- and the hepta-hydrates over KOH or H_2SO_4 in vacuum desiccators. The information on the materials used and their methods of preparation is collected in Table II.

TABLE II

Materials and their Methods of Preparation

No.	Method of preparation; condition	Loss of H_2O on ignition %
1	Solution evaporated in wide mouth flask at 70–80°. Uncovered. Crystals formed largely as crust on surface. Neutral.	0.10 (ungraded)
2	Neutral solution slowly evaporated in covered Pyrex beaker at *ca.* 70°. Beautifully formed large crystals up to 6 mm. along an edge. Crystals neutral.	0.046 (48–100 mesh) 0.028 (10–28 mesh)

[1] Bowen and Greig: Am. J. Sci., **10**, 204 (1925).

TABLE II (continued)

No.	Method of preparation; condition	Loss of H$_2$O on ignition %
3	Faintly acid (H$_2$SO$_4$) solution evaporated as in 2. Large clear crystals. Acidity of crystals 0.001 mols acid per mol of salt.	0.0396 (10–28 mesh)
4	Faintly alkaline (NaOH) solution evaporated as in 2. Large clear crystals. Salt almost neutral to phenolphthalein.	No loss
5	0.02 N acid (H$_2$SO$_4$) solution evaporated as in 2. Large clear crystals. Acidity of crystals 0.00400 mols acid per mol of salt.	0.019 to 0.024 (10–28 mesh)
6	Slightly unsaturated solution at 50° poured into 95% alcohol at 50°; motor stirring. Very fine crystalline powder. Crystals neutral.	<0.05
7	Same, hand stirring. Digested with solution at ca. 70° for 2 hrs. Coarsely crystalline powder. Crystals neutral.	<0.05
8	Slow dehydration of heptahydrate over KOH. Exceedingly fine powder. Neutral.	<0.05
9	Slow dehydration of decahydrate over KOH. Exceedingly fine powder. Neutral.	<0.05
10	Baker's Analyzed anhydrous Na$_2$SO$_4$. Neutral. Small amount of foreign matter.	0.10 (ungraded)

All samples dried below 125°.

Nearly all the preparations contained small amounts of moisture; microscopic examination discovered it as minute inclusions, lying principally along the planes of coalescence of adjoining crystals. This moisture is not completely removed by heating to 200°; in fact, there is reason to believe that some aqueous inclusions persist to a considerably higher temperature.

2. *The Experimental Arrangement.* Two to three gram samples were heated in a Pt thimble crucible alongside a neutral body in a nichrome-wound tube electric furnace at controlled rates. The temperature of the charge and the differential temperature between the charge and the neutral body were measured by means of calibrated Au-Pd vs. Pt-Rh (40% Pd, 10% Rh) thermocouples and a high-sensitivity potentiometer system,

capable of reading directly to 0.5 microvolt. At the temperatures involved these thermocouples have a sensitivity of 50 microvolts per 1° so that 0.01° changes could be easily detected; in general only whole microvolts were recorded.

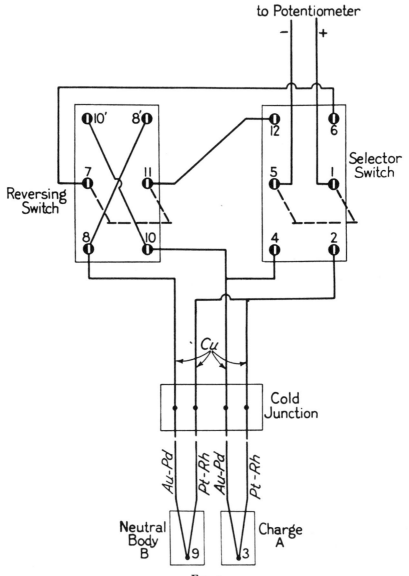

FIG. I
Switching arrangement for thermocouples.

Two thermocouples were connected in a differential arrangement as shown in Fig. 1, one placed in the sample, the other in a neutral body of alundum cement which shows no heat effects in the region under investigation. The selector switch connects the potentiometer either with the thermocouple reading the temperature of the charge, or with the two thermocouples opposed to give the differential temperature. The reversing switch controls the direction of the differential e.m.f., making it possible to balance the differential on the potentiometer in the usual way, whether the charge or the neutral body is warmer. This switch is particularly useful when the direction of the differential changes during the course of a run. All essential parts of the switching arrangement are of copper to guard against stray electromotive forces.

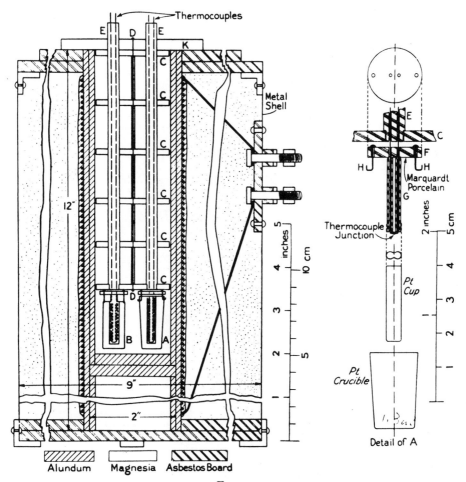

FIG. 2
The furnace and accessories.

The charge and the neutral body with their thermocouples are located in the furnace at the point where the thermal gradient is smallest. The furnace and its accessories are shown in Fig. 2, which is largely self-explanatory. The winding of the furnace consists of 20 ohms of No. 16 (B & S) nichrome type wire and consumes approximately 600 watts when connected directly on the 115 volt line. It is capable of maintaining temperatures up to 1150° with a long life. The baffles C divide the interior of the furnace into a series of chambers which are very effective in eliminating disturbances due to convection of air. The detailed assembly of *A* shows the manner in which the thermocouple is placed in the charge. The small Pt cup serves to keep the couple from being contaminated by the charge, and facilitates its removal at the close of the measurement; it is particularly useful in case the charge is melted in the crucible. In the usual arrangement when a bare couple is placed in the charge it is often impossible to remove it undamaged from the solidified melt. The present apparatus has proved entirely satisfactory not only in this work, but with heating curves of silicates as well.

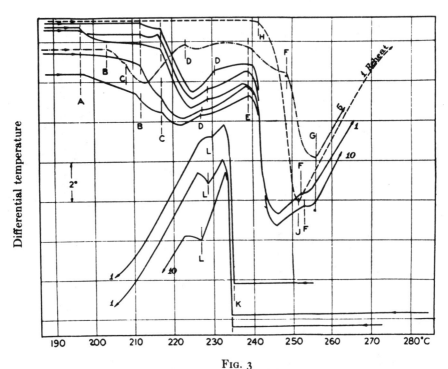

Fig. 3

Heating and cooling curves for neutral Na_2SO_4, preparations 1 and 10. Reheat curve is shown dotted, first heat curve for the slightly acid preparation 5 drawn in for comparison is indicated by the dot and dash curve. The breaks *A*, *B*, *C*, etc., denote the temperatures of the beginning of the various heat effects. Note the grouping and magnitudes of the heat effects. Compare with Figures 4 and 5.

The rate of heating or cooling is suitably controlled by periodically changing the resistance of the circuit and employing storage battery current. With some care it is possible to maintain a sensibly linear rate over periods as long as desired. The rate in ordinary runs was approximately 1° per minute; in some cases rates as low as 0.3° per minute were used successfully, particularly in the reversing experiments.

Readings of the temperature of the charge and of the differential were made alternately every half minute and appear respectively as abscissas and ordinates in the curves which follow. The results are presented in graphical form, for this is the most satisfactory way to show the position of the breaks in the curves and the relative magnitude of the heat effects obtained.

3. *Experimental Results.* In this work three distinct types of heating and cooling curves are to be distinguished for each type of preparation, namely:

(a) The first heating curve of material never previously heated beyond 100° to 125°;

(b) The cooling curve of material previously carried thru the inversions at least once;

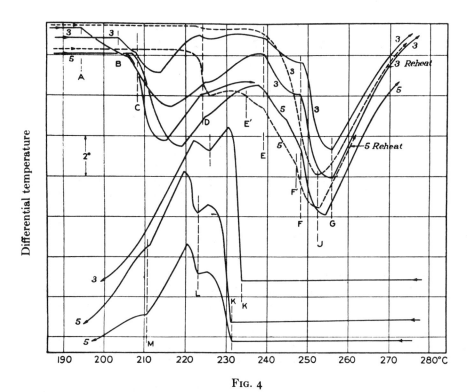

FIG. 4

Heating and cooling curves for the slightly acid Na_2SO_4 preparations 3 and 5. Reheat curves are shown dotted. Compare with Fig. 3. Note the changed contour of the reheat curves, the additional break M on the cooling curve of 5, and the shift in the temperatures of the beginning of the lower inversions A, B, C, and D.

(c) The reheating curve of material which has passed thru the inversion cycle at least once.

Each type of curve is characteristic, as will be seen by reference to the accompanying Figs. 3, 4 and 5. Figs. 3 and 4 present the assembly of heating, cooling and reheat curves for the neutral and the slightly acid preparations respectively, and Fig. 5 shows an assembly of first heat curves for several other types of preparations studied, together with the usual characteristic reheat curve included for reference. On these curves the breaks indicating the temperatures at which heat effects begin, whether on heating or cooling, have been marked and labeled A, B, C . . . etc. These have been collected and classified in Table III for the various preparations.

Table III

Prep.	A	B	C	D	E	F	G	H	J	K	L	M
1	197.5	210.8	217.5	226.8	239.4	253.0	257	242	252	235	230	
	—	211.6	217.0	227.2	238.4					234.6	228.8	
	196.4	212.2	217.2	229.0	236.5	251.6				234.6	227.0	
	196.2	211.8	216	229.2	238							
2		212	217.2	230.5	238	253.0	260 ±	242	251	234.7	228.2	
		212.0	217.0	231.6	237.8	255	260 ±			234.2	226.5	
			216		235	—	260 ±					
3	202	204	217.0	231.0	238.4	—	258			233.8	225.8	
	197		216.5	230.3	238.6	248.6	256			233.2	225.2	
	195		206.4	224.4	238.5	247.8						
	—		206.6	224.4	238.0	248.5						
			208	223.5	239.5							
				*222								
4		202.8	217.8	230.8	238.2	246	253.5	244	*252	234.2	230.4	
		203.5	219.5	230.0	237.8	248.2	254.5		251.6	234.3	228.8	
		—	207	224.6	235.0	247.0	253.0					
			?	223.6			—					
5	<195.0	201.4	206.6	224.0	235.0							
	<195			222.5								
6		209.0	218.6	230.5	238.5	242.5	256	240	251	231.0	223.0	
		206.0	219.0	230	237	243.6	251.5			231.5	223.0	
				230.5	237.2	242.8	251.0			233.7		
7	<195				236.8	246	264	240	251	233.8		211.6
						247.5	264			233.7		210.4
8	—	—	219.5									
9	—	—										
10	200.5	211.2	217	227.5	240	250	254	241	252	234.5	227.0	—
	—	211.5	217	227	240.5	249	253					
	—	211.8	217.5	227.5	238.5	251	256			234.2	226.2	
	197.4	212.0	218.3	226.7	240.5	251.5	258					
	198.0	211.8	217.4		238.4	250.5	258					
		211.9			—	—	—					

* Reheat.

Examination of the diagrams in connection with this table brings out the essential features of the results obtained. The first point of interest is the difference between the behavior of the salt on first heating and on reheating, the latter showing, with *neutral preparations*, merely one large heat effect beginning at a somewhat lower temperature than that of the beginning of the highest heat effect F on the first heating curve. The course of the cooling curve is no less remarkable, two adjacent heat evolutions K and L being registered with the neutral salt. The slightly *acid* preparations 3 and 5 exhibit additional features which will be discussed at a later point.

While the temperatures at which the various heat effects begin in different preparations are not definitely reproducible in all cases, there is a parallelism in the characteristics of the curves which is easily recognized. The recurrence of the heat effects is not accidental and certainly can not be ascribed to experimental uncertainties. The small irregularities in the temperatures of beginning of the various changes in heat capacity are the result of hysteresis accompanying the inversions to which the heat effects owe their origin, and are an essential feature of the results. Some of the inversions do not occur promptly, particularly on heating. The best reproducible temperature in the series, and the most prompt heat effect is that corresponding to the first break K in the cooling curve, occurring generally at 234.5° in the neutral salt. Small excess of NaOH or H_2SO_4 serves to lower this temperature to some extent. This is the temperature recorded by previous workers as representing the inversion in Na_2SO_4. The second break L on the cooling curve has heretofore escaped notice altogether.

The phase which is stable at ordinary temperature is thenardite. This is the phase which always crystallizes from aqueous solutions above 32.5° under ordinary pressure. This fact alone, however, is insufficient to afford proof of stability, because many compounds crystallize first in a metastable modification from solutions, in preference to the stable one, e.g., the calcite and aragonite forms of $CaCO_3$, etc. For this purpose thenardite and "inverted thenardite" in known amounts were sealed in glass tubes with the requisite amounts of water and oscillated in an oil bath at 50° for 18 to 24 hours. In all cases the solubilities were the same and equal to the value of 31.8 weight per cent Na_2SO_4 interpolated from the known solubility curve for this salt, at 50°. Since the solubility is not dependent upon the previous thermal history of the salt used, the phase which ordinarily crystallizes from solution, thenardite, must be the stable phase. The irreversibility noted in thermal analysis is then due to metastability, and furnishes a clear example of pseudo-monotropic behavior. This apparent irreversibility is shown convincingly by the characteristics of the first heating curves as compared with the reheat curves, the breaks A, B, C (and D) being absent upon the latter, particularly with the neutral salt.

No heat effect has been detected on heating beyond 260° and up to 700°; accordingly, Wyrouboff's qualitative estimate of 500° for the highest inversion to hexagonal form[1] is not confirmed.

---

[1] See footnote, p. 1282.

The first heating curves for the *slightly alkaline* preparation 4, and the *neutral* preparations 6, 7, 8 and 9 differ from those of the slowly crystallized neutral preparations partly in the amount of energy associated with the different inversions and partly in the number of recognizable heat effects. This is particularly true of preparation 9 derived from the decahydrate.

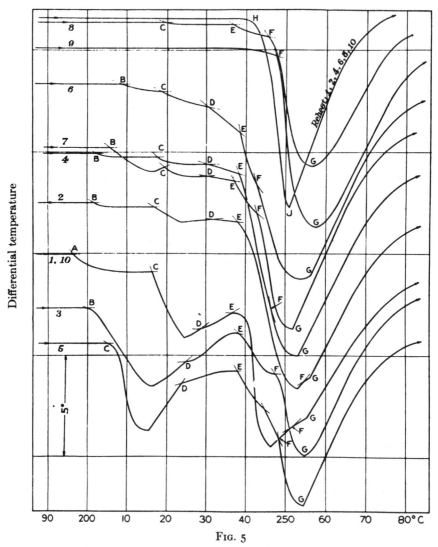

FIG. 5

The first heating curves for various preparations of Na_2SO_4, together with a typical reheat curve. Note the positions of the different breaks, and the change in the magnitude of the heat effects with relation to the amount of occluded water present. (See Table II).

Microscopic examination of these preparations shows that, as prepared, they consist principally of thenardite. (Preparations 8 and 9 were so exceedingly fine-grained that even with the highest power of the microscope the examination was difficult and the result uncertain.)

The behavior of the *slightly acid* preparations 3 and 5, as shown in Fig. 4, differs decidedly from that of the neutral preparations. All the heat effects begin at somewhat lower temperatures than in the neutral salt. The distribution of energy among the various heat effects is of different character,

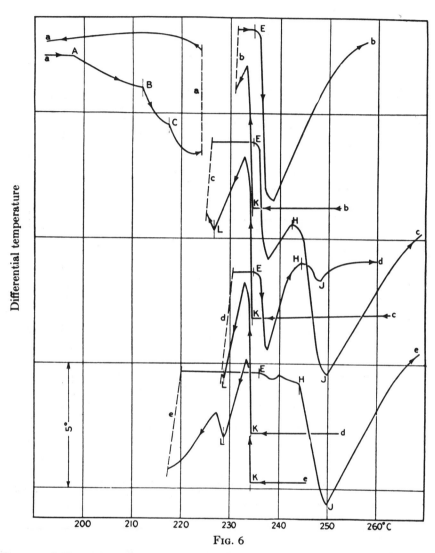

FIG. 6

The reversibility of inversions in neutral Na_2SO_4. Note that A, B, C (and D, not included in the figure) are not reversed, and the manner in which E and F occur in dependence upon how far cooling proceeds before reheating.

particularly the heat effects following the breaks E and F. The reheat curves show a reversal of the heat effect following D in both cases; the cooling curve of 5 also shows a well-defined heat evolution following M, which occurs in preparation 3 only under certain conditions (see reversal experiments). The reheat curve of 5 shows, beside the break D, the breaks E and F.

The first heating curves show six reproducible breaks (Fig. 3). Classified according to the amount of energy associated with the accompanying heat effects the inversions can be arbitrarily divided into two groups, the heat effects corresponding to breaks *A*, *B*, *C* (and *D*) being denoted as belonging

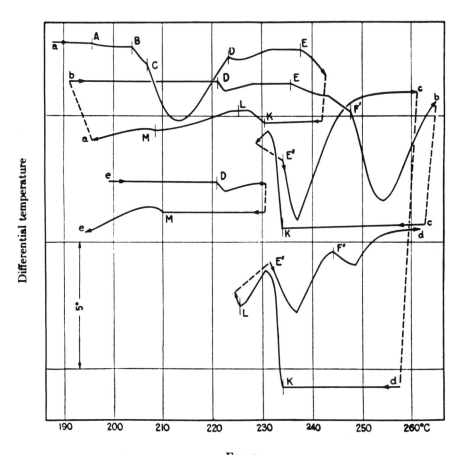

FIG. 7

Reversibility of inversions in the Na_2SO_4 preparation 3 which contained 0.1 per cent free H_2SO_4. Compare with Fig. 4. Note the irreversibility of *A*, *B* and *C*, and the small heat effect accompanying the reversal of *D* with respect to *M*. Also note the manner in which the upper inversions reverse.

to the lower inversions, the remaining two, *E* and *F* as belonging to the upper inversions. This arbitrary classification has some basis in fact, as will be pointed out later.

4. *Reversal Experiments.* In addition to the heat runs already described, attempts were made to reverse the various inversions by quickly changing the direction of heating of the furnace. The results are shown graphically in Figs. 6, 7 and 8, and, because of their importance, are described in some detail in the following tabulation.

4.1. Preparations 1, 2, 4 and 10. Fig. 6; also consult Fig. 3.

Curve a. First heating. Break A at 198°, B at 212°, C at 217.5°. Following break C furnace was reversed and cooling curve recorded. No heat effects on cooling curve. Another similar first heat curve taken thru D at 227°; no breaks on cooling curve; not recorded in the diagram.

Curve b. Cooling curve from 260°. Break K at 234.5°.
Furnace reversed quickly before L takes place. Heating curve shows E at 236°, no other break to 260°.

Curve c. Cooling curve from 260°. Break K at 234.4°, L at 227°. Furnace reversed before heat effect after L is completed. Heating curve break E at 236°, break H (F) at 242.5°, major heat evolution begins at 245°. Heat effect following H (F) greater than that following E.

Curve d. Cooling curve from 260°. K at 234.3°, L at 229°. Furnace reversed immediately after L occurred. Heating curve shows main heat absorption after E at 236°, H (F) at 244.5° followed by small heat absorption.

Curve e. Cooling curve from 260°. K at 234.3°, L at 228.5°. Furnace reversed after heat evolution following L is apparently complete, and the charge has given up most of its accumulated heat. Heating curve shows meager heat effect following E at 236°, major effect after H (F) at 244°.

Conclusions.

1. Breaks A, B, C and D do not reverse.
2. Cooling from high temperature, K takes place sharply at 234.3° to 234.5°; heat evolution proceeds rapidly at almost constant temperature till heat effect is completed.
3. After cooling thru K a sharp reversal takes place with respect to E at 236°. If temperature does not fall to L, effect after H (F) is eliminated. If L occurs, H (F) occurs; the farther the heat effect after L proceeds toward completion the more energy is associated with H (F), the heat effect after E becoming less and less significant.

4.2. Preparation 3. Fig. 7; see also Fig. 4.

According to Table II this preparation contains less than 0.1 per cent H_2SO_4. The cooling curve is of the usual type, the reheat curve shows a faint effect corresponding to break D.

Curve a. First heating thru A at 195.5°, B at 203.5°, C at 207° and D at 223°. Furnace reversed immediately after break E occurs at 237.5°. The cooling curve then shows break K very much lower than usual, at 230.5° followed by a small heat effect. The level portion of this heat evolution indicates that L also takes place. On further cooling the recovery curve is interrupted by a heat evolution beginning at M, 208.5°.

Curve b. Furnace reversed following completion of curve a and heating curve taken. Break D occurs sharply at 221°. Heating continued thru E at 236°, F' (F) at 248°, until completely inverted.

67

Curve c. Cooling curve following on b. K at 234°, furnace again reversed before L is encountered. The following heating curve shows
E at 234°; F is eliminated.

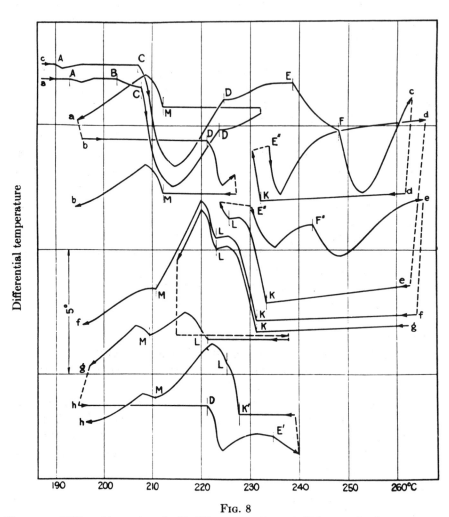

FIG. 8

The reversibility of inversions in Na$_2$SO$_4$ preparation 5 which contained 0.3 per cent free
H$_2$SO$_4$. Note the irreversibility of A, B and C, the magnitude of the heat effect of the
D $\rightleftharpoons$ M inversion and the reversibility conditions of the upper inversions.

Curve d. Cooling curve thru K at 234°, L at 225.5°, furnace reversed
before heat effect following on L is completed. Following heating
curve shows a heat absorption corresponding to E (E″) whose location is indeterminate, and a heat absorption corresponding to
F (F″) at 244°.

Curve e. Heating curve from below 180°. Only D on heating curve at
221°; furnace reversed, cooling curve thru M at 210°.

Conclusions.

    1. Heat effects following on A, B and C do not reverse.

68

2. Effect D, while totally irreversible in the neutral salt is found at least partly reversible, the corresponding break on the cooling curve is at M. Hysteresis range is from 210° to 221°.

3. Heat absorption following E (E', E'') on heating curve corresponds to heat evolution following K on the cooling curve. The reversal is prompt.

4. Break F (F', F'') is lost on heating if cooling is stopped before L takes place.

4.3. Preparation 5. Fig. 8; see also Fig. 4.

This preparation contains about 0.3 per cent H_2SO_4.

Cooling curve shows three heat evolutions beginning at K, L and M. Reheat curve shows three heat absorptions beginning at D, E' and F'. E' and F' occur at lower temperatures than E and F on first heating curve.

Curve a. First heating curve thru A at 193°, B at 202.5°, C at 208° and D at 223.5°. Furnace reversed. M on cooling curve at 212°; no other breaks on further cooling.

Curve b. Heating curve following a thru same range of temperatures. D found at 221°. Furnace reversed. Cooling curve shows M at 212°.

Curve c. Complete first heating curve for comparison.

Curve d. Cooling curve from above 260° thru K at 232°.

Furnace quickly reversed. Heating curve shows heat absorption following on E''. Owing to rapid reversal of heating the position of E'' is indeterminate. Heat absorption proceeds with great speed at 234°. F is lost.

Curve e. Cooling curve from above 260° thru K at 233°, L at 225.5°. Furnace reversed before the heat evolution is completed. Heating curve shows heat effect following on E'' completing at 235°, followed by another heat absorption beginning at F'', 243°.

Curve f. Cooling curve for comparison. K at 231.5°, L at 223°, M at 211°.

Curve g. Cooling curve from above 260° thru K at 231.5°, L at 223°; cooling stopped before M is reached. Slowly heated from 215° to 237°, followed by cooling curve. Heat evolution at 221° corresponding to L; M is at 209.5°.

Curve h. Following on curve g heating curve begins at 195°, shows D at 221° followed by a moderately large heat absorption, then by E (E') at 235°. Furnace reversed quickly. Cooling curve shows K at 228° followed by L at 225°, the two heat effects partly overlapping; M is at 210.5°.

Conclusions.

1. Heat effects beginning at A, B and C do not reverse.

2. Effect beginning at D on heating curve is reversible and begins at M on cooling curve. Hysteresis range extends from 210° to 221°.

3. Heat effect beginning at K on cooling curve is reversible and begins at E on heating curve. Hysteresis range about 2°.

4. F is encountered on heating curves only if L takes place on cooling. Otherwise it is lost.

Several experiments were carried out with the aim of bringing about a reversal of the three heat effects following on A, B and C. Inverted samples were held at various constant temperatures between 200° and 230° for periods ranging from 12 hours up to two weeks, and heat runs were then made to determine if any appreciable inversion took place. The results were in all cases negative.

Other experiments performed with the view of reducing the hysteresis range between D and M, by holding the inverted acid preparations 3 and 5 for 12 hours at constant temperatures ranging between 212° and 220° showed that the speed of inversion is inappreciable over this region. Samples held at 211° gave the break D on heating, likewise samples held at 221° showed M on cooling.

5. *Optical Examination.*[1] Attempts were made to detect changes in the optical properties of the salt corresponding with the breaks obtained by thermal analysis. The studies with a heating microscope showed that the changes proceed slowly thru a crystal over a temperature interval, with a gradual lowering of intensity of the interference colors, with nicols crossed. No sudden changes were noticeable. This shows that the speed of the observed changes is low. Examination of samples held for some time at a particular constant temperature and then quenched gave results which further indicate that the observations of Wyrouboff and Mügge[2] are essentially correct. Thenardite heated to various constant temperatures between 195° and 215° and then rapidly cooled shows increasing quantities of a much less birefringent phase whose indices of refraction are between 1.480 and 1.485, just about equal to the high index of thenardite. This phase is identical with Wyrouboff's γ. Another phase occurs in the same samples, with greater birefringence than thenardite; its high index is about 1.480, low index is less than 1.46, and incidentally, much less than the low index of thenardite (1.47). This coincides in occurrence and appearance with Wyrouboff's β form. It always occurs mixed with thenardite (α), or the γ form; often all three phases are present. Samples of thenardite heated at constant temperatures between 220° and 230° are almost completely converted to Wyrouboff's γ with usually a trace only of the more birefringent modification remaining.

Thenardite heated for several hours above 260° should be, according to the heating curves, completely converted to the high-temperature modification. When such samples containing 2 or 3 mg. of salt are sharply quenched by dropping them into cold mercury and immediately subjected to microscopic examination, they reveal the presence of a large proportion (*ca.* 50

---

[1] I am indebted to Dr. J. W. Greig for much of the microscopic work reported here. He is, however, not to be held responsible for any of the conclusions arrived at in this paper.

[2] See Table I.

per cent) of a highly birefringent modification with high index about 1.480, low index much below 1.465, together with a small amount of an apparently isotropic phase with indices lower than 1.475, the remainder being of the familiar γ. The highly birefringent phase encountered here corresponds in occurrence with Mügge's observation that Na_2SO_4 while cooling from high temperature passes thru a brief stage of higher birefringence than that of the high-temperature form or of the subsequent form which results on further cooling and which becomes inert at room temperature. Altho in general appearance and in indices of refraction this highly birefringent phase closely resembles the β modification, it is significant that the phase occurs in large amounts only in quenched samples. When the hot sample is lifted out of the furnace and cooled in air this phase disappears almost completely. These observations agree with the presence of the breaks K and L on the cooling curves, and also in that the heat effects following K and L, particularly the former, are very prompt. On these considerations it appears that the nearly isotropic phase of index less than 1.475 must be the high temperature (hexagonal) form denoted as δ by Wyrouboff, and that the highly birefringent form occurring in the quenched samples is the same as that observed by Mügge, with its stability range between those of Wyrouboff's γ and δ.

These studies, when considered with the results of thermal analysis make the assumption of pentamorphism in Na_2SO_4 tenable; in fact, it is not possible to correlate all the experimental data on the basis of a smaller number of phases.

Discussion

1. *Theoretical.* For the purpose of clarity it seems desirable to recall here the more important principles involved in the theoretical treatment of the equilibrium conditions governing inversions.

In a polycomponent system of two or more phases coexisting *in equilibrium* the chemical potential μ of each actual component has the same value in all the phases. In a system of one component μ is equal to the molar thermodynamic potential of that component. The slope of μ for a given phase as a function of temperature is given by the molar entropy of that phase,

$$\left(\frac{\partial \mu}{\partial T}\right)_{p,m} = -S;$$

that is, for the coexistence of two phases in equilibrium at the equilibrium temperature

$$\mu_I = \mu_{II}$$

$$\left(\frac{\partial \mu_I}{\partial T}\right)_{p,m} - \left(\frac{\partial \mu_{II}}{\partial T}\right)_{p,m} = S_{II} - S_I$$

At the equilibrium temperature

$$S_{II} - S_I = \frac{\triangle H}{T_i}$$

71

where $\triangle H$ is the change in the heat content of phase II on conversion into phase I, numerically equal to the heat of inversion at constant pressure, and T_i is the equilibrium inversion temperature.

The stable phase is always characterized by the lower value of μ. If phase I is stable at the higher temperature, then, in the event of inversion, below the equilibrium temperature $\mu_{II} < \mu_I$, while above the equilibrium $\mu_{II} > \mu_I$. Below the inversion temperature the reaction II $\rightarrow$ I can not take

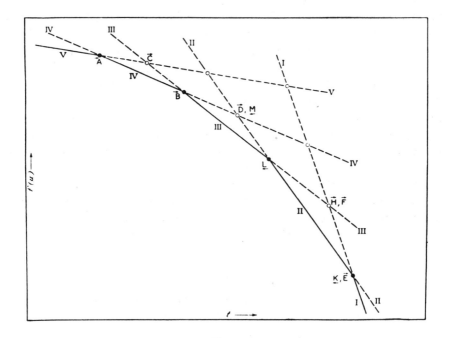

FIG. 9

Equilibrium diagram for pentamorphism applied to the case of Na_2SO_4. See Table IV and the text.

place. At the equilibrium temperature, where $\mu_{II} = \mu_I$ the reaction proceeds infinitely slowly. It reaches a measurable speed only when the value of the potential difference $\mu_{II} - \mu_I$ has attained sufficient magnitude to overcome the forces of restraint within the crystal operative virtually as internal friction. This imparts to the system a kind of inertia or hysteresis, which appears as apparent superheating or undercooling. Whether it is justifiable to speak of superheating and undercooling in the case of inversions remains to be decided by experiment. It is certain that the low to high inversion in quartz recently re-investigated by Gibson[1] is the only known case of apparently well substantiated superheating on inversion.

When a substance is capable of existence in more than two modifications the complexity of behavior increases rapidly with the number of possible phases. The total number of possible stable and metastable inversions is

$$1 + 2 + \ldots\ldots\ldots + (n - 1)$$

[1] R. E. Gibson: J. Phys. Chem., **32**, 1197, 1206 (1928).

where n is the number of possible crystalline phases.[1] The stable inversions number $n-1$. All the rest are metastable.

2. *The system* Na_2SO_4. In describing the experimental results it was pointed out that the facts are best accounted for on the basis of pentamorphism of Na_2SO_4. I shall now attempt to give a justification for this assertion.

In Fig. 9 is given a schematic diagram of thermodynamic potential μ as a function of temperature for five mutually convertible phases, arbitrarily employing a straight line relationship between μ and t. The lines intersect at the theoretical inversion temperatures with no hysteresis. The position of the μ lines for the various phases is fixed in agreement with the equations on p. 1299. In accord with the current convention of petrography, the phases are numbered with Roman numerals beginning with the phase stable at high temperature. The various intersections are labeled A, B, C, etc. to accord with the breaks encountered on the heating and cooling curves in the experiments on reversibility and the optical studies.

The established facts may be interpreted as follows:

1. Reactions beginning at A, B and C on first heating curves do not reverse. Thenardite is stable at low temperatures, hence these phase reactions are pseudo-monotropic. Under a heating microscope, the reactions are seen to be sluggish. Instead of taking place rapidly throughout a crystal at appreciably constant temperature, the formation of a new phase spreads slowly thru the crystal, occupying a considerable temperature interval. Because of the very considerable hysteresis the reactions can proceed simultaneously, the breaks on the heating curves merely marking the temperatures at which each reaction has attained sufficient speed to begin to absorb appreciable quantities of heat.

2. The reaction beginning at D does not reverse in neutral or slightly alkaline salt; in slightly acid salt it reverses at M, with a hysteresis range of about 12°.

3. Reaction beginning at K on cooling curves reverses at E on heating curves. This reaction shows comparatively little hysteresis, and takes place promptly, particularly on cooling. It is interpreted as representing the inversion of the high temperature modification into the next stable form, that is, I $\leftrightarrows$ II.

4. The appearance of F on heating curves depends on whether the preceding cooling curve in reversibility experiments had proceeded as far as L. When the reaction following L goes to completion, E is eliminated, as is seen from the reheat curves of neutral preparations. The form following L on the cooling curve is Wyrouboff's γ; but the simplest tenable assumption is that L must represent the inversion of II $\rightarrow$ III. Reheat curves for the neutral salt show that H (F) represents the inversion of Wyrouboff's γ to the high-temperature form; accordingly we have

---

[1] True monotropic inversions are not considered.

L = II → III, F = III → I, identifying Wyrouboff's γ with the form numbered III in the present scheme.

5. The arbitrary classification of the heat effects into two groups, comprising the lower inversions (A, B, C) and the upper inversions (E, F, K. L). alluded to earlier in the paper (p. 1294) is now justified on the basis that the lower inversions are pseudo-monotropic while the upper inversions are more readily reversible. The inversion $D \leftrightarrows M$ is reversible only in the slightly acidified preparations; it fits into the scheme on the assumption that it represents the metastable transition IV $\leftrightarrows$ II. In neutral preparations the breaks M on the cooling curve, and D and E on the reheat curve are eliminated, and no modification lower than III is formed on cooling. If II is the highly birefringent Mügge form with but a brief temperature interval of existence, then IV is the Wyrouboff β. Thenardite, the ordinary low-temperature modification, becomes form V in this scheme.

TABLE IV

1. *Identification of Phases*

Phase   I   High temperature modification, observed by Mügge, named δ by Wyrouboff.

Phase II   Highly birefringent phase of brief stable temperature interval resulting from the high temperature modification, observed by Mügge.

Phase III   "Altered thenardite" observed by Mügge, named γ by Wyrouboff. Inert at ordinary temperature.

Phase IV   Wyrouboff's β.

Phase V   Thenardite, Wyrouboff's α.

2. *Identification of Inversions*

| Break | Phase Change | Temperature °C. | |
		Heating Curve	Cooling Curve
A	V→IV pseudo-monotropic	197 ± 2 (neutral) 195 ± 2 (acid)	——
B	IV→III pseudo-monotropic	210 ± 2 (neutral) 202 ± 2 (acid)	——
C	V→III pseudo-monotropic	217 ± 2 (neutral) 207 ± 2 (acid)	——
D, M	IV⇆II	230 ± 2 (neutral) 222 ± 1 (acid)	Irreversible in neutral 210 ± 2 (acid)
E, K	II⇆I	238 ± 2 (neutral) 236 ± 2 (acid)	234.5 ± 0.1 (neutral) 233 ± 1 (acid)
F (H)	III→I	244 – 250	——
L	II→III	——	228 (neutral) 225 (acid)

The presence of traces of H_2O in the inclusions, and particularly the presence of small amounts of H_2SO_4, greatly increases the facility with which the inversions take place. It has already been remarked that the amount of energy associated with the various heat effects differs according to the mode of preparation of the salt. The behavior of the slightly acidified preparations 3 and 5 is unique in this respect. The breaks take place at lower temperatures than in neutral salt. Reluctance toward inversion is decreased to such an extent that $D \leftrightarrows M$ can take place. It is plausible that the role played by the occluded water is that of enabling the lower inversions to proceed; it is certain that when this water is eliminated by previous heating the lower inversions do not reverse, and also in preparations with very low H_2O content the lower heat effects on first heating are less energetic (see Fig. 5, preparations 4, 6, 7, 8, 9).

Conclusion and Summary

The close proximity of the inversions, the large hysteresis and the tendency toward pseudo-monotropic behavior of some of the inversions serve to make this unary system of outstanding importance in the elucidation of the theory of solid state. The conclusions deduced from this preliminary investigation are that Na_2SO_4 is capable of existence in five distinct modifications, the scheme of inversions being made evident in Fig. 9. Owing to hysteresis in the inversions the equilibrium temperatures can not be fixed; the same holds true for the stability intervals of the various phases. The essential conclusions are summarized in Table IV, the temperatures given representing values at which each inversion proceeds with sufficient speed to produce an appreciable heat effect.

Acknowledgment

I wish to express my thanks to my colleagues R. E. Gibson, J. W. Greig, E. Posnjak, G. Tunell, and C. J. Ksanda for aid in connection with various phases of this work.

Geophysical Laboratory,
Carnegie Institution of Washington,
May, 1929.

5

Reprinted from *J. Am. Ceram. Soc., 22,* 54–63 (1939)

CRITICAL STUDY OF THE DIFFERENTIAL THERMAL METHOD FOR THE IDENTIFICATION OF THE CLAY MINERALS*

BY F. H. NORTON

ABSTRACT

An apparatus is described for studying clay minerals by the differential thermal method, which has some advantages over previous types of apparatus. Thermal curves are given for many of the pure clay minerals for use as standards of identification. The use of this method quantitatively for the determination of the relative amounts of clay minerals in a natural clay is discussed, and examples are given for a number of high alumina clays.

Introduction

Any one who has worked with clays or soils realizes the difficulty in identifying the clay minerals in the finer fractions. The petrographic method, which is so satisfactory for crystals more than a few microns in size, is useful in the smaller sizes only to a limited extent. The X-ray, on the other hand, is able to identify the finer grained clay minerals if they occur in fairly large proportions, but when small amounts are combined with large amounts of some other mineral, it is difficult to recognize the smaller portion because they are all sheet minerals with many common lines.

This difficulty becomes more pronounced as the crystal size decreases because of the broadness and general indistinctness of the X-ray lines. In looking about for some additional way of identifying the clay minerals, the differential thermal method seemed promising in view of the qualitative work which had already been done with it. This paper, therefore, discusses the thermal method for the identification of the clay minerals, using an improved apparatus which attempts to minimize some of the errors occurring in previous investigations so that quantitative results can be achieved. The various factors influencing the magnitude and position of the absorption peak in a given mineral are also studied.

In Part I, the qualitative identification of minerals will be taken up, including a comparison of the present data with those of other experimenters. In Part II, the differential thermal method will be discussed from the viewpoint of quantitative determinations.

Part I. Qualitative Determination

1. Previous Work

The thermal method of identifying minerals has been investigated by a number of authors. The differential thermal method was suggested in 1887 by Le Chatelier.[1] Later Wallach[2] showed how this method could be used to identify types of clays. Satoh[3] gave a number of thermal curves of Japanese fire clays. In 1933, the fine work of Orcel and Caillère[4] on the bentonite minerals brought out the value of this method more clearly than any one before. Granger[5] discussed the thermal analysis of clay a little later, and Orcel[6] then presented a complete paper, giving thermal curves of most of the known clay minerals as well as curves of many natural clays. Orcel brought out still further the value of this method for identification purposes. In a study of the kaolin minerals, Insley[7] obtained thermal curves of kaolinite and dickite. A thermal curve of talc was obtained by Ewell, Bunting, and Geller,[8] which showed the high temperature stability of this mineral. One of the latest papers on the subject by Jourdain[9] gives thermal curves for a number of the clay minerals. Thilo[10] also shows curves for pyrophyllite.

A study of this literature will indicate continued improvement in the methods to obtain the thermal curves and a greater purity of samples of the clay minerals. Owing to differences in apparatus and methods, however, it is not possible to compare the results of different experimenters more than in a general way, but certain discrepancies are noted which the writer hopes to clear up in this paper. As far as is known, no quantitative measurements by the thermal method have been

* Received September 22, 1938.
[1] Le Chatelier, "De L'Action de la Chaleur sur les Argiles," *Bull. Soc. Min.,* **10,** 204 (1887).
[2] H. Wallach, "Analyse Thermique des Argiles," *Compt. Rend.,* **157,** 48 (1913).
[3] Shinjo Satoh, "Heat Effects on Fire Clays and Their Mixtures," *Science Repts., Tôhoku Imp. Univ.* (Series 3), **1,** 3 (1923).

[4] J. Orcel and S. Caillère, "L'Analyse Thermique Différentielle des Argiles à Montmorillonite (Bentonites)" (Differential Thermal Analysis of Montmorillonite Clays (Bentonites)), *Compt. Rend.,* **197** [15] 774–77 (1933); *Ceram. Abs.,* **13** [6] 160 (1934).
[5] A. Granger, "Thermal Analysis of Clay," *Céramique,* **37** [552] 58 (1934); *Ceram. Abs.,* **14** [1] 23 (1935).
[6] J. Orcel, "L'Emploi de L'Analyse Thermique Différentielle dans la Détermination des Constituants des Argiles, des Latérites, et des Bauxites" (Differential Thermal Analysis for Determination of Constituents of Clays, Laterites, and Bauxites), *Congr. Internat. Mines, Met. Geol. Appl., 7ᵉ Session, Paris, 1935, Geol.,* **1,** 359–73 (1936); *Ceram. Abs.,* **16** [7] 218 (1937).
[7] H. Insley and R. H. Ewell, "Thermal Behavior of the Kaolin Minerals," *Jour. Research Nat. Bur. Stand.,* **14** [5] 615–27 (1935); R.P. 792; *Ceram. Abs.,* **14** [8] 201 (1935).
[8] R. H. Ewell, E. N. Bunting, and R. F. Geller, "Thermal Decomposition of Talc," *Jour. Research Nat. Bur. Stand.,* **15** [5] 551–56 (1935); R.P. 848; *Ceram. Abs.,* **15** [2] 76 (1936).
[9] A. Jourdain, "Studies of the Constituents of Refractory Clays by Means of Thermal Analysis," *Céramique,* **40** [593] 135–41 (1937); *Ceram. Abs.,* **16** [11] 357 (1937).
[10] Erich Thilo and Heinz Schünemann, "Chemical Studies of Silicates: IV, Behavior of Pyrophyllite, $Al_2(Si_4O_{10})(OH)_2$, on Heating and the Existence of a 'Water-Free' Pyrophyllite, $Al_2(Si_4O_{10})O$," *Z. anorg. allgem. Chem.,* **230** [4] 321–35 (1937); *Ceram. Abs.,* **16** [10] 311 (1937).

attempted, although such a possibility was suggested by Orcel.[6] A more detailed comparison will be made later of the previous work in relation to the curves obtained in this paper.

II. Design of the Apparatus

A cross-section of the thermal furnace is shown in Fig. 1. The specimens are held in two sockets drilled into a heavy nickel block, which tends to minimize the thermal gradients. The neutral body and the test specimen are placed laterally in the furnace rather than longitudinally, because this assures the same heat input to each throughout the whole run and also gives the uniformity of the curves at the start.

recorded on this same bromide paper by a signal lamp, which is flashed every time the potentiometer reading comes to an even 50° interval.

The rate of temperature rise of the furnace is controlled by a liquid rheostat in which dilute sulfuric acid flows through a capillary tube to a battery jar into which are suspended two carefully shaped graphite electrodes, so that the current is increased at such a rate that a close approximation to the desired temperature rise of 12°C per minute is obtained. The main difficulty with this simple equipment is that changes in room temperature affect both the resistance of the acid and also the heat loss from the furnace, so that certain corrections have to be made in the acid flow rate from

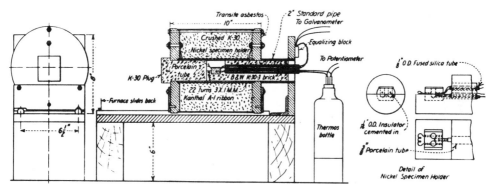

FIG. 1.—Thermal analysis furnace.

There are no other special features of the furnace except the Kanthal winding, which allows running the furnace continuously above 1000°[*] without danger of burn-outs; the sliding furnace also allows the specimen block to be exposed for changing specimens without disturbing the thermocouple connections.

Contrary to most of the previous work, the temperature of the specimen is measured by a separate thermocouple, imbedded in the center of the nickel block, which is connected directly to the potentiometer with the cold junction in ice. This eliminates certain complications of the temperature measurement and gives a temperature which can not depart more than a few degrees from the temperature of the sample. In fact, a temperature taken in the block itself is more suitable for a base than the temperature of the neutral body, which may show variations owing to shifts in the couple position.

The differential temperature is read with a recording galvanometer of such a resistance that no shunt is required. A continuous record is produced on a strip of bromide paper, a method superior to the point-by-point method, which may lose small but important characteristics of the curve. The actual temperatures are

---

[*] Centigrade temperature will be used throughout this paper.

time to time which introduces some unevenness in the temperature rise. Small variations in rate, however, do not seem to affect the area appreciably under the various peaks.

III. Samples Used

Table I shows the generally accepted relation between the clay minerals, although there is no uniformity of opinion on just which one should or should not be included. In Table II there are shown the chemical

TABLE I
CLAY MINERALS

Kaolin	Montmorillonite	Micaceous	Hydrated alumina
Kaolinite-anauxite	Montmorillonite-nontronite	[*]Muscovite-sericite	Gibbsite
Dickite	Beidellite	Potash clay	Diaspore
Nacrite	Talc[*]		
Halloysite	Magnesium clay		
Allophane	Pyrophyllite[*]		

[*] While these minerals are not usually considered to be clays because of their large crystal size in nature, they do acquire claylike properties when finely ground.

analyses and the origin of the various clay minerals tested. In view of the limitations of space, only the purest mineral obtainable of each type is listed.

although in most cases several samples have been tested; any great deviation will be mentioned in connection with the curves. The samples were air-dried, ground in a mortar until they passed a 150-mesh screen, and then were tamped lightly into the test chamber; the junction of the thermocouple was exactly centered. This tamping was done carefully so that as far as possible various minerals were tamped approximately to the same density. Between 0.3 and 0.4 gram of material is necessary for this purpose, the small amount be-

for about one second; this produces a horizontal black line across the film corresponding with this temperature. The time is noted, and the potentiometer is set ahead to the millivolt reading corresponding to 100° and so on until the last reading of 1000° is reached, after which the differential galvanometer circuit is opened so that a zero reading can be obtained on the film. The flow of acid through the capillary tube is controlled if necessary to maintain a time interval of about 4 minutes for each 50° interval.

TABLE II

	(1)	(2)	(3)	(4)	(5)	(6)	(7)	(8)	(9)	(10)	(11)	(12)	(13)	(14)	(15)	(16)	(17)
SiO_2	44.70	48.80	46.35	44.75		27.61	61.34	65.35	53.68	47.28	41.38	50.20		55.64	50.10	1.6	9.7
Al_2O_3	38.64	35.18	39.59	39.49		32.29	0.71	29.25	0.60	20.27	9.84	16.19		16.18	25.12	65.0	90.2
FeO															1.52		
Fe_2O_3	0.96	1.24	0.11	0.53		0.23		0.10	1.12	8.68	27.47	4.13		2.68	5.12	None	
TiO_2	0.22	0.61									Tr.	0.20			0.50	1.1	
CaO	0.24	0.22		0.13		0.02	Tr.	None	0.52	2.75	Tr.	2.18		1.80	0.35	None	
MgO	0.08			0.19		0.10	32.32	0.05	25.34	0.70	Tr.	4.12		8.88	3.93		
BaO																	
K_2O	0.14	0.40		{0.10				{0.05	0.07	Tr.	Tr.	0.16		5.16	6.93		
Na_2O	0.62	0.25							3.00	0.97		0.17		0.04	0.05		
P_2O_5						1.31											
F			0.15														
SO_3						0.12											
CO_2						0.72											
H_2O-	0.64	1.16		0.61				0.08	7.28			12.10	15.58	{9.72	1.90		
H_2O+	13.88	12.81	13.93	14.40		18.05	5.81	0.25	8.24	19.72	9.25	7.57			7.18	33.0	

(1) **Kaolinite** (Ione, Amador County, Calif.).
(2) **Anauxite** (Newman pit, near Ione, Calif.).
(3) **Dickite** (National Belle mine, Red Mountain, Ouray, Colo.).
(4) **Nacrite** (Brand, Saxony).
(5) **Halloysite** (Anamosa, Iowa).
(6) **Allophane** (Morehead, Ky.).
(7) **Talc** (Manchuria, very pure sample).
(8) **Pyrophyllite** (More County, N. C.).
(9) **Magnesium clay** (Hector County, Calif.).
(10) **Beidellite** (Beidell, Colo.).
(11) **Nontronite** (Sandy Ridge, N. C.).
(12) **Montmorillonite** (Smith County, Miss.).
(13) **Muscovite** (Keystone, S. Dak.).
(14) **Potash clay** (High Bridge, Ky.).
(15) **Illite, colloidal fraction** (Maquoketa shale, near Gilead, Ill.).
(16) **Gibbsite** (Dutch Guiana).
(17) **Diaspore** (Rolla, Mo.).

Analyses (3) and (7), see footnotes 7 and 8.
Analyses (1), (2), (4), (6), (8), (9), (10), (11), (12), (14), Dr. Ross, U. S. Geol. Survey.
Analyses (16) and (17), Babcock and Wilcox Co.
Analysis (15), *Amer. Mineralogist*, **22**, 818–23 (1937).

ing desirable in cases where the clay can be obtained only in small quantities. In some cases, the clay mineral is dried at 110°C to constant weight before the test, and (as will be shown later) this drying materially affects the initial part of the curve for many of the minerals.

IV. Method of Conducting the Test

After the sample has been placed in its socket in the nickel block, a nickel cover is put on top of the block to protect both the neutral body, which is calcined alumina, and the specimen from the direct radiation of the furnace walls. The furnace is then slid over the specimen. The liquid rheostat is adjusted so that the ends of the electrodes are just touching the acid, and a shunt resistance regulates the initial current to five amperes. As soon as the switch is closed to pass this current through the furnace, the acid is started flowing through the capillary tube and the recording galvanometer is started. The potentiometer is then set at a millivolt reading corresponding to 50°, and the needle of the potentiometer galvanometer is watched until it reaches the zero mark, at which time the signal switch is closed

The method described could be considerably simplified if a program controller maintained the temperature definitely on schedule and an automatic contactor flashed the 50° signals. It is hoped to construct such a device in the future.

V. The Kaolin Minerals

The thermal curves for the kaolin minerals are shown in Fig 2. In all of these thermal curves, the first vertical line represents 50°, with 50° intervals up to 1000°. All of these minerals are characterized by a strong exothermic peak in the region of 980°, and, as far as the writer has been able to determine, no minerals outside this group have this characteristic. The minerals in this class were of high purity and certainty of composition so that the results are much more definite than in the other classes. The thermal curves for nacrite, dickite, and kaolinite show only a small deviation from a straight line in the initial portion owing to absorbed moisture, but the beginning of an endothermic effect is found in all of them at about 475°. The nacrite shows a comparatively long peak running from 610° to 670°. The dickite, on the other hand, shows a rather long range of slight heat absorption and then a rather sudden peak at 610°. It is possible, therefore, to distinguish definitely these three kaolin minerals by the thermal curves, because a mixture of kaolinite and dickite would give a double peak rather than the broad peak of the nacrite.

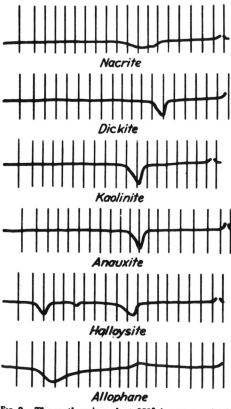

Nacrite

Dickite

Kaolinite

Anauxite

Halloysite

Allophane

FIG. 2.—The exothermic peak at 980° is not completely shown on these curves.

Allowing for differences in heating rate and size of sample, the kaolinite curves of Orcel and Jourdain agree with that shown here except for the large, low temperature absorption peak, due probably to the fine grain size of their samples.

The anauxite gives a curve substantially the same as the kaolinite except for a slightly smaller area under the endothermic peak, which might be expected from the lower content of combined water. It must be concluded, therefore, that the thermal method can not distinguish between anauxite and kaolinite, and a chemical analysis must be made to assign values to the kaolinite-anauxite series.

The halloysite is readily distinguished from the kaolin minerals named because of the two low temperature peaks, which are sharp and distinct, the first at 150°

and the smaller at 325°. These peaks do not show to any extent if the material has been previously heated to 110°. It is advisable, when identifying clay minerals, to run them in the air-dried condition and, if possible, with the same moisture they contained in the ground. Undoubtedly in the past, some so-called halloysite has really been fine-grained kaolinite, and in a mixture of the two there is sometimes great difficulty in distinguishing between them. It will also be noticed that the main absorption peak at 580 °C in halloysite comes at a lower temperature than for the kaolinite, as might be expected from the finer grain size. A fine-grained kaolin will show the first absorption peak but not the second. The halloysite curves of Orcel and Jourdain indicate only one broad low temperature peak rather than two separate ones.

The mineral, allophane, is of such fine grain size as to be practically amorphous, and therefore the large endothermic effect at low temperatures, which shows a peak at 180° but extends over a long range, is more or less to be expected. The low exothermic peak at 600° is difficult to explain at present because the usual exothermic peak due to the amorphous-alumina–gamma-alumina change occurs in its usual position. Because this material is noncrystalline, it is perhaps not justified to group it with the kaolin minerals, but at present this seems to be the most logical location.

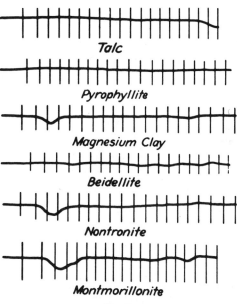

Talc

Pyrophyllite

Magnesium Clay

Beidellite

Nontronite

Montmorillonite

FIG. 3

In conclusion, the kaolin minerals as a class can be determined definitely by the exothermic peak at 980 °C, and the various members of the group can be distinguished by the endothermic effects which, except for anauxite, are unique for each one of the minerals.

VI. The Montmorillonite Minerals

These minerals are believed to have a symmetrical sheet structure similar to pyrophyllite, but owing to the minute and imperfect crystallization, there is still some doubt as to the exact crystal structure. Contrary to the case of the kaolin minerals, many minerals of this group are difficult to obtain in the pure form, and because of the fine grain size the X-ray does not always indicate when small admixtures occur. Only by studying a great number of these minerals from different

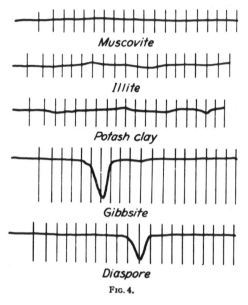

Muscovite

Illite

Potash clay

Gibbsite

Diaspore

Fig. 4.

localities can it be definitely determined which peaks are characteristic of the pure mineral studied and which are due to admixtures. While the minerals listed in this group are considered to be of a high degree of purity, there is no certainty that all admixtures have been excluded. Curves for the minerals of this group are shown in Fig. 3.

The specimen of montmorillonite is characterized by a strong double peak at low temperatures, which is quite characteristic of this mineral. The endothermic peak at 870° has also occurred in a number of samples of montmorillonite The other endothermic peak at 690° seems to be rather variable between different samples, but it is probably characteristic of this mineral. The curves of Jourdain and Orcel agree perfectly with this curve, except that the 690° peak is larger, corresponding exactly with the curve for Wyoming bentonite (Fig. 5).

Nontronite is a mineral believed to be more or less isomorphous with montmorillonite, and it is therefore not surprising to find that the thermal curve is somewhat similar with a low endothermic peak at 870°. It can be distinguished, however, at least in this

specimen, from the montmorillonite by the initial peak which is much less broad than the double peak of the montmorillonite. The curve agrees perfectly with that of Orcel for this mineral.

The curve for the beidellite shows a single low temperature peak. Because the sample had been dried at 110°C, this peak is not as prominent as it would have been in air-dried clay. The endothermic peak is also found at 575° and 870° and an exothermic peak at 900°, which seems to be the distinguishing feature of this mineral. There is also the small endothermic peak at 690° which occurs in the montmorillonite. This curve agrees with that of Orcel except that the 690° peak is absent in his curve; in Jourdain's curve the 870° peak is absent.

The magnesium clay shows a single low peak at 175° and an endothermic peak at 800°; there is also a small sharp peak at 275°. This peak has been noticed on at least eight records of minerals containing fairly large quantities of magnesium, such as talc and some of the bentonites. It is a peak which would probably be missed if a point-by-point method were used, and it might be thought accidental if it occurred on only one record. Because, however, it occurs at the same temperature on all of these runs, it must correspond to some definite change in the structure, which at present is unknown.

The mineral, pyrophyllite, shows no low temperature effect as would be expected from the large crystal size. On the other hand, the endothermic peak at about 625° is so broad and shallow that it is difficult to identify this mineral by thermal methods. It usually occurs in large crystal sizes, however, and it is readily determined by the petrographic method. Neither the Jourdain nor the Orcel curves for this mineral agrees with that of the writer, and Orcel states that the results for this mineral are variable. There can be little question of the purity of the sample used by the writer, because of the large and perfect crystals.

The talc gives a distinct endothermic peak at 990°, which is characteristic of this mineral. The small endothermic peak, shown at 600°, can not be due to the impurities because this particular sample of talc was of great purity, but some other pure samples of talc have shown only the point at 990°. The small peak at 275° is also present on this curve. The relatively high temperature of the main endothermic peak indicates the great strength of the OH bonds in this crystal. This curve agrees with those of Orcel and Ewell.

VII. Micaceous Clay Minerals

These minerals, like the preceding, can not be obtained in as pure form as the kaolin minerals, and there is still a great deal of work to be done in this class. In Fig. 4, a curve of muscovite is given; although muscovite is not usually considered as a clay mineral, it is included because of its close association. The curve shows a broad absorption peak at about 600°, which is so shallow as to make identification of this mineral by the thermal method not too certain. Several specimens of muscovite from different sources have shown identical curves. On the other hand, Insley* stated in a private communication that a muscovite from Laurel, Md., showed a broad but distinct endothermic peak at 840°.

The potash clay shows a rather complex curve with a double absorption peak at a low temperature, due to surface water and endothermic peaks at 690° and 900°, which are fairly sharp and distinct. The exothermic peak at 490° is still to be explained, because it would hardly seem to be due to organic matter in this material.

The illite gives a curve somewhat analogous to the potash clay but with fewer low temperature effects, which would indicate a less active surface or a larger grain size. The endothermic peaks at 600° and 880°, although small, are sufficiently distinct for identification purposes. The 600° peak corresponds closely with that of muscovite.

VIII. Hydrous Alumina Minerals

Figure 4 also shows the curve for gibbsite, indicating a large heat absorption at the comparatively low temperature of 350°. This large effect would be expected because the mineral contains 33% water, which is twice that of the kaolin minerals. It is interesting to observe the double peak in the mineral, which

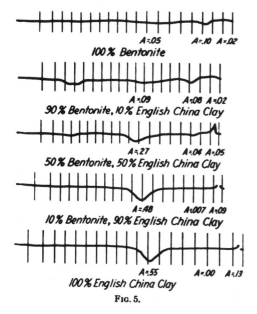

A=.05   A=.10 A=.02
100% Bentonite

A=.09   A=.08 A=.02
90% Bentonite, 10% English China Clay

A=.27   A=.04 A=.05
50% Bentonite, 50% English China Clay

A=.48   A=.007 A=.09
10% Bentonite, 90% English China Clay

A=.55   A=.00 A=.13
100% English China Clay

Fig. 5.

may be due to some admixture, because some other gibbsites contained in high alumina clays show only a single peak at this point. It was impossible to obtain a pure crystalline gibbsite, which would make the theory of an admixture possible. The small endothermic peak at 550° and the small exothermic peak at 980° indicate the presence of a few per cent of kaolinite.

* Herbert Insley, National Bureau of Standards.

The diaspore is a crystalline material, but it has some silica admixture. It seems difficult, in fact, to obtain a sample of diaspore which does not have about 10% of silica. Is this silica free quartz, or does it occur in the diaspore structure? It is assumed here that it is thermally inert material. Diaspore gives a single definite peak at about 550°, but it shows no exothermic peak at 980° and it need not be confused with the kaolin minerals.

Both of these minerals give results in good agreement with Jourdain and Orcel.

Part II.   Quantitative Determination

I.   Measurement of Heat Effect

If it is desired to determine precisely the amount of a hydrous mineral in a clay by the thermal method, it is necessasry to use some definite measure of the heat effect. The measure need not be absolute, because it is difficult to make such a measurement, but it should be relatively accurate as compared with the standard samples. As a basis for this measurement, the height of the peak on the thermal curve may be used, which is the maximum temperature difference between the test sample and the neutral body. It has been found, however, that the height of the peak varies considerably with small changes in heating rate; therefore, in this work the area under the peak has been found to be satisfactory for measuring the heat effect because this quantity is not greatly affected by small changes in heating rate. In defining the area under the peak, certain limits must be set; otherwise the area will be somewhat indefinite. In general, a straight line is drawn from a tangent to the curve on both sides of the peak, and the area under this line is determined. There are some cases, however, where one peak is disturbed by adjacent peaks, in which case it is necessary to resolve the curve into its elementary peaks, each one representing a particular heat affect. In most cases, this can be done readily with considerable precision.

II.   Reproducibility

In measuring the heat effect, certain variables enter into the measurement which must be taken into account for quantitative work. One of the most important variables is the physical property of the sample when packed in the test cavity. With care, it is possible to pack the sample in with sufficient reproducibility to check the heat effect within 2%, but there are cases where the properties of the mineral itself change, such as density, specific heat, and thermal conductivity as various types of clay are used. An example is shown by kaolin which, when packed in the test cavity in the usual manner, has a density of 0.70 g. per cu. cm., whereas a hard Missouri flint clay packed in the same manner has a density of 1.19. Both of these materials are essentially pure kaolinite, and yet one will naturally give a much greater heat effect than the other. At present, a method of eliminating this variable has not been determined, but it happens to a serious extent only in the case mentioned. The change in heat effect with the variation in physical properties of a single mineral are shown in Table III.

The heating rate, as explained before, influences both the temperature at which the peak comes and also the height of the peak, but as far as has been determined, it affects only slightly the area. It may be concluded, therefore, that any reasonable control of temperature rise will give little difficulty in quantitative work.

There is also a number of other factors which may affect the reproducibility of the result, but all of these effects can be minimized by careful technique. Among these are the position of the thermocouple, which must

TABLE III

Materials	Area under curve (sq. in.)
Pure kaolinite crystals	0.45
No. 1 English china clay	0.55
Hard Mo. flint clay	0.64

be accurately centered in the sample, and the calibration of the couple, which must be checked. For example, in some of the first runs, there was difficulty in obtaining reproducible results because the moisture from the sample passed out through the thermocouple tube and condensed in the cooler portions in such a way that several errors in the temperature resulted. It should also be remembered that the resistance of the couple current changes with the temperature and

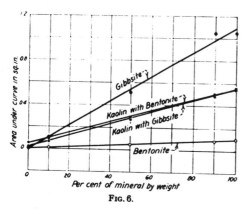

Fig. 6.

that the relation between e.m.f. and temperature is not linear. These effects compensate each other to some extent, giving an average sensitivity of 1.5° per mm. deflection on the record.

III. Individuality of Separate Heat Effects in Mixtures of Minerals

To ascertain the proportionality between the amount of minerals in the mixture and their respective heat effects, a series of mixtures was made of kaolin and montmorillonite, giving the curves in Fig. 5. If the area under each endothermic peak is plotted against the percentage weight of the mineral, a curve will result, such as is shown in Fig. 6, indicating proportionality between the weight of mineral and the heat effect, i.e.,

each mineral acts independently of the other. The maximum deviation of the points from a straight line is 5% and the average 2%. It is interesting to note that, in many cases, amounts as small as 1% of a

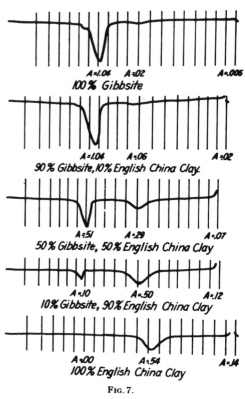

Fig. 7.

mineral can be detected by this method. A similar series of runs was made with the kaolin-gibbsite mixtures (shown in Fig. 7), and when these areas are plotted, a straight line is obtained as shown in Fig. 6.

IV. Disturbing Factors

In measuring the heat effect quantitatively, there are factors entering in which appear to influence the results in addition to those variables listed under reproducibility. For example, the particle size has a distinct influence on the heating curve, the finer fractions giving up their heat more rapidly than the coarser fractions, which is expected from the physical nature of the crystal. To show this more clearly, a sample of Edgar plastic kaolin from Florida was fractionated in the centrifuge to give a series of monodispersed fractions as shown in Table IV. The heating curves for this set of fractions are shown in Fig. 8. The initial part of the endothermic curve occurs at the same temperature, and the peak comes at about the same temperature in all cases, but the finer the particle the more rapidly

TABLE IV

INFLUENCE OF PARTICLE SIZE

Edgar plastic kaolin	Completion of endothermic reaction (°C)	Endothermic maximum (°C)
10–44μ	670	600
0.5–1.0	650	605
0.25–0.5	630	605
0.1–0.25	615	600
Less than 0.1	610	600

the curve comes back to the zero line. Incidentally, this method might be used for a measure of particle size.

Particular attention is called to the finest fraction which gives a curve distinctly different from the others. The fine fraction curve gives a normal endothermic

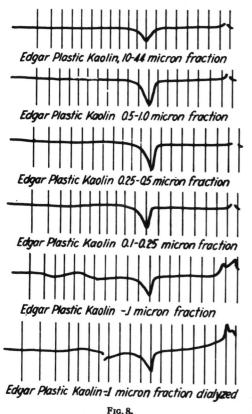

Edgar Plastic Kaolin, 10-44 micron fraction

Edgar Plastic Kaolin 0.5-1.0 micron fraction

Edgar Plastic Kaolin 0.25-0.5 micron fraction

Edgar Plastic Kaolin 0.1-0.25 micron fraction

Edgar Plastic Kaolin -.1 micron fraction

Edgar Plastic Kaolin -.1 micron fraction dialyzed

FIG. 8.

peak, but other characteristics, such as the sudden drop in temperature at 375° and the large double exothermic peak at 900° to 1000°, are peculiar and difficult to explain. Thinking that some of this peculiarity may have been due to adsorbed ions in the fine fraction, the sample was dialyzed and run again,

producing a curve of somewhat different but quite similar nature. This curve can be explained either by the fact that there is a mineral in the fine fraction for which there is no prototype in the pure specimen or

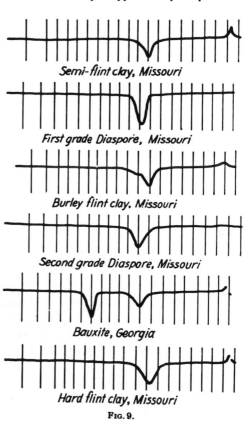

Semi-flint clay, Missouri

First grade Diaspore, Missouri

Burley flint clay, Missouri

Second grade Diaspore, Missouri

Bauxite, Georgia

Hard flint clay, Missouri

FIG. 9.

that the greatly increased surface area gives effects quite different from the same mineral in a coarser state of aggregation.

It is found, of course, that different clay specimens shrink to different amounts on heating in the test. Some of the montmorillonite samples shrink to a great degree so that a large space exists around the sample preventing any contact with the walls of the cavity. On the other hand, the purer kaolins and high aluminous materials show little shrinkage up to 1000°C, and a few, such as dickite, show an actual expansion. It would be expected that this shrinkage, which changes the density, specific heat, and thermal conductivity of the specimen as well as its ability to give up heat to the walls of the cavity, would influence the heating curve.

There are also conditions where the peaks of two minerals will somewhat overlap, which makes the determination of the area a little difficult. This, however, can be taken care of in most cases by analyzing the curve into its individual components by graphical methods. There are cases, however, where the overlapping is so close that a careful analysis of the curve is impossible.

V. Selection of Peaks for Each Mineral

In Table V, the important peaks for each mineral are listed, which are most useful for quantitative work, together with their areas. The measurement of the amount of any one mineral in a clay is carried out by

TABLE V

Mineral	Selected peaks (°C)	Area* (sq. in.)
Nacrite	625	0.35
Dickite	700	0.42
Hard flint clay	610	0.64
English china clay	610	0.55
Kaolinite	610	0.45
Anauxite	610	0.37
Halloysite	575	0.33
	310	0.03
Allophane	190	1.22
Montmorillonite	875	0.08
Nontronite	860 (ex.)	0.07
Beidellite	700	0.03
	575	0.03
Magnesium clay	800	0.07
Pyrophyllite		0.16
Talc	990	0.17
Muscovite		0.10
Illite	600	0.15
	890	0.01
Potash clay	900	0.08
Gibbsite	360	1.05
Diaspore (corrected)	550	0.61

* These areas can be converted to degrees times seconds by multiplying by 18,300.

first determining which mineral it is and then by measuring the area under the selected peak. This area is then divided by the area given in Table V for the pure material and multiplied by 100, which gives the percentage of the mineral in the mixture.

VI. Examples of Quantitative Analysis of Clays

To bring out this method of computation more clearly, a number of high alumina clays has been analyzed, both thermally and chemically. These clays are particularly suited to thermal analysis because the minerals contained are kaolinite, gibbsite, or diaspore.

In Fig. 9, the thermal curves of the various clays are quite characteristic, giving distinct heat-absorption peaks for kaolin, diaspore, and gibbsite. Even when the peaks for diaspore and kaolin occur together, it is possible to resolve the separate peaks readily to determine their relative areas. Of particular interest is the distinct difference in the thermal curve for the Georgia bauxite and the Missouri diaspore, although the chemical analyses of the two are nearly identical. It would, perhaps, be well to confine the term "bauxite" to mixtures of kaolin and gibbsite and the word "diaspore" to the pure mineral or to its mixture with kaolin.

In Table VI, the areas under the various peaks for these clays are tabulated, and these areas are then

TABLE VI

Material	Area of kaolinite peak (sq. in.)	Area of diaspore peak (sq. in.)	Area of gibbsite peak (sq. in.)	Kaolinite by area (%)	Diaspore by area (%)	Gibbsite by area (%)	Free silica and inert material (%)	Al_2O_3/SiO_2 by area	SiO_2 (%)	Al_2O_3 (%)	TiO_2 (%)	Fe_2O_3 (%)	Combined H_2O (%)	Al_2O_3/SiO_2 (%) (by analysis)
Semiflint (Mo.)	0.45			82			18	0.61	54.5	30.2	1.9	2.2	10.2	0.55
First-grade diaspore (Mo.)		0.55			90		10	7.4	10.9	72.4	3.2	1.1	13.5	6.63
Bauxite (Ga.)	0.37		0.40	67		38		1.95	32.0	65.0				2.02
Burly flint clay (Mo.)	0.39	0.16		71	26		3	1.32	33.8	49.4	2.6	1.8	12.0	1.46
Second-grade diaspore (Mo.)	0.10	0.32		18	53		28	1.41	29.2	53.3	2.7	1.9	12.0	1.82
Hard flint clay (Mo.)	0.64			100*				0.84	46.0	36.0	3.3	1.3	13.5	0.78

* Based on 0.64 area for compacted kaolinite.

converted to percentages by dividing by the area of the peak for the pure material and multiplying by 100. The ratio of alumina to silica is then computed from these percentages, considering the inert matter to be principally silica. This, however, is not entirely correct because it does contain some iron and titanium minerals, and it is not known how much of them is in the crystal lattice of the clay minerals. The Georgia bauxite gives a total kaolinite and diaspore percentage of 105%, which is simply an error in the quantitative measurements.

The chemical analyses of these clays are also given in Table VI, and the alumina-silica ratio is computed from them. In general, it will be found that a surprisingly good agreement exists between the two.

VII. Conclusions

It would seem that the thermal method will be found particularly valuable to identify clay minerals when they occur either in such a finely crystalline state that the petrographic method is of little value or where they occur in mixtures in which the X-ray identification is difficult. The thermal method, of course, will not supplant either of these other methods, but in conjunction with them, it will permit increased ability to determine the clay minerals with certainty. An example of this is the identification of clay minerals in the reaction product of feldspar treated in a pressure chamber. The minerals were finely crystalline and a number of them occurred together so that the X-ray method did not give a very clear picture. The thermal method, however, confirmed the results from the X-ray qualitatively, and gave more or less quantitative results as to the amounts of the various minerals.

The thermal method should also be of considerable value as a control of raw material, particularly in connection with the high alumina clays which are becoming so important in the refractories industry. A thermal curve can be obtained much more quickly and at less expense than a chemical analysis, and in some ways it will give more definite information as to the type of clay. It is believed that a thermal analysis apparatus should be a part of every research and control laboratory dealing with clays.

Acknowledgments

The author wishes to express his thanks especially to Dr. Ross of the U. S. Geological Survey for his interest in this problem, which could not have been carried out without his coöperation in supplying the pure minerals. Thanks are also due to Professor Kerr of Columbia University, who was able to supply the rare mineral, nacrite, for this work, and to Dr. Insley of the Bureau of Standards, who was helpful with advice and suggestions on this problem.

Division of Ceramics
Department of Metallurgy
Massachusetts Institute of Technology
Cambridge, Massachusetts

Reprinted from *Anal. Chem.*, **21**(6), 683–688 (1949)

DIFFERENTIAL THERMAL ANALYSIS

MARJORIE J. VOLD

University of Southern California, Los Angeles, Calif.

Equations are derived which make possible the calculation of heats of transformation from differential heating curves, independent of external calibrations. The rate of restoration of a thermal steady state after a transformation is employed to establish the relation between the differential temperature and the heat absorption producing it. Valid results are obtained for such widely divergent processes as the melting of stearic acid and the vaporization of water, using exactly the same rapid and convenient experimental procedure. A fully automatic self-recording differential calorimeter is described, and the factors influencing the design finally adopted are discussed.

DIFFERENTIAL thermal analysis is essentially a refinement of the classical procedure of studying phase transformations by means of time-temperature records during uniform heating or cooling of the system. It appears to have been first employed by Le Chatelier (*6*) in 1887 but has become popular again only recently. Examples of its employment are given by Kracek (*5*) for sodium sulfate, Partridge, Hicks, and Smith (*8*) for sodium polyphosphates, Hendricks, Nelson, and Alexander (*4*), Grim and Rowland (*3*), Speil *et al.* (*10*) and Norton (*7*) for clays, and Vold (*11-13*) for soaps.

The experimental procedure consists of heating or cooling the sample side by side with an inert reference material in the same furnace, and measuring both the sample temperature and the temperature difference between sample and reference material as a function of time. When a phase change occurs involving absorption or evolution of heat, the temperature difference between reference and sample begins to increase; after the transformation is complete the temperature difference declines again. Thus each transformation produces a peak in the curve of temperature difference against time, from which it should be possible to derive information about the transformation temperature, heat of transformation, and rate of transformation.

This paper describes the construction and operation of a fully automatic differential calorimeter and presents an analysis of the course of the curve of differential temperature against time from which heats of transformation may be calculated.

DIFFERENTIAL HEATING CURVES

In order to analyze the differential heating curve, it is convenient to write down a formal expression for the rate at which heat is transferred into and out of the sample or reference cell.

$$\frac{dq_s}{dt} = K_s(T_w - T_s) + \sigma(T_r - T_s) + \alpha_s(T_e - T_s) \quad (1a)$$

$$\frac{dq_r}{dt} = K_r(T_w - T_r) + \sigma(T_s - T_r) + \alpha_r(T_e - T_r) \quad (1b)$$

Here dq/dt is the rate at which heat is received by the reference material (subscript r) and sample material (subscript s), respectively. K_r and K_s are heat transfer coefficients between the materials and the furnace wall. They are made as nearly identical as possible by choice of reference material and design of cell and furnace. Sigma (σ) is the heat transfer coefficient between the cells, and alpha (α_r and α_s) is the heat loss (chiefly along the thermocouple wires) to the outside environment. T_w, T_r, T_s, and T_e are the temperatures of the furnace wall, reference and sample materials, and external environment, respectively.

For experimental arrangements in which the furnace is a metal block with the materials contained in wells drilled in the block, a linear dependence of the rate of heat transfer on the temperature difference as given in the equations is certainly justified. When the materials are contained in metal capsules or cells suspended in air, and a portion of the transfer is by convection currents in the air, the validity of this assumption is open to question. White (*15*) has shown that heat transfer by convection, when the air flow in cylindrical space (up the walls and down the center of the cylinder) is not turbulent, is proportional to the square of the temperature difference between the walls and the center of the cylinder, but when the diameter of the cylinder is not small compared to its length he shows experimentally that the rate of heat transfer increases less rapidly with temperature difference, and approaches a first-power law.

Next use can be made of the identity

$$\frac{dq}{dt} = \frac{dH}{dt} = \frac{dH}{dt}\frac{dT}{dt} \quad (2)$$

In the case of the reference cell, dH/dt is simple C_r, the heat capacity of the cell plus that of its contents. For the sample it is convenient to segregate the portion of the increased heat content arising from phase change, writing

$$\frac{dq_s}{dt} = C_s \frac{dT_s}{dt} + \Delta H \frac{df}{dt} \quad (3)$$

Here C_s is the heat capacity of the cell plus its contents, while ΔH is the heat of the transformation and df/dt is its time rate of occurrence under the conditions of the experiment, f being the fraction of the sample transformed at any time t.

Every effort is made to have the two materials located symmetrically within the furnace, so that one can write

$$K_s = K_r - \delta K \quad (4a)$$

$$\alpha_s = \alpha_r - \delta \alpha \quad (4b)$$

with the assurance that δK and $\delta \alpha$ are small. The various equations can then be combined to yield an expression for the rate of change of the differential temperature $(T_r - T_s)$ with time which is

$$\frac{d(T_r - T_s)}{dt} = -\frac{K_r + \alpha_r + 2\sigma}{C_s}(T_w - T_s) + \left(1 - \frac{C_r}{C_s}\right)\frac{dT_r}{dt} + \frac{\Delta H}{C_s}\frac{df}{dt} - \frac{1}{C_s}[\delta K(T_w - T_s) - \delta\alpha(T_s - T_s)] \quad (5)$$

At times when $df/dt = 0$—i.e., when the sample is not undergoing a transformation— Equation 5 can be integrated directly, subject to the assumptions that dT_r/dt, $(T_w - T_s)$, $(T_s - T_s)$, and C_r and C_s are independent of time and temperature. In the apparatus here described dT_r/dt is controlled at a constant value. $(T_w - T_s)$ is at least slowly varying. $(T_s - T_s)$ appears with the coefficient $\delta\alpha$ (small), so that its rise with increasing time can be safely neglected.

It is convenient at this point to introduce a more compact nomenclature—i.e., $y = T_r - T_s;$ $A = (K_r + \alpha_r + 2\sigma)/C_s;$ y_1 a given value of y at $t = t_1$, serving as a boundary condition, and

$$y_s = \left[(C_s - C_r)\frac{dT_r}{dt} - \delta K(T_w - T_s) + \delta\alpha(T_s - T_s)\right] \Big/ (K_r + \alpha_r + 2\sigma) \quad (6)$$

With these changes the integrated form of Equation 5 for $df/dt = 0$ is

$$y = y_s(1 - e^{-A(t-t_1)}) + y_1 e^{-A(t-t_1)} \quad (7)$$

It is apparent that y_s is a steady state value of the differential temperature achieved at a sufficiently long time after the initial condition $y = y_1$ at $t = t_1$. At the outset of an experiment $y_1 = 0$ at $t = t_1 = 0$. The differential temperature rises to a value y_w dependent primarily on the difference in heat capacity of the sample and reference materials, the heating rate, and the heat transfer coefficients. After a transformation is complete, the differential temperature again approaches y_s according to the same equation. The constant, A, can thus be evaluated from a plot of $\log (y - y_s)$ versus t. In the new nomenclature, Equation 5 may be written

$$\frac{\Delta H}{C_s}\frac{df}{dt} = \frac{dy}{dt} + A(y - y_s) \quad (8)$$

Graphical or numerical integration of Equation 8 over the period of time during which the transformation is occurring then yields a value for the heat of the transformation.

Any time interval wide enough to include the whole transformation may be taken. When $df/dt = 0$ before the transformation has begun $dy/dt = 0$ and $y = y_s$. When $df/dt = 0$ after the transformation, dy/dt is equal in value and opposite in sign to $A(y - y_s)$. The integral of the second term on the right-hand side of Equation 8 is simply the area under the peak, while that of the first term is $(y_2 - y_1)$ where y_1 and y_2 are the values of y at the beginning and end of the time period chosen.

In practice, in analysing a differential heating curve, it is convenient to plot $(y - y_s)$ against time, beginning at the top of a peak. The points lie on a curve which becomes linear at the end of the transformation and thus yields a value of the time at which the transformation is over. The temperature of the sample at

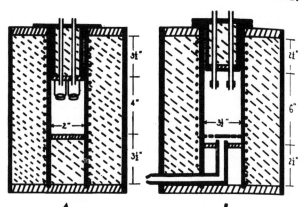

Figure 1. Differential Calorimeter Furnaces

■ Metal (mild steel for A, brass for B)
■ Transite
■ Asbestos magnesia packing
C, C'. Glass tubes for entry of thermocouple leads
D. Glass cell support for sample and reference cells
O. Heating element

this time is of phase significance as the "liquidus" point in a binary system if it can be shown by experiment to be independent of the heating rate.

DIFFERENTIAL COOLING CURVES

If the rate of cooling is controlled, the analysis of differential cooling curves is identical to that given for differential heating curves. If the calorimeter is allowed to cool at its natural rate which is measured experimentally, an entirely similar, though numerically much more complex, analysis can be carried through. The complexity arises from the fact that y_s is now time-dependent. In many cases the rate of transformation on cooling is very small, so that the differential cooling curve exhibits a wide shallow dip whose area beneath the varying base line, y_s, is hard to determine with any accuracy. For these reasons, differential heating curves are much more convenient in the study of phase transformations in most cases.

RATE OF TRANSFORMATION

Equation 8 can be used to calculate df/dt at any time (and hence any T_s) after ΔH has been obtained. At low heating rates, the values probably have little significance, as they are determined primarily by the rate at which heat is made available to the sample. From differential cooling curves, or at high heating rates, values of df/dt may be an interesting property of the system under study.

INHERENT LIMITATIONS OF DIFFERENTIAL THERMAL ANALYSIS

Although sensitive and convenient, this method contains two factors militating against its development into a technique of high precision. One is the assumption of a constant value of the heat capacity of the sample. The second is the assumption that the sample temperature is uniform throughout at each time instant.

The heat capacity of the sample is that of the cell plus that of the transformed portion of the sample plus that of the untransformed portion. The relative proportions of transformed and untransformed sample change during the heating. In practice, if the heat capacity of the cell is deliberately made large, this fluctuation is minor, but sensitivity is sacrificed, because a given heat effect then produces a smaller differential temperature. A solution of Equation 8, taking the variation of C_s into account, has been worked out by successive approximations, but it is too cumbersome a procedure for routine use.

Precise analysis of the effect of the existing temperature gradient in the sample on the calculated heat effect has not been accomplished. Judging from the results obtained, it is not serious enough to vitiate the method. Its chief effect is in the determination of the transformation temperatures rather than on the calculation of heat effects. The differential temperature begins to rise when the outside of the cylindrical sample reaches its transformation point. The center of the cylinder has been shown to be as much as 3 to 4° cooler at a heating rate of 1.5° per minute and dependent on the heating rate and the thermal conductivity of the sample. Reduction of the heating rate (with accompanying loss of sensitivity), measurement of the sample temperature at its outside surface nearest the furnace wall, and various extrapolation procedures all reduce the error involved but can never render it entirely negligible.

APPARATUS

The essential parts of a differential calorimeter are the furnace, devices for containing the sample and reference materials, the accessories for controlling and measuring the heating rate, and the accessories for measuring the temperature difference between sample and reference material.

Considerations underlying the design of a furnace are adequate insulation to prevent undue dissipation of the heat input to the room and erratic temperature gradients within the active space, adequately sized heating elements arranged to minimise the gradients within the active space, and a heat capacity low enough for rapid response in the furnace temperature to changes in heat input, but high enough so that fluctuations in the heat input do not cause rapid fluctuations of the rate of temperature rise about its mean value. In addition, the heat transfer coefficient ($K + 2\sigma$) has to be considered. The smaller this quantity, the larger will be the peak obtained for a given transformation, but the temperature difference will decline more slowly to its steady value, so that successive transformations will be more difficult to resolve.

On the other hand the rate of the transformations probably depends to a certain extent on the rate at which heat can be supplied, so large values of $K(T_w - T_s)$ are desirable. If a metal block is used, the high value of ($K + 2\sigma$) must be compensated by using relatively larger samples and a high heating rate to achieve the same sensitivity. The most desirable type of furnace is believed therefore to consist of an air oven, suitably insulated. The designs given in Figure 1 have proved very satisfactory.

Furnace A consists of two concentric steel tubes 12 inches tall of 0.125-inch wall thickness, the outer 8 inches in diameter, the inner 2 inches. The inner is wrapped with a single layer of 0.125-inch asbestos cloth upon which are wound 20 feet of chromel A resistance wire (No. 22). The coil is more closely wound (4.75 turns per inch) around the upper and lower portions than around the central space (3.125 turns per inch). The space between the two tubes is packed with asbestos magnesia insulation, as is the lower third of the inner tube. The base of the furnace is 0.5-inch Transite board and the top is 3/16 inch Transite board. The central space is closed by a sliding plug of 0.125-inch steel tubing closed at top and bottom by 0.125-inch Transite and filled with asbestos magnesia insulation. Pyrex tubes 6 mm. in outside diameter carry the thermocouple leads through this plug to the measuring instruments, while the glass supports for the sample containers are suspended from its bottom. The lead wires are covered with glass fiber sleeving and completely fill the inlet tubes, so that air currents through these tubes are minimised.

The most desirable shape for the sample is that of a thin cylindrical shell whose inside diameter is just large enough to accommodate the necessary thermocouples. In the present application it was desired also to keep air out and vapor in; consequently a closed cell is necessary. Furthermore, the cell should have as small a heat capacity as possible. Figure 2 shows two designs, both of which have proved satisfactory. Cell A proved very difficult to machine, and the small diameter of the thermocouple well made it difficult to insert and remove the thermocouples without mechanical damage to the junctions. Furthermore, the glass supports are awkward to handle and crack frequently when samples are quenched by immersing the cells in freezing mixtures. Cell B, though larger and heavier, has an improved design of support. The first cells were constructed

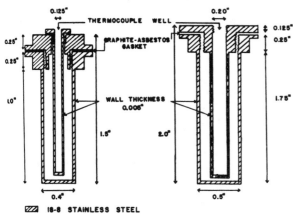

Figure 2. Differential Calorimeter Cells
Made of 18-8 stainless steel

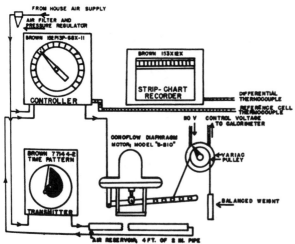

Figure 3. Control Apparatus for Differential Calorimeter

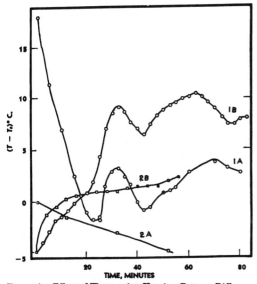

Figure 4. Effect of Fluctuating Heating Rate on Differential Heating Curves

1A. Fluctuating difference between actual and desired temperature during heating $(T - T_c)$
1B. Corresponding curve of temperature difference between reference and sample cells $(T_r - T_s)$ versus time. In both cases average heating rate is 1.5°/min.

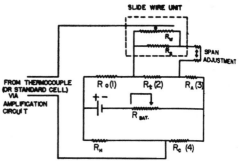

Figure 5. Wiring Diagram of Strip Chart Potentiometer

R_A (3), 253.4 ohms; R_c (4), 509.5 ohms; R_H, 5.00 ohms; slide wire unit, 20 ohms. R_0 (1) was originally 2.474 ohms but was made variable by placing a 20- to 150-ohm resistance in parallel with it. $\frac{1}{2}R_t$ (2) was originally 1.474 ohms but was made variable at from 0.3 to 1.5 ohms, the lower figure giving maximum sensitivity.

of aluminum alloy 24ST, but later stainless steel was employed because the threads machined in the aluminum alloy wore out very rapidly.

The control and measuring assembly is shown in Figure 3.

The heating rate is measured and controlled by means of the Brown Instrument Co. pneumatic time-temperature pattern controller. The desired time-temperature pattern is plotted on an aluminum disk in polar coordinates and the disk is then cut evenly along the curve to give a control cam. This is mounted on the shaft of an electrically driven clock motor. A wheel rolling along the curved surface of the cam controls the position of a pointer on the scale of the circular chart temperature recorder. The furnace temperature is measured by means of a single-junction iron-constantan thermocouple. When this temperature differs from the control temperature as indicated by the pointer, an e.m.f. proportional to the difference is applied to open a valve by which air pressure is transmitted to a Conoflow air-operated motor, which drives the control spindle of an autotransformer (Variac model VMT5, 0 to 130 volts) governing the current passing through the heating elements of the furnace. The drive is adjusted so that 30 volts are applied to the heating element when the applied air pressure is zero, and 130 volts when the applied air pressure reaches its maximum (14 pounds). The range of this model is from 0° to 400° C.

A control knob enables the experimenter to apply a given fraction of the total available air pressure for a given percentage deviation between recording pen and control pointer, based on the full scale. This is known as "per cent throttling." Thus 20% throttling applies full air pressure for an 80° deviation, 10% for 8°, etc. Owing to lag in the furnace, low throttling gives considerable fluctuation of the temperature about its controlled value. With high throttling, the applied current is generally not enough to maintain the desired rate of heating, so that a further control is necessary. This is designated "automatic reset," and provides a continuous, independent increase in the applied air pressure, again proportional to the deviation between actual and control temperature but applied at a controllable rate to overcome the fluctuations that would result from a sudden increase in heating current. The operation of these instruments has been described in greater detail (14).

Selection of the proper control settings for the given furnace is essential. Figure 4 shows two sets of curves for the deviation of

actual and control temperatures, and for fluctuation of the differential temperature between sample and reference cells (in the absence of phase changes) for different values of throttling range and reset control. In set 1 the heating rate, dT_r/dt, oscillates, and the differential temperature, y, oscillates also (as it should according to Equation 7). The corresponding peaks in the curve of differential temperature against time simulate peaks due to phase changes. The better choice of control variable gives rise to the curves in set 2, where the $y - t$ curve is almost a straight line, after time enough has elapsed for the temperature difference to reach its steady state value.

The temperature difference between sample and reference cells is measured by means of a three-junction iron-constantan thermopile, and recorded continuously on the Brown strip-chart potentiometer. The thermocouple junctions are formed of No. 30 wire, spot welded, and insulated from each other by fiber glass sleeving further impregnated with Dow Corning silicone varnish 996. They have been found durable over periods of 2 to 3 months' almost continuous use, and fail ultimately because of breakdown in the insulation.

The potentiometer as supplied has a pen travel of 12 inches (50 scale divisions) per 5 millivolts. As the rated sensitivity is such that full voltage is applied to the balancing motor for an e.m.f. across the thermocouples of 20 microvolts, it is practicable to increase the sensitivity about fivefold by altering the resistance ratio in the arms of the bridge. The wiring diagram for the instrument is given in Figure 5, showing this change and the addition of a further auxiliary variable resistance to vary the position of zero e.m.f. on the scale. With this arrangement both positive and negative values of the temperature difference can be recorded; high values of the sensitivity are used for transformations having small heat effect and lower values for transformations in the same sample having larger heat effects.

Experimental Results. To test both the instrument and the equations described above, five runs were made on a specially pure sample of stearic acid (9) and one on Baker's c.p. analyzed benzoic acid.

The powdered sample was tamped firmly into its calorimeter cell and hung in position from the plug of the colorimeter furnace. The reference cell was then filled with sufficient white mineral oil (Nujol) to give nearly the same calculated heat capacity, and assembly of the calorimeter was completed. After about 20 minutes the initial temperature difference between the cells (due to handling) had disappeared, and the automatic heating and recording had begun. For these relatively large heats of fusion the instrument was operated at a sensitivity of 0.131° per scale

89

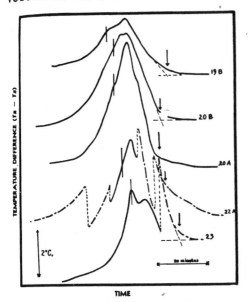

Figure 6. Differential Heating Curves of Stearic and Benzoic Acids

Pantograph reductions of automatic records. Arrows show time at which melting begins. Vertical full lines show time at which calculation indicates that transformation is complete. Vertical dotted lines correspond to breaks in automatic record produced by deliberate test changes in sensitivity of recorder.

division for the measurement of temperature difference as compared with its maximum of 0.089° per scale division. However, in two cases the highest sensitivity was employed at the beginning of the run, and decreased sensitivity later as required to keep the recording pen on the scale.

Pantograph reductions of some of the automatically recorded curves of temperature difference *vs.* time are given in Figure 6.

Temperatures of Transition. The differential temperature, y, begins to rise as soon as the outside of the sample reaches a transition point, (T_m). The inside of the sample at the same instant, T_s, however, is below T_m by an amount dependent on the rate of heating and the thermal conductivity of the sample. The difference has been measured for the reference cell filled with Nujol and found to be of the order of 5° for a heating rate of 1.5° per minute. The thermal conductivity of the powdered sample, however, is not necessarily even approximately equal to that of Nujol nor reproducible from run to run, so no reliable estimate of the difference $(T_s - T_m)$ can be formed to serve as a systematic correction. By the time that the differential temperature is rising rapidly the sample temperature must be somewhat above T_m in order to keep up a steady flow of heat to the transforming portion. The best estimate of T_m is therefore obtained by extrapolating the steeply rising portion of the curve backward to its intersection with the initial base line. As there is a degree of arbitrary choice involved in such an extrapolation, the values obtained are uncertain to about 3° to 4°, with a tendency to be low rather than high. The arrows on Figure 6 show the times at which the sample temperature was taken to be equal to T_m. The values obtained average 4° lower than the known melting points of the samples, as seen in Table I.

Heats of Transition. Heats of transformation were calculated from Equation 8.

Table I. Temperatures and Heats of Fusion of Stearic and Benzoic Acids[a]

Run	Sample Weight, G.	dT_r/dt, °/Min.	T_m, (Extr.) Obsd., °	ΔH calcd., Cal./G.
20 A	1.523	1.95	68	50.5
20 B	1.523	1.45	64	49.4
22	1.300	1.90	67	49.6
23	1.300	1.75	63	47.3
44[b]	1.077	1.70	67	48.0
Mean			65	49.2
19 C	1.306	1.40	118	31.4

[a] M.P. of stearic acid, determined directly, was 69°. ΔH_f (8) is 47.6 cal./g. For benzoic acid accepted m.p. is 122° and ΔH_f is 33.9 cal./g. (1).
[b] Run made using model B furnace and cells of Figures 1 and 2. All others with model A furnace and cells.

The value of A was first determined by plotting $\log (y - y_s)$ against time. Two such plots are shown in Figure 7. The points fall on good straight lines after the transformation is over, at the times indicated on the figure. At the end of the transformation, so determined, the base line y_s, was assumed to have its final value—i.e., c_s equal to the heat capacity of the cell plus that of the melted sample. At the beginning of the transformation—i.e., the point selected as T_m—y_s was assumed to have its initial value (c_s equal to that of the cell plus that of the unmelted sample). The base line under the peak was taken as a straight line between these two points.

Figure 8 shows a typical run replotted to show the actual magnitude of T_r, T_s, and y. The base line and area taken in evaluating ΔH are shown. The results, given in Table I, are obviously not of high precision, but when it is remembered that they are absolutely independent of any empirical calibrations, they give an assurance of validity to the theoretical analysis.

In one experiment, employing technical stearic acid which melted visually over a range from 56° to 60° C., two runs were made at heating rates of 0.5° and 1.5° per minute. Heat effects of 48.5 and 47.4 calories per gram were obtained, respectively.

To guard against the remote possibility that the agreement between observed and theoretical values might be fortuitous, the same method was applied to an experiment in which powdered,

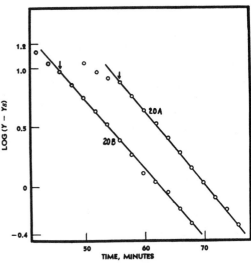

Figure 7. Plot for Determining Constant A

A governs rate at which a thermal steady state is re-established after a transformation occurs. Arrows mark time at which transformation is finished; subsequent points lie on straight lines

688

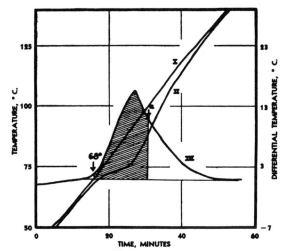

Figure 8. Total and Differential Heating Curves for Stearic Acid and Nujol

I. Heating curve for Nujol cell. II. Heating curve for stearic acid cell. III. Differential heating curve. At 68° melting begins. *a* is time at which calculation shows that melting is complete. Shaded area is that taken into account in calculating ΔH.

oven-dried (150°) silica was used as reference material and the sample cell was packed with powdered glass and filled with water, which was allowed to vaporize through a hole drilled in the side

of the cell as it was heated at a rate of 1.5° per minute. A value of 500 calories per gram for the heat of vaporization of water was obtained. This type of experiment is of potential value in studying the thermal dehydration of clay minerals and similar problems.

LITERATURE CITED

(1) Andrews, D. H., Lynn, G., and Johnston, J., *J. Am. Chem. Soc.*, **48**, 1274 (1926).
(2) Garner, W. D., Madden, F. C., and Rushbrooke, J. E., *J. Chem. Soc.*, **1926**, 2941.
(3) Grim, R. E., and Rowland, R. A., *Am. Mineral.*, **27**, 746-61, 801-18 (1942).
(4) Hendricks, S. B., Nelson, R. A., and Alexander, L. T., *J. Am. Chem. Soc.*, **62**, 1457 (1940).
(5) Kracek, F. C., *J. Phys. Chem.*, **33**, 1281 (1929).
(6) Le Chatelier, H., *Z. physik. Chem.*, **1**, 396 (1887).
(7) Norton, F. H., *J. Am. Ceram. Soc.*, **22**, 54 (1939).
(8) Partridge, E. P., Hicks, V., and Smith, G. W., *J. Am. Chem. Soc.*, **63**, 454 (1941).
(9) Philipson, J. M., Heldman, M. J., Lyon, L. L., and Vold R. D., *Oil & Soap*, **21**, 315 (1944).
(10) Speil, S., Berkenhamer, L. H., Pask, J. A., and Davies, B., U. S. Bur. Mines, *Tech. Paper 664* (1945).
(11) Vold, R. D., *J. Am. Chem. Soc.*, **63**, 2915 (1941).
(12) Vold, R. D., Grandine, J. D., 2nd, and Vold, M. J., *J. Colloid Sci.*, **3**, 339 (1948).
(13) Vold, R. D., and Vold, M. J., *J. Phys. Chem.*, **49**, 32 (1945).
(14) Werey, R. B., "Instrumentation and Control in the Oil Refining Industry," Philadelphia, Pa., Brown Instrument Co., 1941.
(15) White, W. P., "Modern Calorimeter," pp. 74-6, New York, Reinhold Publishing Corp., 1928.

RECEIVED June 30, 1948. Presented before the Division of Physical and Inorganic Chemistry at the 113th Meeting of the AMERICAN CHEMICAL SOCIETY, Chicago, Ill. Work supported in part by a grant from the Office of Naval Research, Contract No. N6-onr-238-TO2; NRO67057.

7

Reprinted from *J. Am. Ceram. Soc.*, **34**(7), 221–224 (1951)

Temperature Distribution During Mineral Inversion and Its Significance in Differential Thermal Analysis

by **HAROLD T. SMYTH**

New Jersey Ceramic Research Station, Rutgers University, New Brunswick, New Jersey

Temperature distributions have been calculated in a thin slab of material for which the temperature of the outside faces was raised at a uniform rate. Calculations cover the entire period during which the material is transformed from one form to another with an endothermic reaction. Application of these calculations to the interpretation of thermal analysis data is discussed.

I. Introduction

IN A usual form of differential thermal analysis apparatus[1] a metal block is placed in an electrically heated furnace. The block has in it two cavities, equal in size and shape, and, as far as is possible, symmetrically situated with respect to the block and the furnace. One cavity contains a material whose thermal properties are being investigated; and the other, a reference material, such as alumina, whose thermal properties are reasonably well known and which does not itself show any polymorphic inversions or chemical reactions within the temperature range covered.

Located at the center of each cavity is a thermocouple. Proper electrical connection of these thermocouples enables the differential temperature between them to be recorded by a sensitive potentiometer. As the temperature of the furnace is raised at a uniform rate, this differential temperature is plotted against time. On the same chart is plotted another temperature which may be the temperature of the furnace atmosphere, the temperature of the metal block, or, more usually, the temperature of the center of the reference material.

If at some temperature a transformation, such as a polymorphic inversion, a loss of combined water, or any other chemical reaction, occurs in the material being investigated, and if this transformation requires heat or evolves heat, there will be a "bump" in the differential temperature recording. The bump will show on one or the other side of zero, depending on whether the transformation is endothermic or exothermic.

Received May 22, 1950; revised copy received September 15, 1950.

The author is research professor at Rutgers University.

[1] F. H. Norton, Refractories, 3d ed. Ch. V, pp. 76–93. McGraw-Hill Book Co., Inc., New York, 1949. 782 pp.; *Ceram. Abstracts*, **1950**, July, 145a.

For reasons which will become clear in the analysis to follow, these bumps or peaks are not always instantaneous, but may spread over quite a range of temperature. It was for the purpose of helping to interpret the exact meaning of the shape and location of these peaks that the calculations were undertaken. Only endothermic reactions were considered. The shape of exothermic peaks may be quite different since enough heat may be generated to trigger a more or less instantaneous reaction.

For reasons of mathematical simplicity the shape of the cavity was taken as the space between two infinite parallel planes with the thermocouple located midway between the planes. The temperatures of these two bounding planes were equal and were assumed to be increasing at a uniform rate. At each instant the boundary surfaces of the test specimen and reference specimen were equal. Although these shapes are physically unrealizable, the qualitative conclusions arrived at will apply to cylindrical or other more practical shapes.

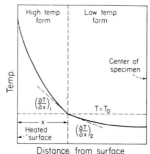

Fig. 1. Temperature gradients close to boundary between low- and high-temperature forms.

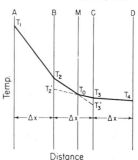

Fig 2. Diagram used in computing the movement of the phase boundary as inversion proceeds.

II. Temperature Distribution in Reference Specimen

In both samples the heat flow is one dimensional, being always perpendicular to the boundary planes. If distance in the direction of flow is denoted by x, then the temperature distribution in the reference sample and in the test sample, until the temperature of transformation is reached, is governed by the well-known heat-flow equation

$$\frac{\partial^2 T}{\partial x^2} = \frac{1}{a}\frac{\partial T}{\partial t} \qquad (1)$$

where T is the temperature, t is the time, and a is the diffusivity of the material. If the zero of x is taken midway between the boundary walls, and if the walls are brought up at a uniform rate of α degrees per second, then the material will sooner or later arrive at a state where the temperature distribution is given by

$$T = T_c + \alpha t + \frac{1}{2}\frac{1}{a}\alpha x^2 \qquad (2)$$

where T_c is a constant of the dimensions of a temperature. It is easy to verify that the distribution (2) actually satisfies the heat-flow equation (1), and that it represents a state in which the boundary walls (or, indeed, any point in the sample) are increasing uniformly in temperature.

It is assumed in the calculations that heating in both the reference and test samples has been going on for a time sufficiently long to allow the steady-state condition of equation (2) to be established in both samples.

Using equation (2), it is very simple to calculate the temperature distribution in the reference sample at any instant.

III. Temperature Distribution in the Test Specimen

It is assumed that at a certain temperature T_0 a reaction in the test specimen takes place which requires, at that temperature, L cal. of heat to convert 1 gm. of the specimen from the original state to the completely reacted state. This reaction may be a polymorphic inversion or any chemical reaction which takes place at a definite temperature.

Solutions of equation (1) which would satisfy the boundary conditions at the surface of the sample and the conditions at the interface of the unreacted and the reacted material could not be found. It was therefore necessary to use an approximate method based on that given by Schack[2] and attributed to Schmidt. In this method the slab is divided by equi-

distant planes separated from each other by a distance of Δx cm., and the distribution is calculated at successive time intervals, each of which is called Δt seconds. If Δx and Δt are so chosen that they satisfy the relation

$$\Delta t = \frac{(\Delta x)^2}{2a} \qquad (3)$$

the temperature at any one of these dividing planes at any instant will be the mean of the temperatures of its two immediate neighbors at an instant Δt seconds earlier. Thus, if the distribution at any instant in a uniform material and the boundary conditions are given, the distribution at any succeeding instant can be calculated.

The procedure used in calculating the activity at the interface between the reacted and unreacted material is given. For simplification it was assumed that the density ρ, specific heat C, and thermal conductivity k of the unreacted material were the same as those of the reacted material. As already stated L cal. of heat is required to convert 1 gm. of the unreacted material to the completely reacted state at temperature T_0, the temperature at which the reaction takes place.

In Fig. 1 there is shown the temperature distribution close to the boundary between the unreacted and the reacted material if a chemical reaction is involved, or between the high and low forms of the material if a polymorphic inversion is being considered.

The heat flowing into the boundary layer from the left side is

$$- k \left(\frac{\partial T}{\partial x}\right)_1 \text{ cal./sq. cm./second}$$

and the heat flowing out to the right side is

$$- k \left(\frac{\partial T}{\partial x}\right)_2 \text{ cal./sq. cm./second}$$

The net gain in time Δt seconds is therefore

$$\left[- k \left(\frac{\partial T}{\partial x}\right)_1 + k \left(\frac{\partial T}{\partial x}\right)_2\right] \Delta t \text{ cal./sq. cm.}$$

This is enough to convert a layer of thickness δx from the low to the high form where

$$\delta x = \frac{1}{\rho L}\left[- k \left(\frac{\partial T}{\partial x}\right)_1 + k \left(\frac{\partial T}{\partial x}\right)_2\right] \Delta t \qquad (4)$$

Now Δt has been chosen such that

$$\Delta t = \frac{(\Delta x)^2}{2a} \qquad (3)$$

[2] A. Schack, Industrial Heat Transfer, translated by Hans Goldschmidt and Everett P. Partridge. John Wiley & Sons, Inc., New York, 1933. 371 pp.; *Ceram. Abstracts*, **13** [3] 64 (1934).

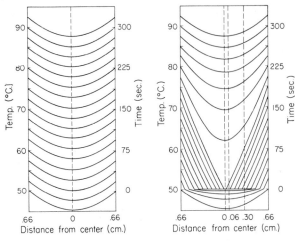

Fig. 3. Temperature distributions in the reference material at intervals of 18.75 seconds.

Fig. 4. Temperature distributions in the test material at intervals of 18.75 seconds.

Hence

$$\delta x = \frac{1}{\rho L}\left[-k\left(\frac{\partial T}{\partial x}\right)_1 + k\left(\frac{\partial T}{\partial x}\right)_2\right]\frac{(\Delta x)^2}{2a} \quad (5)$$

or

$$\frac{\delta x}{\Delta x} = \frac{1}{\rho L}\left[-k\left(\frac{\partial T}{\partial x}\right)_1 + k\left(\frac{\partial T}{\partial x}\right)_2\right]\frac{\Delta x}{2a} \quad (6)$$

But

$$a = \frac{k}{\rho C} \quad (7)$$

Therefore

$$\frac{\delta x}{\Delta x} = \frac{C}{2L}\left[-\left(\frac{\partial T}{\partial x}\right)_1 + \left(\frac{\partial T}{\partial x}\right)_2\right]\Delta x \quad (8)$$

This gives the distance the boundary moves in time Δt expressed as a fraction of Δx, the distance between neighboring planes.

The actual method of evaluating the right side during the calculations is given. In Fig. 2 are shown four of the equally spaced planes, A, B, C, and D, by means of which the sample is divided for calculation purposes. These have been so chosen that the reaction interface M is between B and C. It is assumed that, by previous calculations, at a certain instant of time the temperatures T_1, T_2, T_3, and T_4 at A, B, C, and D, respectively, are known, as well as the position of M between B and C. That is, the distances BM and MC are known. A method for calculating the temperatures and the position of M at the succeeding instant of time Δt seconds later is then needed.

Equation (8) gives the movement of the interface in time Δt expressed as a fraction Δx which, in the practical calculations, is the most convenient way to express it. It is necessary, however, to modify this to express $\left(\frac{\partial T}{\partial x}\right)_1$ and $\left(\frac{\partial T}{\partial x}\right)_2$ as ratios of differences, rather than in their differential form.

For $\left(\frac{\partial T}{\partial x}\right)_1$, $\frac{T_0 - T_2}{BM}$ is used.

For $\left(\frac{\partial T}{\partial x}\right)_2$, $\frac{T_3 - T_0}{MC}$ is used.

Therefore equation (8) can be rewritten in the form

$$\frac{\delta x}{\Delta x} = \frac{C}{2L}\left(\frac{T_2 - T_0}{BM} - \frac{T_0 - T_3}{MC}\right)\Delta x \quad (9)$$

In the calculations it is more convenient to express BM and MC as fractions of Δx.

Therefore $\frac{BM}{\Delta x} = f$ and $\frac{MC}{\Delta x} = g$ is used, the quantities f and g being known. One can therefore write for equation (9)

$$\frac{\delta x}{\Delta x} = \frac{C}{2L}\left(\frac{T_2 - T_0}{f} - \frac{T_0 - T_3}{g}\right) \quad (10)$$

This shows how, if the position of M at any instant of time and the temperatures, T_2 and T_3, are known, the position of M at Δt seconds later can be calculated.

The temperature of B at the succeeding instant of time was taken to be the mean between T_1 and the temperature of C if the slope is extrapolated at B on to C as shown at T_3' on the drawing. Similarly the temperature at C was taken as the mean between T_2' and T_4. The temperatures at the other planes were calculated as in the regular Schmidt[2] method, the outside boundaries being brought up at the chosen uniform rate.

IV. Numerical Values Chosen

The numerical values chosen were as follows:

$$\Delta x = 0.06 \text{ cm.}$$
$$k = 0.003 \text{ cal./sq. cm./second/°C./cm.}$$
$$\rho = 2.0 \text{ gm./cu. cm.}$$
$$C = 0.25 \text{ cal./gm./°C.}$$
$$\Delta t = 0.3 \text{ second}$$
$$L = 15 \text{ cal./gm.}$$
$$T_0 = 50°C.$$

Rate of rise of outside $= 8°C./minute$
Total thickness of slab $= 22 \Delta x = 1.32$ cm.

The reference specimen was taken as identical with the test specimen except that it showed no inversion.

The physical properties of the test specimen above the inversion were taken as the same as those below the inversion.

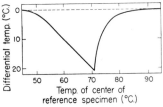

Fig. 5. Differential temperature plotted against the temperature at the center of the reference material.

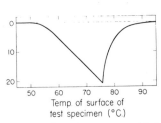

Fig. 6. Differential temperature plotted against the surface temperature of the test or the reference sample.

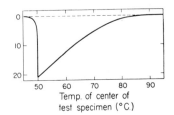

Fig. 7. Differential temperature plotted against the temperature at the center of the test specimen.

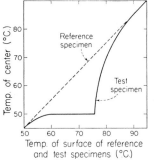

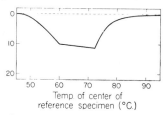

Fig. 8. Temperature at the center of the reference and the test specimens plotted against the surface temperature of either.

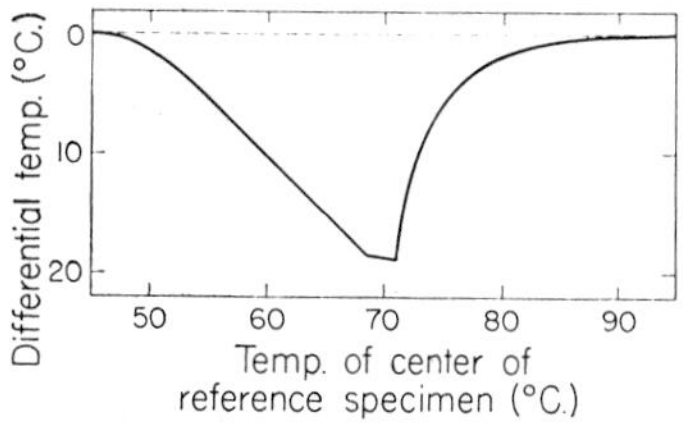

Fig. 9. Effect of having thermocouple 0.06 cm. from center of test specimen.

Fig. 10. Effect of having thermocouple 0.30 cm. from center of test specimen.

V. Results

Temperatures were calculated for the reference specimen and the test specimen at intervals of 0.3 second, starting at a time when the test specimen was completely below the inversion temperature and continuing until it had, within the accuracy of the calculations, recovered from the effects of the inversion and resumed its original parabolic temperature distribution. Selected distribution curves are shown for the reference specimen in Fig. 3 and for the test specimen in Fig. 4.

The distribution curves for the reference sample are at all instants of time identical parabolas, as one can see from equation (2). The test specimen curve starts out as a parabola; but as the outer layers reach the inversion temperature, so much heat is required to change them from the low to the high form that the heat supply to the interior is seriously interrupted. The rate of heating at the center, therefore, slows up quite a while before the material at the center has reached the inversion temperature.

As soon as the rate of heating at the center starts slowing up, the pen tracing the differential temperature will start to deviate from its zero line. At this point neither the center of the test specimen nor the center of the reference specimen is at the inversion temperature, and one must be rather careful not to attach too much importance to this point of initial deviation.

It will be seen in Fig. 4 that when the inversion is complete the distribution curve comes to a sharp point at the center. Since such a sharp point corresponds to a very high value of the second derivative (corresponding roughly to the curvature), one would correctly deduce from equation (1) that there would be a rapid rise in temperature at the center. This rapid rise at the center gradually slows down until, after more than 300 seconds, the test specimen has regained its original parabolic distribution and has caught up with the reference specimen bringing the differential temperature back to zero.

In order to illustrate more clearly how these calculations relate to actual differential thermal analysis curves several curves were plotted.

In Fig. 5 the differential temperature is plotted against the temperature at the center of the reference specimen. The curve departs from zero some distance below the inversion temperature (50°C.) and reaches its peak some 20°C. above the inversion temperature.

In Fig. 6 the differential temperature is plotted against the temperature of the surface of the test (or reference) specimen. This latter would correspond to the temperature of the metal block in which the sample cavities are located. This curve actually starts its deviation from the zero line at

the inversion temperature; and if this point of initial deviation could be accurately determined, it would have useful significance on such a curve.

In Fig. 7 the differential temperature is plotted against the temperature at the center of the test specimen itself. If by proper instrumental ingenuity this arrangement were used, the location of the peak of the curve would be a good indication of the inversion temperature.

As an additional illustration of how heating progresses at the center of the specimens, Fig. 8 was plotted showing the temperatures at the center of the reference and test samples plotted against the surface temperature of either. The center of the reference sample heats uniformly, always lagging 4.84°C. below the temperature of the surface. The increase in the temperature of the center of the test specimen slows down and finally effectively stops at the inversion temperature until the inversion is complete, at which time it heats very rapidly finally catching up with the reference sample. The vertical distance between the two curves at any point gives the differential temperature.

VI. Effect of Lack of Symmetry

If the thermocouple in the test sample had been 0.06 cm. from the center, instead of at the center, the differential thermal analysis curve would be that shown in Fig. 9. If, in a rather extreme case, it had been 0.30 cm. from the center, the curve would be that shown in Fig. 10. The inversion occurs now, not at the peak of the curve, but at the first break. From there on the differential rises slowly until the material is completely inverted, after which it drops fast. White[3] presented somewhat similar considerations semi-quantitatively in connection with discussions of the precision of melting-point determinations.

VII. Conclusions

It is shown that some rather simple mathematical calculations on the mechanism can be of help in the interpretation of differential thermal analysis curves where an endothermic reaction or polymorphic inversion is involved.

If the differential temperature is plotted against the surface temperature of the sample, the point of initial departure from the straight line corresponds to the inversion temperature.

If the differential temperature is plotted against the temperature of the center of the sample while the outside is being heated uniformly, the peak of the curve corresponds to the inversion temperature.

If the differential temperature is plotted against the temperature of the center of the reference specimen, neither the point of initial departure from the straight line nor the peak of the curve corresponds to the inversion temperature.

Acknowledgment

This work was carried out in connection with a project for the investigation of high-temperature thermal and other physical properties of ceramic materials sponsored by the Office of Naval Research. The author would like to express his thanks to Robert B. Sosman, Myril C. Shaw, and Harriet R. Wisely for reading the manuscript and offering useful suggestions.

[3] W. P. White, "Melting Point Determination," *Am. J. Sci.,* **28,** 453 (1909).

8

Reprinted from *Am. Mineralogist*, **36**, 615–621 (1951)

THE DIFFERENTIAL THERMAL ANALYZER AS A MICRO-CALORIMETER*

MARK WITTELS, *Massachusetts Institute of Technology, Cambridge, Mass.*

ABSTRACT

Calorimetric measurements of very small magnitude are made possible by a sensitive method of differential thermal analysis. The increased sensitivity of the technique is due to the use of a low-mass radiation-type furnace operated in a vacuum, and at high heating rates. Calibration of the instrument was accomplished by employing the well-known reaction of calcite dissociation as a standard.

INTRODUCTION

The technique of differential thermal analysis has been largely confined to the investigation of qualitative thermal characteristics of various mineral species. In particular, the clay minerals have been widely studied, and thermal analyses for these minerals are familiar to mineralogists, ceramists, and soil scientists.

Hitherto, the differential thermal analyzer has not been developed into a precise calorimetric tool. This paper describes a successful calibration of a differential thermal analyzer, so that it is able not only to detect, but also to measure heat changes as small as ten millicalories in magnitude. This device might well be termed a micro-calorimeter, and, as such, should be of interest to all physical scientists.

APPARATUS

The apparatus used in this investigation was essentially the same as that described by Whitehead and Breger.[1] The controller and recorder were unchanged, but the mass of the entire furnace housing was reduced so that its design is that shown in Fig. 1. As a result of these changes, the furnace was brought under closer temperature control, especially at high heating rates, and thermal changes, as recorded by the thermographic curves, became highly magnified.

THEORY OF METHOD

Cohn,[2] Macgee,[3] and Shorter[4] have described calibrations of differential thermal analyzers for use on ceramic bodies. All of these calibrations,

---

* An investigation as part of a Ph.D. thesis by the author.

[1] Whitehead, W. L., and Breger, I. A., Vacuum differential thermal analysis: *Sci.*, **111**, 279–281 (1950).

[2] Cohn, W. H., The problem of heat economy in the ceramic industry: *J. Am. Cer. Soc.*, **7**, 475–488 (1924).

[3] Macgee, A. E., The heat required to fire ceramic bodies: *J. Am. Cer. Soc.*, **9**, 206–247 (1926).

[4] Shorter, A. J., The measurement of heat required in firing clays: *Trans. Brit. Cer. Soc.*, **47**, 1–22 (1948).

however, are on a macro scale, and entail a comparatively long and cumbersome procedure.

The method of calibration described here is simple. Briefly, it involved heating carefully weighed samples of a known reactive substance at a

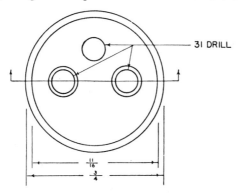

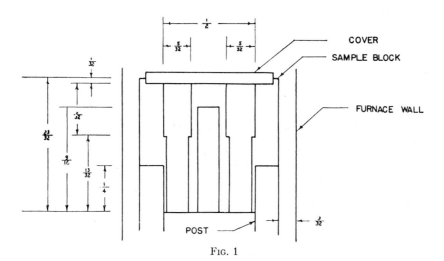

Fig. 1

constant rate, through its reaction temperature range, and then relating the energy changes with the corresponding responses (areas) recorded by the thermographic curves.

The theoretical basis upon which this calibration was developed, was formulated by Speil[5] and is briefly reviewed here with the aid of Fig. 2,

[5] Speil, S., Applications of thermal analysis to clays and aluminous materials: *U. S. Dept. of Int.—Bur. of Mines*, **R.I. 3764** (1944).

97

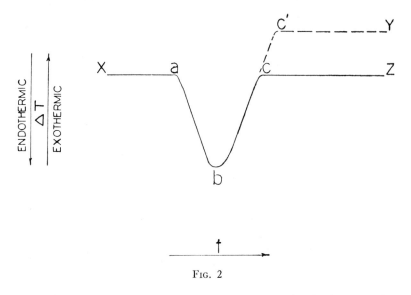

FIG. 2

a sample thermographic curve showing a reaction beginning at point *a*, and continuing towards point *c*.

$$(\text{Area}) \; abc \propto \int_a^c \Delta T dt = \frac{M(\Delta H)}{gk} \tag{1}$$

where,

M = mass of the reacting specimen
ΔH = specific heat of reaction
g = geometrical shape constant
k = thermal conductivity of the reacting specimen.

Equation (1) neglects the temperature gradient in the sample as well as some insignificant differential terms, and therefore is a close approximation. Transposing terms we get

$$(\text{Area}) \; abc \propto gk \int_a^c \Delta T dt = M\Delta H = Q \tag{2}$$

or,

$$(\text{Area}) \; abc \propto Q \tag{3}$$

where

Q = heat of reaction.

As an approximation, then, the area is a linear function of the heat of reaction as expressed by equation (3).

CALIBRATION

The reactive substance chosen for the calibration was $CaCO_3$ (calcite)

98

which follows the reaction

$$CaCO_3 + Q \rightarrow CaO + CO_2 \uparrow$$

beginning at a temperature of about 630° C. when heated statically. The values[6] for Q in the above reaction are very large and permitted samples smaller than the thermocouple head to be used. Later, the significance of these minute samples will be shown.

The dissociation of the $CaCO_3$ samples at a heating rate of 30° C per minute in an evacuated furnace produced the thermographic curves shown in Fig. 3, and the relation between the area and the heat of reaction, as described above, gave the linear plot of the same figure. The validity of equation (3) is demonstrated by the experimental evidence displayed by the linear plot. Table 1 is a summation of the experimental and calculated data of the calibration.

TABLE 1

Reactive Sample			Thermographic Response (Area in Sq. In.)
Sample	Weight (milligrams)	Heat of Reaction (millicals.)	
A	0.30	123	0.294
B	0.40	165	0.363
C	1.00	410	0.900
D	1.50	614	1.385
E	2.10	853	1.910
F	2.50	1015	2.283
G	3.00	1215	2.721

Measurements of areas were made with a planimeter and checked by means of a simple grid. The measurement of the areas, however, was the crux of the entire calibration since the area is not always clearly defined. In Fig. 2 the thermographic base-line X-Z is not linear in the general case. It is obvious that when a large sample is tested in the differential thermal analyzer, a major reaction is usually accompanied by a sharp change in the thermal conductivity and the specific heat capacity of the specimen. This results in a thermal gradient between the differential thermocouples that gives rise to a non-linear base-line. In Fig. 2 point c is often displaced towards c', and the curve continues toward Y. In addition, slow heating rates (12° C. per minute) decrease the slope of ab

[6] Smyth, F. H., and Adams, L. H., The system, calcium oxide-carbon dioxide: *Jour. Am. Chem. Soc.*, 45, 1167–1184 (1923).

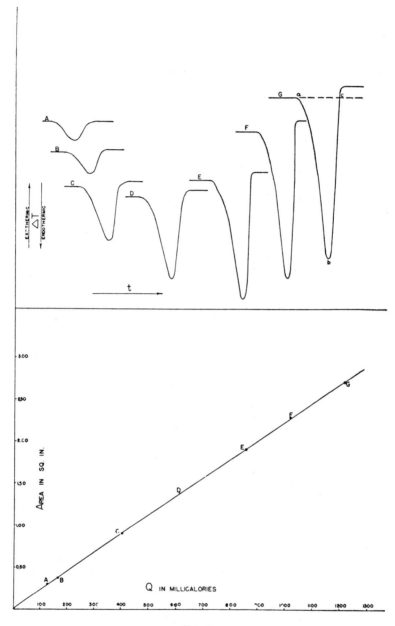

FIG. 3

100

so that point *a* becomes unidentifiable. The work of Norton[7] and that of Berkelhammer[8] emphasize this fact clearly.

The difficulties of the non-linear base-line are largely obviated by reducing the mass of the furnace housing, employing high heating rates, (30° C. per minute), and testing small samples with large ΔH. The reaction point *a*, Fig. 2, is sharply defined as the slope *ab* steepens, and the base-line deviation is reduced to a small magnitude. The method of area measurement adopted overcame this slight deviation as shown in Fig. 3, curve *G*. The base-line is merely extended from point *a*, the reaction origin, to point *c*, where the reaction is completed for all intents and purposes. The area measured is that enclosed by *abc*.

A standard procedure of analysis was established for this investigation and strictly adhered to. The use of uniform thermocouples, precise centering within the furnace wall, careful particle sizing, and uniform packing of samples are the important rules to be followed.

DISCUSSION

The results of this experiment justify the belief that a low-mass radiation-type furnace can render important service in investigations of this nature. The sensitivity of the instrument described was found to be 30 to 100 times that of instruments described in the recent literature,[5,7,8,9,10] and it should be mentioned that no amplification was needed to achieve these magnified thermographic curves.

The apparatus described here is an excellent tool in the range 300° C– 1100° C. Below 300° C it cannot be used as a quantitative instrument because the mass of the furnace housing is too large to permit sensitive response and control at temperatures this low, especially at high heating rates which have been shown to be desirable in the higher temperature ranges. It is suggested that further diminution of the furnace housing to one having lower inertia may make calorimetric studies over a wider temperature range possible. The possible applications of such a quantitative calorimetric tool are numerous.

[7] Norton, F. H., Critical study of the differential thermal method for the identification of the clay minerals: *J. Am. Cer. Soc.*, **22**, 54–63 (1939).

[8] Berkelhammer, L. H., Differential thermal analysis of quartz: *U. S. Dept. of Int.— Bur. of Mines*, **R. I. 3763** (1944).

[9] Faust, G. T., Thermal analysis studies on carbonates, I. Aragonite and calcite: *Am. Mineral.*, **35**, 207–224 (1950).

[10] Beck, C. W., An amplifier for differential thermal analysis: *Am. Mineral.*, **35**, 508–524 (1950).

Acknowledgments

The author is indebted to Professor M. J. Buerger for critically review-ing the manuscript. To Professors W. L. Whitehead and H. W. Fairbairn go the writer's gratitude for helpful advice and full use of the equip-ment.

A special note of thanks is due Mr. John Solo who gave much technical assistance.

Manuscript received Nov. 3, 1950

Reprinted from *J. Am. Ceram. Soc.*, 35(3), 76–82 (1952)

Apparatus for Differential Thermal Analysis Under Controlled Partial Pressures of H₂O, CO₂, or Other Gases

by ROBERT L. STONE

Engineering Experiment Station, The Ohio State University, Columbus, Ohio

An apparatus was designed so that an atmosphere (a continuously flowing atmosphere) containing any partial pressure of the active vapor such as H_2O, CO_2, or SO_3, could be maintained around the particles of a mineral powder while the differential thermal analysis was being run. The sample holder and its attendant parts were mounted on a movable vertical shaft so that the assembly could be lowered into or raised out of the furnace. The gas was carried to the powders under very slight pressure through $1/4$-in.-diameter porcelain tubes, called the injector, which fit snugly into the sample holder cavities. The bottoms of the cavities were the ends of two 4-hole porcelain tubes which carry the thermocouples and also support the sample holder. Two types of sample holders were used, (1) Inconel tubes, $1/4$ in. in outside diameter and $3/4$ in. long with a wall thickness of 0.008 in., and (2) a solid Inconel block $7/8$ in. in diameter and $3/4$ in. long with two symmetrically placed holes 0.234 in. in diameter drilled parallel to the length. These sample holders accommodate about 0.15 gm. of clay or 0.20 gm. of alumina as the reference material. There were four 28-gauge thermocouple wires in each cavity. In the sample cell, two wires were one leg of the differential couple and the other two wires were dummies to balance the cell. In the reference cavity were the other leg of the differential couple and the measuring couple. All four wires in each cell were fused into a single bead and the wires were spread as far apart as possible, forming a sturdy unit. The furnace was $4^{1}/_{2}$ in. square and 5 in. long with an Alundum tube core, $1^{1}/_{16}$ in. in outside diameter, on which 45 ohms of 28-gauge Nichrome V wire was noninductively wound. A motor-driven Powerstat was used to control the temperature rate at 16°C. per minute.

I. Introduction

FUNDAMENTALLY, the differential thermal method of analyzing powdered materials, such as clays, is a method in which the temperature and the temperature interval at which endothermic and exothermic chemical reactions take place are used for mineral identification. Each mineral has characteristic intensities and temperatures of these reactions which serve as a means of qualitative and semiquantitative determination. The differential thermal method is applicable to any process where heat is either absorbed or evolved, but it was not until 1887 that Le Châtelier[1] demonstrated that it could be applied to heat-absorbing reactions like the dehydration of clays.

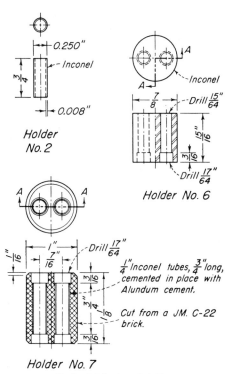

Fig. 1. Details of sample holders.

This paper deals with the design of apparatus for differential thermal analysis in which the composition of the atmosphere around the particles of powder being studied can be controlled both before and during a reaction. The equipment presented in this paper is far from perfect, but it is hoped that the presentation of the fundamental principles will serve as a guide to those interested in this field.

II. Review of Literature

The literature on differential thermal analysis of clay minerals is abundant, and several papers, such as the one by Kerr, Kulp, and Hamilton[2] on reference clay specimens, give excellent reviews of literature. However, there are only

[1] H. Le Châtelier, "Action of Heat on Clays," *Bull. soc. franç. minéral.*, **10**, 204–11 (1887); *Z. physik. Chem.*, **1**, 396 (1887).
[2] P. F. Kerr, J. L. Kulp, and P. K. Hamilton, "Differential Thermal Analysis of Reference Clay Mineral Specimens," A.P.I. Project 49, Preliminary Report No. 3, Columbia Univ., May, 1949.

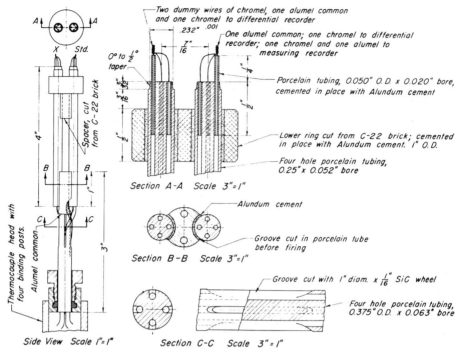

Fig. 2. Detail of thermocouple assembly No. 8.

two published papers dealing with the application of atmosphere control to differential thermal analysis.

In 1950 Saunders and Giedroyc[3] applied an improvised atmosphere control to thermal analysis of powdered materials. The sample holder was machined from a block of lightly sintered Alundum cement. Two cavities, $5/16$ in. in diameter and $1/2$ in. deep, into which the standard and the unknown were packed, were drilled into the block before sintering. This assembly was then placed on top of a hollow tubular shape cut from a silica brick. On top of the crucible was placed another cylinder cut from a silica brick. This entire unit was placed in a 2-in.-diameter fused quartz tube through which the desired gas, nitrogen, was circulated. The amount of gas flowing through the specimens depended entirely upon the permeability of the packed specimen compared with the permeability of the silica brick, as well as on the contact space between the silica parts and the quartz tube. Quite satisfactory results were obtained by using pure gases with this apparatus, but no attempt was made to explain the temperature effects in the light of thermodynamics.

Rowland and Lewis[4] published a paper in 1951 on furnace atmosphere control in differential thermal analysis. They used the conventional horizontal-tube furnace through which a long porcelain combustion tube was inserted. The conventional nickel-block specimen holder assembly was placed in this tube, and the ends of the tube were closed with cylinders of insulating firebrick. The desired gas was introduced into the combustion tube through glass tubing inserted

[3] H. L. Saunders and V. Giedroyc, "Differential Thermal Analysis in Controlled Atmosphere," *Trans. Brit. Ceram. Soc.*, **49**, 365–74 (1950).

[4] R. A. Rowland and D. R. Lewis, "Furnace Atmosphere Control in Differential Thermal Analysis," *Am. Mineral.*, **36** [1 and 2] 80–91 (1951).

through one end of the furnace. Nitrogen and carbon dioxide were employed in studying carbonaceous clays and carbonates, respectively. The results obtained were similar to those obtained by Saunders and Giedroyc. Rowland and Lewis did, however, explain the phenomena observed with carbonates by stating the van't Hoff equation, $\ln K = -(\Delta H/RT) + C$.

Neither of these groups of researchers have employed a flowing atmosphere (that is, atmosphere in motion) where the partial pressure of a given vapor can be kept reasonably constant within the specimen while a reaction is taking place. Saunders came close, but the gas was forced through the specimen only insofar as the resistance to flow within the specimen before and during a gas-evolving reaction was related to flow resistance in the surrounding media. In Rowland's method the controlled atmosphere was around the specimen holder but not dynamically controlled within the powdered specimen.

III. The Apparatus Developed

Many of the equipment components used in the apparatus were designed by the author and fabricated in the shops of the Engineering Experiment Station of The Ohio State University. The recording and power-input components were purchased as commercial units.

The furnace, sample holder, gas injecting system, and thermocouple assembly constitute the new development. The dimensions of these parts were partially dictated by materials on hand or easily available. Choice of other dimensions, materials, and design was based on the function to be accomplished by the component, and these are stated under the headings given in the following paragraphs. A vertical arrangement was selected as being more applicable for the purposes and more versatile than a horizontal arrangement.

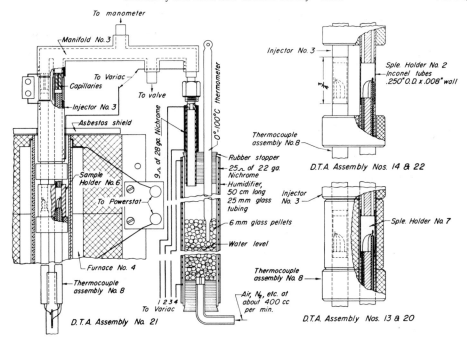

Fig. 3. Principal differential thermal analysis assemblies.

(1) Sample Holders

The primary function of the sample holder is to provide a container for the specimens, but it must meet specific requirements as a container. Because of the prerequisite that a constant gas flow be maintained through the powdered specimen, a thin-walled metal tube seemed practical since the gas could be injected at one end of the tube, passed through the powdered specimen, and then discharged into the open air. Inconel metal was chosen because it resists corrosion in air up to at least 1200°C. and has no Curie point between 25° and 1200°C. Inconel tubing $1/4$ in. in outside diameter with a wall thickness of 0.008 was used. This sample holder with air as the insulator is referred to as holder No. 2, and the details are given in Fig. 1. A diatomaceous earth-insulated holder was made by cementing Inconel tubes of the same size as holder No. 2 inside a cylindrical shape cut from a C22 brick. This holder is designated as holder No. 7, and the details are given in Fig. 1.

It was fully realized after the first run (in this paper the tests are referred to as runs) that some sensitivity might have to be sacrificed in order to restrain the natural consequences due to differences in the diffusivity and specific heat equations of the reference and the sample. Accordingly a sample holder in the form of an Inconel block with 0.234-in. cavities was made for use when the great sensitivity of the thin-walled tubes was not required. This holder is designated as holder No. 6 in Fig. 1.

A cavity of the size provided by these holders will accommodate about 0.15 gm. of the clay minerals or 0.20 gm. of the alumina standard.

(2) Thermocouple Assemblies

The thermocouple assembly as interpreted here includes the thermocouples and the spaghetti or porcelain insulators which carry the thermocouple wires.

The porcelain spaghetti in this apparatus not only carries the wires but also supports the sample holders, carries the

gases away from the samples, and forms the bottom of the cavity. Refractory porcelain tubing $1/4$ in. in outside diameter with 4 holes 0.050 in. in diameter was used. Two 4-in. lengths of this tubing were cemented to a spacer cut from a C22 brick as shown in Fig. 2. The center-to-center distance of $7/16$ in. was dictated by the size ($7/8$ in. in diameter) of the Inconel bar that was available for holder No. 6. The top ends of these tubes (and the lower ends of the injector tubes to be discussed later) were ground to fit snugly inside the Inconel sample holders.

The most satisfactory arrangement of the thermocouples was the three-couple, four-wire type with the temperature-measuring couple placed in the reference cell beside the standard-of-comparison half of the differential couple. In order to balance the heat capacity in the two cells, a second, or dummy, couple was placed in the sample cell. The four wires in each cell were fused into one bead, thus making a four-wire couple of considerable rigidity.

It was presumed that an appreciable temperature gradient would exist between the center of the specimen and the Inconel tube (sample holder No. 2), that the temperature at the walls of the tubes would be highest, and that it would be here that the reactions would start first. Hence, the thermocouple bead was placed at a point $1/32$ in. from the tube walls, and the wires were spread as far apart as possible, as shown in the drawings of assembly No. 8, Fig. 2. This arrangement proved to be very satisfactory.

A ring of C22 material was cemented to the tube assembly $3/16$ in. from the top to assure centering of the unit in the furnace. The design and placing of this centering ring is shown in the details of assembly No. 8, Fig. 2, and its position in the furnace is shown in Fig. 3. A similar centering ring was cemented to the injector tubes (injector No. 3) for the same purpose.

(3) Components for Injecting the Gases

The function of these parts was to carry the gases from the

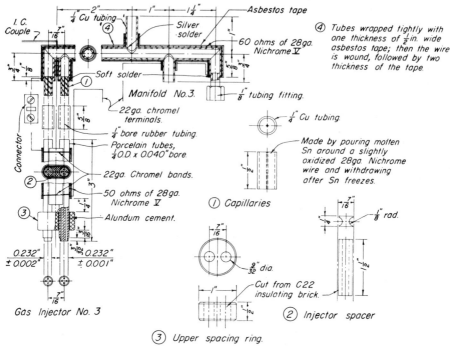

Fig. 4.   Details of injector No. 3 and manifold No. 3.

generating source to the sample holders; hence, part of the unit had to be in the furnace in contact with the two sample holders and part of it had to be outside the furnace connected with the gas-generating source.

The part of the unit that carried the gas into the sample holders was called the injector and the part that distributed the gas to the injector and connected to the gas source was called the manifold.

Drawings for injector No. 3 and manifold No. 3 are shown in Fig. 4. Refractory porcelain tubing, $1/4$ in. in outside diameter, was selected as the material for the injector. Two 4-in. lengths of this tubing were cemented with Alundum cement to a spacer cut from a C22 brick. The center-to-center distance of the tubes was $7/16$ in. to correspond with the spacing of the sample holder cavities. The lower ends were ground to fit snugly into the sample holders. A snug fit was necessary to be sure that the resistance to flow of the gas through the samples was less than that of the connection; otherwise, the gas would be lost completely at the connection and would not be forced through the powder. Some loss was expected when the furnace was hot because the coefficients of expansion of the metal and the porcelain were about 11.5 and 4.5×10^{-6}, respectively, hence the metal tubes would tend to expand away from the porcelain.

To take care of this loss, experience showed that the gas flow into the injector should be four times that desired through the powder. A gas flow of about 100 cm.³ per minute into each injector tube gave a flow of about 0.40 cc. per second through the loose powder in the cavity. If the powder was assumed to be 50% porous, the 0.40 cc. per second corresponds to 3 cm. per second; that is, in a sample cavity 1 cm. long and $1/4$ in. in diameter, the gas was changed about three times per second.

Two identical capillaries were soldered into the manifold at the head ends of the injector tubes, shown as part (1)

in Fig. 4. In this way about 2 to 3 cm. of Hg pressure could be maintained in the manifold so that the pressure drops, hence the gas flow through the two capillaries, were identical for all practical purposes.

When dry gases (i.e., free of H_2O vapor) were used, no problem of condensation of vapor existed; but when the gas contained a partial pressure of water vapor exceeding the vapor pressure of water at the room temperature, it was necessary to heat the manifold and injector. This was accomplished by winding these parts with 28-gauge Nichrome V wire of the resistances shown in Fig. 4. These resistances gave temperatures, as indicated by the thermocouple, in the order of 150°C. with 110 volts. A small Powerstat was used to give lower temperature as desired.

The manifold was connected to a mercury manometer to measure the pressure within it. A short length of rubber tubing and a pinchcock also were connected to the manifold and used to accurately regulate the manometer pressure.

(4) Sources of Gas Supply

The apparatus was designed so that any one of three gases could be used during a run, and one gas could be changed almost instantly to a gas of another composition. This was done by use of the four-way distributing valve shown in Fig. 5 and as part J in Fig. 6. For example, the three gases might be dry air, steam, and CO_2, or any combination of them. The setup in Fig. 7 shows steam, generated in the boiler I, as one of the gases; air was brought to the valve at O; and a mixture of $N_2 + CO_2$ was brought to the valve at P. The N_2 and CO_2 were metered through two flowmeters and mixed in the glass mixer Q. The flow of all three gases was preadjusted so that the desired gas could be selected instantly by turning the distributing valve II. A fourth outlet on the valve discharged the gases not being used to the atmosphere.

Atmospheres containing water vapor were prepared with three different setups. Atmosphere containing 5.1 mm. of water vapor was prepared by passing compressed air over an Ehrlenmeyer flask half filled with water at room temperature. The air, partially saturated, was then passed through a U tube filled with granules of $CaCl_2$ which removed all water in excess of 5.1 mm. partial pressure.

Water vapor pressure of 1 atmosphere, or about 750 mm., was obtained by using steam from the steam generator I in Fig. 7. When partial pressures of water vapor between those limits were required, the basic gas (N_2, CO_2, or air) was bubbled through the humidifier shown in Figs. 3 and 5. Any desired water temperature between room temperature (about 25.5°C.) and 98°C. was obtained by adjusting the current through the Nichrome winding on the humidifier as shown in Fig. 5.

(5) Furnace

Several small furnaces were made, but the most successful one was designated as furnace No. 4. The detail drawings of this furnace are shown in Fig. 8. It is also shown in Fig. 7 as part D. Its relation to the sample holder assemblies is shown in Fig. 3.

Furnace No. 4 is of very simple construction, inexpensive, and yet very effective. Two halves of a 9-in. straight brick of J-M C22 form the insulation. The muffle is a thin-walled Alundum tube on which the resistance wire is wound. Several metals have been used. About 40 ohms of Chromel A or Nichrome V wire (both 28 gauge) were used up to 1200°C. Platinum-rhodium (20% Rh) alloy wire (5 mils diameter) has also been used up to 1400°C.)

If the winding burned out, the furnace could be taken apart, a new winding put on the core, and be back in service in about 1 hour.

One important feature of the construction was the air space between the winding and the insulation. This air space made possible very rapid heating and cooling schedules; the cooling time was only about 1 hour. The core and winding constituted the only heat reservoir, and because of their small weight, the temperature response of the furnace was

Fig. 6. Close-up of differential thermal analysis assembly No. 8. (A) electrical connections to manifold No. 3; (B) manifold No. 3; (C) capillaries; (D) upper centering ring; (E) holder No. 6; (F) lower centering ring; (G) thermocouple assembly No. 8; (H) thermocouple head; (I) pinchcock for pressure control; (J) selector valve; (K) boiler needle valve; (L) steam generator; (M) humidifier which substitutes for L.

[*Editors' Note:* Figures 5 and 7 have been omitted because the original halftones were unavailable.]

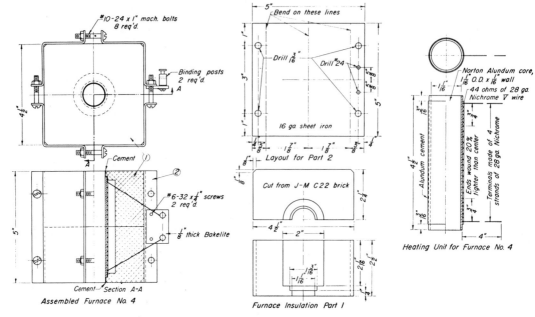

Fig. 8.   Details of furnace No. 4.

rapid.   The temperature almost kept up with the voltage input, even on a 20 °C.-per-minute schedule, when the Inconel tube sample holder was used.   This state of equilibrium made possible abrupt halts in temperature rise—the abruptness being controlled more by the heat characteristics of the sample holders than by the temperature lag of the furnace.

The power input source to this furnace was a 0- to 270-volt Powerstat.   The starting voltage for sample holder No. 2 (the small Inconel tubes) was 37 volts.   For sample holder No. 6 (the Inconel block), the required starting potential was 54 volts.   In both cases, a rate of voltage increase of 1.5 volts per minute gave a temperature rise of approximately 15 °C. per minute.

(6)   Recorders and Controllers

The make of instruments is not an important consideration, but the specifications cannot be treated so lightly.   The temperature rate controller must be of a type which will provide a *continuous* power input to the furnace at a rate to produce the desired temperature rise.   On-off types or types which increase the input in relatively large increments are not satisfactory.

Any reliable type of instrument for measuring the reference and differential temperatures can be used.   An Elektronik recorder made by the Brown Instrument Company was selected for the differential recorder.   A range of −1.0 to 0 to +1.0 mv. was selected, based upon the calculation of the temperature differential that would be obtained with the α- to β-quartz inversion.   This is a rather weak reaction, and it was felt that this inversion should show at least a 2-cm. peak in the thermal curve.   Calculations based on specific heat data showed that the quartz reaction for a 0.20-gm. quartz sample with the Inconel tube sample holder would give a maximum temperature differential of 5.1 °C.   This corresponds to about 0.20 mv. for a Chromel-Alumel couple.   With a recorder having a 10-in. span and a 2.0-mv. total range, the 0.20 mv. would correspond to 1.0 in. or 2.5 cm.   Hence, the theoretical maximum peak height for the assumed sample weights corresponds closely to the desired 2.0 cm.

An actual run with approximately the assumed conditions gave a peak height of 2.1 cm.

(7)   Method of Mounting the Parts

A general view of the entire apparatus is shown in Fig. 7.

With vertical arrangements, like that by Kerr, Kulp, and Hamilton,[2] the sample holders were stationary and the furnaces were on vertical tracks so that they could be lowered over the sample holders.   In the present apparatus, it was easier to make the furnace stationary and mount the array of parts constituting the differential thermal analysis assembly on a single movable shaft.

This setup is shown in Fig. 6.   (The movable shaft is designated *G* in Fig. 7.)   The manifold-injector assembly (parts *A, B, C, D, I, N,* and *X* in Fig. 6) were fastened to the shaft with one clamp.   The thermocouple assembly, including parts *F, G,* and *H,* were fastened to the shaft with another clamp.   The clamps could be adjusted with reference to length and rotation of the jaws, hence the assemblies of components could be aligned perfectly and held rigidly in alignment by tightening the clamp thumbscrews.   The aligned parts and the shaft on which they were mounted could then be raised or lowered as desired.

IV.   Procedure

(1)   Data for Calculating Heats of Reaction

In order to plot reaction temperatures according to the van't Hoff equation, several values are required for $1/T$ and ln p, where ln p is the natural log of the pressure in millimeters of the gas evolved and $1/T$ is the reciprocal of the absolute temperature in degrees Kelvin.   In order to obtain these data, the decomposition partial pressures of water vs. decomposition temperatures,* for example, were obtained

---

* The reference temperature at which the recorder first indicated a temperature differential due to the reaction was recorded as the desired decomposition temperature.

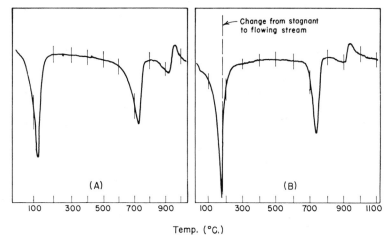

Fig. 9. Differential thermal analysis curves of a Na-montmorillonite showing the effect of dry air vs. steam on the 730° endotherm. (A) Na-montmorillonite, Belle Fourche, S. D. in flowing air, In $p_{H_2O} = 1.63$, differential thermal analysis assembly No. 22; (B) Na-montmorillonite, Belle Fourche, S. D. in stagnant air and flowing steam, differential thermal analysis assembly No. 14.

for four or more partial pressures of water vapor. After obtaining and plotting four or more values for T and In p_{H_2O}, the heats of reaction could be calculated by substituting two values of $1/T$ and the corresponding values for In p_{H_2O} in a modified form of the van't Hoff equation.

(2) Technique of Loading and Unloading Sample Holders

The technique of loading and unloading the sample holders was very simple and required less than 5 minutes. If one started with the differential thermal analysis assembly in operating position as shown in Fig. 7, the connector nut on the manifold was unscrewed from the distributing valve, then the shaft G and all the parts fastened to it were raised vertically up through the furnace until the upper surface of the lower centering ring, F in Fig. 6, was just above the top of the furnace. The clamp holding the shaft to the lattice frame was tightened and the manifold-injector assembly was removed by loosening the thumbscrew of its clamp, raising the assembly a half inch or so, then swinging it to one side out of the way. The sample holder, E in Fig. 6, was raised vertically from the thermocouple assembly. Powders that did not sinter were easily blown clear of the thermocouple and out of the sample holder. Samples which sintered were cracked like a nut with a pair of pliers, and then removed from the thermocouples without any damage to the wires. This technique was first developed by McConnell and Earley.[5]

After cleaning, the sample holder was replaced and the desired quantities of powders were poured into the cavities with a tiny spatula. The thermocouple assembly was tapped lightly with the spatula to settle the powders around the thermocouples. The process of adding powder and settling by tapping was repeated once if necessary or until the surface of the powder was about $1/8$ in. below the surface of the holder. The quantities of powders were about 0.15 gm. of clays, 0.30 gm. of powdered quartz, and 0.20 gm. of calcined c.p. alumina.

The principal reason for not packing the sample powders was to keep the mass as porous as possible to permit free

movement of the flowing gas stream. Test materials were crushed and screened through 65 mesh. This particle size and technique of settling the powders by tapping gave excellent reproducibility of differential thermal analysis curves.

After loading the sample holder, the manifold-injector assembly was swung back into vertical position and lowered until the ends of the injector tubes were properly seated in the holder. The clamp thumbscrew was tightened and the whole assembly on the shaft was lowered to its operating position in the furnace. The connector nut on the manifold was tightened on the distributing valve, and the pressure on the gas was adjusted to about 2.5 cm of Hg by turning the pinchcock, I of Fig. 6. The apparatus was now ready for a run.

It might be pointed out that a certain amount of grief was avoided at the completion of a run by raising the manifold-injector assembly about $1/16$ in. out of the sample holder immediately after shutting off the furnace. If this was not done, the metallic sample holder contracted tightly around the ground ends of the injector tubes with the danger that the tubes would be cracked.

V. Type of Results Obtainable

This paper was written for the purpose of describing the equipment; and the data on only one mineral, a sodium-montmorillonite, are given here as an illustration. More data on other minerals will be forthcoming in the near future.

The differential thermal analysis curves for the Belle Fourche, S. D., montmorillonite, A.P.I., No. 27, are shown in Fig. 9. These data were all obtained with the differential thermal analysis assembly No. 21 which is shown in Fig. 3. Partial pressures of water vapor from 5.0 mm. and 760 mm. were used in the montmorillonite study. There is one point on the differential thermal analysis curves of particular interest, namely, the temperature at which the 730° endothermic peak started, which was about 300°C. with dry air and about 600°C. in steam.

This serves only as an example. Excellent results have been obtained on dissociation of carbonates and on oxidation-reduction of iron oxides, in addition to the dehydration studies.

[5] Duncan McConnell and J. W. Earley, "Apparatus for Differential Thermal Analysis," *J. Am. Ceram. Soc.*, **34** [6] 183–87 (1951).

10

Reprinted from *J. Am. Ceram. Soc.*, **38**(8), 281–284 (1955)

A Theory of Differential Thermal Analysis and New Methods of Measurement and Interpretation

by S. L. BOERSMA   Delft, Holland

A new-type nickel-block sample holder for use in differential thermal analysis is described. A simple machine to aid in the interpretation of complex curves is suggested.

I. Introduction

A WELL-KNOWN method of mineral analysis is to heat a sample in a furnace at a certain rate of temperature rise and to determine the temperature lag or lead that results from the heat of transformation at certain temperatures specific for the mineral. A discussion of this method is given by Kronig and Snoodijk.[1]

An apparatus for differential thermal analysis built at the Laboratorium voor Grondmechanica has been described by de Bruyn and Marel.[2] Since the completion of this apparatus several thousand samples have been investigated. The results of these measurements have been analyzed by Marel, who showed that differential thermal analysis cannot be used for precise quantitative analysis. In this paper a theory of differential thermal analysis is given that accounts for the shortcomings found by Marel, and methods are outlined for improving measurement and interpretation.

II. Theory of Differential Thermal Analysis

(1) Nickel Sample Holder

The following symbols are used:

T = temperature.
θ = differential temperature.
r = radius.
a = radius of cavity filled with sample.
t = time.
c = specific heat of sample material.
λ = thermal conductivity of sample material.
ρ = specific density of sample material.
q = heat of transformation per unit volume.

In a block of material with an infinitely high thermal conductivity (e.g. nickel) there are two identical cylindrical or spherical cavities. Cavity No. 1 is filled with an inert material having the same thermal properties as the sample. When the nickel block is heated (in practice at a uniform rate),* the surface temperature of both cavities is the same during the entire process because of the high conductivity of nickel. Cavity No. 2 contains the sample.

The temperature T_1 in cavity No. 1 is given by

$$\rho c \,(\partial T_1/\partial t) - \lambda \operatorname{div} \operatorname{grad} T_1 = 0$$

Received November 12, 1954.

The work described in this paper was undertaken at the request of E. C. W. A. Geuze, director of the Laboratorium voor Grondmechanica (Soil Mechanics Laboratory) at Delft.

The author is a consulting engineer for this laboratory.

¹ R. Kronig and F. Snoodijk, "Determination of Heats of Transformation in Ceramic Materials," *Appl. Sci. Research*, **A3** [1] 27–30 (1951); *Ceram. Abstr.*, **1953**, January, p. 17a.
² C. M. A. de Bruyn and v. d. Marel, *Geol. en Mijnbouw*, pp. 69–83, March 1954.

* It can be shown that a uniform heating rate becomes necessary when there is a difference between the thermal properties of sample and reference material.

The temperature T_2 in cavity No. 2 is given by

$$\rho c \,(\partial T_2/\partial t) - \lambda \operatorname{div} \operatorname{grad} T_2 = \partial q/\partial t$$

$\partial q/\partial t$ = heat of reaction eventually produced (per unit time).

The differential temperature θ is found from the difference of the two linear equations

$$\rho c \,\frac{\partial \theta}{\partial t} - \lambda \operatorname{div} \operatorname{grad} \theta = \frac{\partial q}{\partial t} \tag{1}$$

with the boundary condition $\theta = 0$ for $r = a$ and $\theta = 0$ for $t = t_1$. It is assumed that $t = t_1$ is before the beginning of a reaction and that $t = t_2$ is far enough after its completion for the temperature difference θ to become zero again.

The integration of equation (1) with respect to time gives

$$\rho c(\theta_2 - \theta_1) - \lambda \int_{t_1}^{t_2} \operatorname{div} \operatorname{grad} \theta \, dt = q \tag{2}$$

q = entire heat of reaction per unit volume.

The temperature difference caused by the reaction will have died away at the moment t_2, so the first term vanishes.

After interchanging the independent time and place operations in the second term, equation (2) becomes

$$-\lambda \operatorname{div} \operatorname{grad} \int_{t_1}^{t_2} \theta \, dt = q \tag{3}$$

Integrating equation (3) over the volume V within a sphere of radius r gives

$$-\lambda \int_V dV \operatorname{div} \operatorname{grad} \int_t \theta \, dt = \int_V q \, dV$$

or

$$-\lambda \int_S dS \operatorname{grad} \int_t \theta \, dt = qV \tag{4}$$

where the left part is transformed by application of the Gauss theorem.

Because of radial symmetry, $\operatorname{grad} \int_{t_1}^{t_2} \theta \, dt$ is everywhere perpendicular to the surface S and of the same magnitude; therefore

$$-S\lambda \operatorname{grad} \int_t \theta \, dt = qV$$

$$-S\lambda \frac{d}{dr} \int_t \theta \, dt = qV \tag{5}$$

According to the boundary condition $\theta = 0$ for $r = a$ and so $\int_t \theta \, dt = 0$. The temperature integral in the center follows from equation (5):

$$\int_{t_1}^{t_2} \theta_0 \, dt = -\frac{q}{\lambda} \int_a^0 \frac{V}{S} \, dr = \frac{q}{\lambda} \int_0^a \frac{V}{S} \, dr \tag{6}$$

For a cylindrical sample $V/S = r/2$, so the peak area in the cylindrical case

$$\int_{t_1}^{t_2} \theta_0 \, dt = \frac{qa^2}{4\lambda} \tag{7}$$

For a sphere $V/S = r/3$, so for a spherical sample

$$\int_{t_1}^{t_2} \theta_0 \, dt = qa^2/6\lambda \tag{7a}$$

In the one-dimensional case of a flat plate

$$\int_{t_1}^{t_2} \theta_0 \, dt = qa^2/2\lambda \tag{7b}$$

(2) Peak Area in a Ceramic Sample Holder

For the sample holder some investigators use a ceramic material instead of nickel. The thermal conductivity, λ_c, here is of the same order of magnitude as the conductivity, λ_s, of the sample itself. The boundary condition is $\theta = 0 = \int \theta \, dt$ for $r = \infty$. For an infinitely large ceramic sample holder, equation (6) becomes

$$\int_{t_1}^{t_2} \theta \, dt = - \frac{q V_s}{\lambda_c} \int_{\infty}^{a} \frac{dr}{S} - \frac{q}{\lambda_s} \int_{a}^{0} \frac{V}{S} \, dr \qquad (6a)$$

V_s = entire sample volume.
$q V_s$ = all the heat produced that is carried away by the ceramic sample holder.

For the case of an infinitely large ceramic block, equation (6a) has no finite solution in the one- and two-dimensional case; there is a solution only for a spherical sample (radius a, $V = (4\pi/3)r^3$, $V_s = (4\pi/3)a^3$, $S = 4\pi r^2$):

$$\text{Peak area} \int \theta \, dt = \frac{qa^2}{6}\left(\frac{2}{\lambda_c} + \frac{1}{\lambda_s}\right) \qquad (8)$$

If λ_c becomes infinite, equation (8) reduces to equation (7a) for the nickel block. In a ceramic block, λ_c is of the same order of magnitude as λ_s of the sample material. Thus, according to equation (8), the peak area becomes three times as large as with a nickel block. Although a ceramic block gives the greater peak area, the method is not attractive because the temperature fields of several samples in the same block can penetrate into each other, thus causing mutual interference.

(3) Thermocouple Interference

The temperature in the sample center is measured by means of a thermocouple. Part of the heat produced in the sample is carried away by the thermocouple wires; too low a temperature therefore is measured. The error from this effect is quite large. Figure 1 shows the sample-filled cavity in the nickel block. The thermojunction consists of a sphere of radius r_0. During a reaction there is a temperature gradient in the leads over the lengths l. It is assumed that at the distance l (slightly larger than a) the wires have attained the temperature of the nickel block. The area of the cross section of the leads will be A. If θ_0 is the junction temperature, the thermocouple leads will carry away an amount of heat

$$Q = \int_{t_1}^{t_2} \frac{A \lambda_p}{l} \theta_0 \, dt$$

λ_p = heat conductivity of wire material (platinum).

This heat no longer passes through the surface of integration in equations (4) and (5); Q therefore must be subtracted there:

$$- S \lambda \frac{d}{dr} \int \theta \, dt = qV - Q \qquad (9)$$

The correct peak area is now

$$\int_{t_1}^{t_2} \theta_0 \, dt = \frac{1}{\lambda} \int_{a}^{r_0} \frac{qV - Q}{S} \, dr \qquad (10)$$

For spherical samples ($S = 4\pi r^2$ and $V = (4\pi/3)(r^3 - r_0^3)$) the peak area is

$$\int_{t_1}^{t_2} \theta_0 \, dt = \frac{qa^2}{6\lambda} \cdot \frac{\alpha}{1 + (\Lambda/\lambda)} \qquad (11)$$

$\alpha = 1 - (r_0^2/a^2)[3 - 2(r_0/a)]$
$\Lambda = \lambda_p(r_0/l)(A/4\pi r_0^2)[1 - (r_0/a)]$

Equation (11) differs from the elementary expression (equation (7a)) by the factor $\alpha/1 + (\Lambda/\lambda)$. Here α (very nearly unity) comes from the altered geometry and Λ from the heat leakage through the thermocouple leads.

It follows from equation (11) that with low-conductivity samples $[(\lambda/\Lambda) \ll 1]$ the peak area will become independent

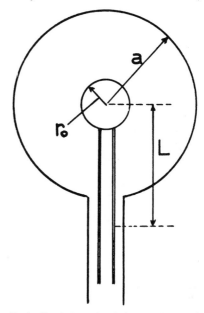

Fig. 1. Heat leakage through thermocouple wires of lengths l.

of the sample conductivity λ, whereas a high sample conductivity will cause an inverse proportional relationship between peak area and thermal conductivity.

For cylindrical samples an expression very similar to equation (11) exists:

$$\int_{t_1}^{t_2} \theta_0 \, dt = \frac{qa^2}{4\lambda} \cdot \frac{1 - \frac{r_0^2}{a^2}\left(1 + 2 \ln \frac{a}{r_0}\right)}{1 + \frac{A}{l} \frac{\lambda_p}{\lambda_s} \frac{\ln (a/r_0)}{2\pi h}} \qquad (11a)$$

The differential thermal analysis apparatus in use at the Laboratorium voor Grondmechanica has the following constants: $l \cong a = 0.4$ cm.; $r_0 = 0.12$ cm.; $A = 4 \times 10^{-3}$ cm.; $\lambda_p = 72$ joules per m. sec. °C.

Most sample materials have thermal conductivities of about 0.3 joule per m. sec. °C., which, with the foregoing constants, makes λ/Λ about unity (equation (11)). According, therefore, to equation (11), the heat leakage through the thermocouple leads reduces the peak area to less than 50% of its theoretical value. The peak area has become sensitive for the thermocouple geometry; in practice, changes in the calibration factor of about 20% have been observed when exchanging the thermocouple.

III. Shortcomings of Existing Differential Thermal Analysis

When used to obtain quantitative results, differential thermal analysis has several defects. The peak area is related not only to the amount of unknown material but is influenced by the following:

(1) Sample volume. The same sample will, when pressed into a smaller volume, give other readings because of the geometrical factor a^2 in equations (7) and (11).

(2) Sample conductivity. According to equations (7), (8), and (11) the peak area depends on the thermal conductivity of the sample material. This conductivity varies with the composition and density and is mostly unknown.

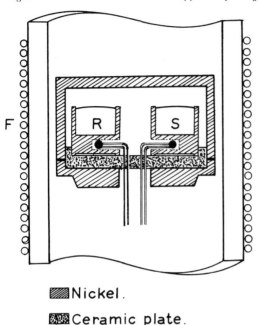

▨ **Nickel.**

▨ **Ceramic plate.**

Fig. 2.   A new sample holder for differential thermal analysis.   S, sample; R, reference material;   F, furnace.

(3)   Thermocouple interference.   As shown by equation (11) the thermocouple has (by its heat conduction) a very large effect on the measured peak area (approximately halves it).

(4)   Sample shrinkage.   Some samples sinter at the higher temperatures and shrink.   An air gap between sample and nickel block is produced that completely upsets the temperature distribution, resulting in a measured peak area that is much too high.

(5)   Loss of material.   In some reactions volatile products (e.g. water) are formed that evaporate.   The sample mass is no longer constant.   The escaping gas can alter the porosity and thermal conductivity, so the linear theory of this paper is no longer valid.   When mass or conductivity is altered, the temperature may not return to the original value after the completion of the reaction; there is no longer a definite peak area.   These shortcomings can be overcome by a new meas-uring technique (new sample holder) and a new method of diagram interpretation (curve synthesis).

IV.   A New Sample Holder

In normal differential thermal analysis the heat of a reac-tion is measured by the temperature difference it produces in the sample material itself.   Thus the sample is used for two entirely different purposes: (*a*) a producer of heat and (*b*) heat measuring resistance in which the flow of heat develops a temperature difference to be measured.

A much better solution is to separate these functions.   This can be done by leading the heat of reaction outside the sample through a special piece of material on which the temperature difference can be measured.   The peak area then is solely de-pendent on the produced heat of reaction and on the calibra-tion factor of the instrument which no longer contains any sample properties (e.g., volume and conductivity).   An ex-ample of such an apparatus is shown in Fig. 2.   The sample *S* and reference material *R* are held in small nickel containers placed on a ceramic base plate.   When a reaction takes place, the heat of reaction is led from the container through the ceramic plate, and the container takes on a temperature that is solely dependent on the heat of reaction and on the specific heat conduction between the container and the surrounding nickel dome.   The heat passes for the greater part through the ceramic plate; the remainder is transferred by radiation and convection.   These properties are a calibration constant for the apparatus, independent of the kind or even the amount of sample material.   The temperature peak area easily can be shown to be

$$\int_{t_1}^{t_2} \theta \, dt = \frac{mq}{G} \tag{12}$$

m = mass of sample.
q = heat of reaction per unit mass.
G = heat-transfer coefficient between small nickel container and surrounding nickel.

The measurement has now become purely caloric: mq, the total heat of reaction, is measured.   There is no need to dilute samples with Al_2O_3 or even to use samples of standardized weight or volume.   Measurements made by this method con-firmed equation (12) quantitatively.   Samples of $CuSO_4$ of widely different packing density produced the same (pre-dicted) peak area.

V.   Curve Synthesis

The shortcomings of differential thermal analysis mentioned in section III have been overcome with the new apparatus.   A number of substances, however, do not produce clear, well-defined temperature peaks (e.g. illite), perhaps because there is no sharp reaction temperature or the properties of the sample vary considerably with temperature.   These curves cannot be characterized by a mere number (peak area), which is a great difficulty in quantitative analysis.   It seems much better to make use of all the data the curve can give.   This can be done in the following manner:

From a number of pure minerals (*A*, *B*, *C*, *D*) are taken the differential thermal analysis curves (*a*, *b*, *c*, *d*).   To analyze an unknown sample its curve *x* is taken and compared with a synthetic curve $x' = \alpha a + \beta b + \gamma c + \delta d$ made up out of known fractions of the pure mineral curves *a*, *b*, *c*, *d*.   The coef-ficients α, β, γ, and δ that give the best match between the curves *x* and *x'* represent the relative concentrations of the minerals *A*, *B*, *C*, and *D* in the sample.

A synthesis of a composite curve made up out of known fractions of pure mineral curves can easily be made with the apparatus shown in Fig. 3 (not yet completed when this paper was written).   The curves obtained with the differen-tial thermal analysis recorder are cut out of opaque sheet ma-terial (e.g. black paper) to form templates.   The templates move simultaneously in front of an illuminated slit, S; they

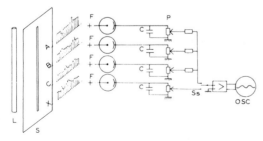

Fig. 3.   Curve synthesis for differential thermal analysis.   L, lamp; S, slit; A, B, C, and x, moving templates; F, photocells; C, integrating condensers; P, calibrated potentiometers; Ss, selector switch; and OSC, cathode-ray oscillograph.

can be bent around a glass cylinder that rotates coaxially with respect to the lamp, L. The slit passes an amount of light that is partially intercepted by the templates. Behind each template is a photoelectric cell, F, that generates a current of the same wave form as the template to which it belongs. Known and adjustable fractions (α, β, etc. of the generated signals) are taken from the calibrated potentiometers, P, and fed into a cathode-ray oscillograph, OSC. The oscillating selector switch, Ss, puts alternatively the unknown sample curve x and the synthetic curve $x' = \alpha a + \beta b + \gamma c + \delta d$ on the screen. By manipulating the α, β, γ, and δ potentiometers the two curves are made to cover each other as much as possible. When this is done, the composition of the sample can be read quantitatively from the potentiometer dials α, β, γ, and δ. Because with differential thermal analysis the temperature-time integral curve rather than the temperature itself is significant, it is better to manipulate some form of integral curve on the screen. This is done readily by means of the electrically integrating condensers, C.

VI. Summary

A theory dealing with the calorimetric shortcomings of the differential thermal analysis of minerals has been given. These shortcomings can be overcome by the use of a new type of sample holder. To improve the interpretation of complex curves, a simple machine is suggested that performs the continuous synthesis of differential thermal analysis curves of widely varying compositions.

11

Copyright © 1957 by the American Chemical Society

Reprinted from *Anal. Chem.*, **29**, 271–275 (Feb. 1957)

Analytical Applications of a Differential Thermal Analysis Apparatus

PAUL D. GARN and STEWARD S. FLASCHEN

Bell Telephone Laboratories, Inc., Murray Hill, N.J.

▶ An apparatus for differential thermal analysis is described, which is useful as a tool in the study of inorganic materials. In it have been incorporated as many advantages as possible of systems previously set up. Thermograms of magnesium carbonates and talcs indicate transition and decomposition temperatures. A phase diagram of the potassium niobate–potassium tantalate system was determined from differential thermal analysis data. Thermograms of "potassium maleate" prepared from different starting materials are shown. Identification of the magnesium carbonates and talcs sourcewise aids in setting up firing schedules in the production of ceramics. The phase diagram of the potassium niobate–potassium tantalate system is of interest in the study of ceramics.

DIFFERENTIAL thermal analysis has been used for many years for detecting phase transitions, principally in minerals. Excellent reviews of differential thermal analysis apparatus and techniques have been prepared by Grim (*6*) and by Smothers, Chiang, and Wilson (*12*).

In addition to its normal use in determining temperatures of phase transformations, differential thermal analysis is useful as a control tool or as a routine tool for comparing similar but not identical materials. As a control tool it may be used to distinguish raw materials quickly and easily in those cases in which the treatment of the material must be modified if slight changes in the material are encountered. As a comparison tool, differential thermal analysis may be used in some cases to test materials that

yield anomalous results by other tests. Determination of transition temperature of samples with systematically varied compositions yields the data necessary to establish the phase diagram of the system.

DESIGN OF APPARATUS

The apparatus used in this work is unique only in that an attempt has been made to incorporate as many advantages as possible of systems previously set up.

The high temperature furnace (Figure 1) was adapted from the design used by Coffeen (*4*).

It has a platinum heating element wound on refractory alumina tubing. A thin cylindrical platinum shield is placed inside the furnace tube. The shield is grounded, with a platinum wire at some convenient point, in order to eliminate noise of thermionic origin. The furnace is mounted on ball-bearing slides and is moved horizontally to enclose or expose the sample holder. To prepare the furnace for use, a platinum–platinum–10% rhodium thermocouple in a $^3/_{32}$-inch insulating tube is placed in the control thermocouple well and the wires are led to the vertical hole in the core and through the ceramic tube to outside terminals. A differential thermocouple consisting of platinum–10% rhodium with a joining wire of platinum, palladium–20% gold, or palladium–10% gold is inserted from the side into two of the sample wells and the wires are led to outside terminals through the four-hole tube. The sample is placed in one of the wells and aluminum oxide as reference material is placed in the other. The furnace is then rolled into place and is ready for use. The furnace thermo-

couple wires are 0.015 inch and the differential thermocouple wires are 0.005 inch in diameter.

A block diagram of the system built by Leeds & Northrup for controlling the temperature and recording the furnace temperature and differential temperature is shown in Figure 2.

The strip chart recorder is an X_1, X_2 Speedomax Model 69955, which gives a continuous plot of the furnace temperature from the control thermocouple and the differential temperature from the differential thermocouple on a single chart.

The program controller consists essentially of a motor-driven slide-wire so designed that full travel takes place in 2.5 hours. A similar slide-wire is mounted on the shaft of the furnace temperature potentiometer. The signals from the two slide-wires are compared by the control unit, L. & N. Model 10864. The control unit introduces proportional band, reset, and rate time action to control sensitivity and eliminate overshoot and "hunting." The unbalance signal controls a motor-driven Powerstat, located behind the panel, advancing it or backing it off as required to maintain a heating or cooling rate of 10° C. per minute. The stabilized direct current voltage amplifier, L. & N. Model 9835-B, amplifies the differential temperature signal before it is sent to the recorder. It provides six ranges from 25-0-25 μv. to 1-0-1 mv., corresponding for the platinum–10% rhodium *vs.* palladium–20% gold thermocouple to ranges of about 0.7°-0°-0.7° to 27°-0°-27° C.

The equipment is designed to heat the furnace to a preselected temperature from 0° to 1500° C. at a constant rate of 10° C. per minute. When the desired temperature is reached by the furnace, the program unit operates to maintain that temperature or to cool the furnace

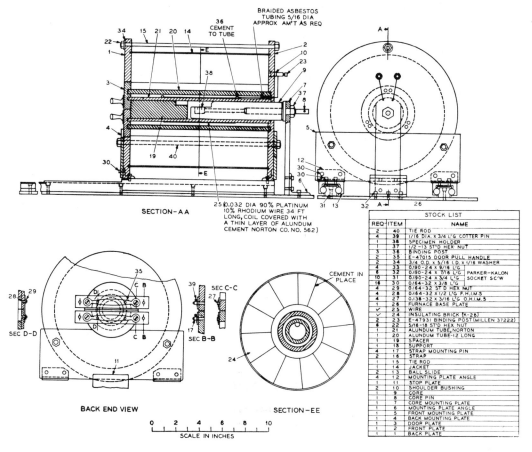

Figure 1.   Furnace for differential thermal analysis

Detail No.	Detail	Material	Suppliers
1	Back plate	Transite	Johns-Manville Co
2	Front plate	Transite	Johns-Manville Co.
3	Door plate	Transite	Johns-Manville Co.
4–5	Mounting plate	Steel	
6	Mounting plate angle	Aluminum	
7	Core mounting plate	Aluminum	
8	Core pin	Steel	
9	Core	Refractory alumina	
13	Ball slide		Grant Pulley & Hardware Co., Flushing, N. Y.
20	Refractory tube 2½-inch I.D.	Alundum	Norton Co.
	Refractory tube ³/₈-inch wall	Alundum	Norton Co.
21	Refractory tube 1½-inch I.D.	Alundum	Norton Co.
	Refractory tube ¼-inch wall	Alundum	Norton Co.
24	Insulating brick	K-28	Babcock & Wilcox

at the same rate, depending on the position of a selector switch.

In addition to the functions described, the equipment may be modified easily to give a heating or cooling rate of 0.5° C. per minute. This, with the use of all four holes in the sample holder, will permit the accurate determination of the temperature of a transition by a simple procedure suggested by Keith and Tuttle (9).

MATERIALS AND APPARATUS

The magnesium carbonates were: lots 2 and 3, Merck heavy grade, Maryland and California, respectively, and lot 6, Baker heavy grade.

The Montana talc was No. 486 Montana talc, supplied by Whittaker, Clark, and Daniels, Inc. The Sierramic and Yellowstone talcs were California

and Montana talcs, respectively, supplied by the Sierra Talc Co.

The potassium niobate and potassium tantalate were prepared from Merck C.P. potassium carbonate, Amend C.P. (99+%) niobium pentoxide, and Amend C.P. (99+%) tantalum pentoxide.

The sample holder, shown in Figure 3, is a platinum cylinder with four wells, the centers of which are equally spaced about the center of the block at a

115

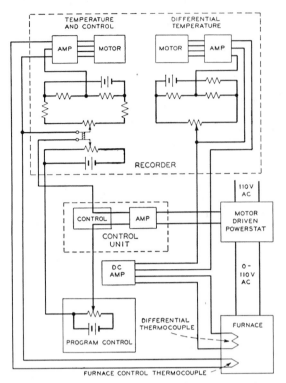

Figure 2.   Block diagram of differential thermal analysis equipment

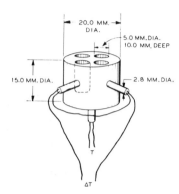

Figure 3.   Sample block showing method of introducing thermocouples for furnace temperature, *T*, and differential temperature, Δ*T*

radius of 5 mm.   The sample holder weighs 82 grams.   For most applications only two holes are used.   A $^{7}/_{64}$-inch hole is drilled through the side of the block into each well at a depth of 5 mm. to accommodate a thermocouple.   A well is drilled from the center of the base of the cylinder to accommodate a control thermocouple.

The platinum cups used in some work are nominally 10 mm. high by 6 to 8 mm. wide.   The wall thickness is 15 mils and the weight is 1.8 grams.

EVALUATION OF EQUIPMENT AND METHODS

The first results obtained with the differential thermal analysis apparatus

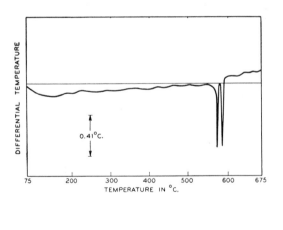

Figure 5.   Comparison of sample holders for potassium sulfate transition (170-mg. sample)

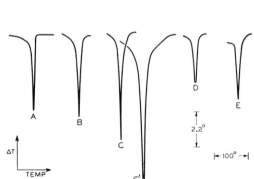

Figure 4.   Differential thermogram of mixture of quartz and potassium sulfate

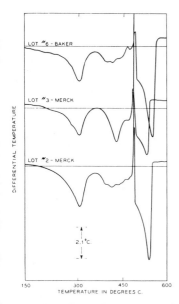

Figure 6. Differential thermograms of commercial magnesium carbonates

Figure 5 shows thermograms of the potassium sulfate transition at 583° C. with a variety of sample holders.

Thermogram *A* was obtained using the platinum block (82 grams) previously described. The sample weight was 170 mg. *B* was obtained by using platinum cups (15-mil wall, 1.8 grams) supported in an insulating firebrick core. The sample consisted of 170 mg. of potassium sulfate mixed with enough aluminum oxide to fill the cup. *C* was obtained in a similar manner, except that the cups were set into loops of platinum wire. *C'* was obtained by the same method, but with the cup filled with potassium sulfate. *D* was obtained using 170 mg. of potassium sulfate and a small amount of aluminum oxide in fused quartz cups set in wire loops. The final thermogram, *E*, was obtained with the same sample as in *B* and *C*, but the cups were fabricated from 1.5-mil platinum sheet.

The ratios of total height to half widths of the deflections are about 25,

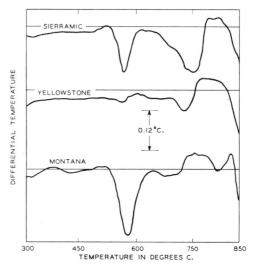

Figure 7. Differential thermograms of Montana, Yellowstone, and Sierramic talcs using platinum cups

were thermograms of known transitions, which were used in order to become familiar with the range and sensitivity of the apparatus. A typical thermogram is shown in Figure 4. The sample consisted of a mixture of quartz and potassium sulfate. The reference material was aluminum oxide. The scale on the differential temperature is 100-0-100 μv. The deflections are due to α-β transitions in quartz (SiO_2) and potassium sulfate, respectively. The thermogram shows the temperatures of transition at 571° and 580° C., in good agreement with Silverman and associates (*11*), who give the transition points at 573° and 583° C., respectively. The transition temperatures were determined by the intersection method (*9*). The magnitude of the deflection corresponds to a differential temperature of approximately 0.7° C. The differential thermocouple was made from 90% platinum–10% rhodium and 60% palladium–40% gold. The higher gold content of the alloy provides an increase in sensitivity at some sacrifice in useful temperature range.

Another method of handling of the sample uses a core- and wire-supported pair of metal cups with the sample and reference materials. Some investigators (*8*) claim that this method permits greater sensitivity because there is no large heat reservoir in contact with the sample. Other investigators (*7*) have used ceramic blocks, claiming the same increased sensitivity because of the low heat conductivity of the block material.

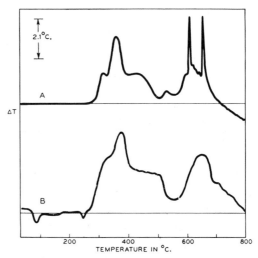

Figure 8.   Thermograms of potassium maleate

A.   Maleic anhydride
B.   Diethyl maleate

117

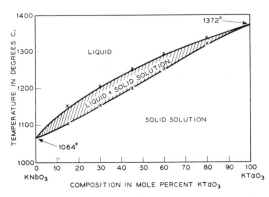

Figure 9. Phase equilibrium relationships at the solidus for system KNbO₃–KTaO₃

23, 39, 11, and 14, for A, B, C, D, and E, respectively. The comparatively wide deflections in D and E indicate that the sample holder is not at a uniform temperature. This effect is probably due to poor heat transfer around the cup because of the poor thermal conductivity of quartz in one case and the very thin cross section of the platinum in the other case. The resulting nonhomogeneity of temperature could permit a portion of the sample to begin its phase transformation well ahead of portions in contact with other parts of the cup. The effect on the difference signal would be a lowering and broadening of the peak.

The conclusion obtainable from these curves is that the use of cups does permit a gain in sensitivity. The cup walls should be thick enough to provide for thermal equilibrium about the cup. There was no indication of error in the measurement of the transition temperature.

EXPERIMENTAL RESULTS

A series of differential thermograms of magnesium carbonate and talc was obtained in order to determine whether or not the method showed promise as an aid in quality control of steatite ceramics. The platinum block was used as the sample holder. The method shows distinct differences in the thermal history of both the magnesium carbonate and the talc. Beck (3) shows thermograms for nesquehonite and lansfordite which are similar in gross structure to the thermograms of magnesium carbonates in Figure 6.

Thermograms of three samples of talc are shown in Figure 7. In each case a strong endothermic reaction begins at about 850° C. The magnitude of the reaction is about the same in each case, but differences due to impurities make it possible to distinguish one

from the other. The Montana and Sierramic talcs show a small broad endothermic reaction in the neighborhood of 570° C. The Sierramic talc also shows a fairly large endothermic effect in the neighborhood of 700° C. This reaction occurs to a lesser extent in the Yellowstone talc, but not in acid-washed samples of either talc.

In order to permit a precise interpretation of the thermograms of Figures 6 and 7, the data obtained would have to be correlated with weight loss and x-ray data. There are, however, sufficient differences in the three thermograms to enable identification of samples having a similar mineralogical composition. The differential thermocouple for the magnesium carbonate and talc studies was of 90% platinum–10% rhodium *vs.* 80% palladium–20% gold.

As part of a study of polyester resins, one observer (13) attempted to determine the degree of completeness of precipitation of maleate as potassium maleate. The amount of maleate precipitated was determined gravimetrically (2) and polarographically (5). An anomaly was observed, in that with certain starting materials the polarographic method showed less than one half the quantity of maleate that was indicated by the gravimetric method, while with other starting materials the results were in agreement. Differential thermograms of representative samples were obtained. The two thermograms shown in Figure 8 indicate that the precipitates obtained with the two starting materials are quite different. Further study of this phenomenon has been started.

Potassium niobate was investigated because of its interest as a ferroelectric. The compound is normally prepared from potassium carbonate and niobium pentoxide. A phase diagram of the potassium carbonate–niobium pentoxide system has been prepared by Reisman

and Holtzberg (10). The melting point of potassium niobate has been reported over a range of temperatures from 1020° to 1125° C. Preliminary investigation indicates that the variable results are due to nonstoichiometric preparation of the potassium niobate. A carefully prepared stoichiometric system yielded a melting point of 1064° $\pm\, 2°$ C.

In Figure 9, the phase equilibrium relationships as determined by differential thermal analysis are shown for the binary system between the isostructural ferroelectric compounds $KNbO_3$ and $KTaO_3$. As the mixed system showed strong supercooling tendencies (evidenced by large exotherms in solid solution mixtures slightly below the solidus), data determination was limited to the heating cycle only. The solidus temperature is marked by initial endothermic departure of the differential signal from its base line; the cessation of the endothermic effect marks the liquidus temperature. The extrapolation of the mixed system to the $KTaO_3$ end member is shown to coincide closely with the melting point established by independent methods at 1372° C. The data for this phase diagram were obtained using wire-supported platinum cups and a platinum *vs.* platinum–10% rhodium differential thermocouple.

REFERENCES

(1) Ahrens, L. H., *Geochim. et Cosmochim.* **2**, 168–9 (1952).
(2) Am. Soc. Testing Materials, Philadelphia, Pa., "Official and Tentative Methods," ASTM D 563-52.
(3) Beck, C. W., *Am. Mineralogist* **35**, 985–1013 (1950).
(4) Coffeen, W. W., Metal & Thermit Corp., Rahway, N. J., private communication.
(5) Garn, P. D., Halline, E. W., ANAL. CHEM. **27**, 1563–5 (1955).
(6) Grim, R. E., *Ann. N. Y. Acad. Sci.* **53**, 1031–53 (1951).
(7) Grimshaw, R. W., Heaton, E., Roberts, A. L., *Trans. Brit. Ceram. Soc.* **44**, 76–92 (1945).
(8) Gruver, R. M., *J. Am. Ceram. Soc.* **31**, 323–8 (1948).
(9) Keith, M. L., Tuttle, O. F., *Am. J. Sci., Bowen Vol.* Pt. 1, 203–80 (1952).
(10) Reisman, A., Holtzberg, F., *J. Am. Chem. Soc.* **77**, 2115–9 (1955).
(11) Silverman, A., Insley, H., Morey, G. W., Rossini, F. D., "Data on Chemicals for Ceramics Use," *Natl. Research Council Bull.* **18**, 94, 102 (1949).
(12) Smothers, W. J., Chiang, Y., Wilson, A., "Bibliography of Differential Thermal Analysis," Univ. Arkansas Inst. Sci. and Technol. Research Ser., No. **21** (November 1951).
(13) Vincent, S. M., Bell Telephone Laboratories, Inc., Murray Hill, N. J., unpublished measurements.

RECEIVED for review July 28, 1955. Accepted October 4, 1956. Division of Analytical Chemistry, 126th Meeting ACS, New York, N. Y., September 1954.

12

Reprinted from *J. Phys. Chem.*, **61**, 917–921 (July 1957)

DIFFERENTIAL THERMAL ANALYSIS OF INORGANIC HYDRATES[1]

By Hans J. Borchardt and Farrington Daniels

Contribution from the Department of Chemistry, University of Wisconsin, Madison, Wisconsin

Received January 8, 1957

Differential thermographs are presented for $CuSO_4 \cdot 5H_2O$, $CoCl_2 \cdot 6H_2O$, $MnCl_2 \cdot 4H_2O$, $SrCl_2 \cdot 6H_2O$ and $BaBr_2 \cdot 2H_2O$. Two kinds of anomalous peaks were observed. One originates from the formation and subsequent vaporization of liquid water when a hydrate decomposes. This situation has sometimes led to the erroneous interpretation of data in the literature. The other type of anomalous peak is due to the sudden change in the thermal conductivity of a sample when liquid forms. X-Ray measurements were used to determine the origin of the peaks in the differential thermographs. The X-ray data are given for the compound $CuSO_4 \cdot CuO$ and new lines are given for $SrCl_2 \cdot 2H_2O$.

Introduction

Differential thermal analysis (DTA) is a useful tool for studying the processes which occur in a material on heating. The difference between the temperature of a sample of powder and that of a standard powder is recorded as the two are heated simultaneously. Endothermic processes such as melting or dehydration are shown by minima or "peaks" in the curve in which temperature differences are plotted against the temperature. Certain conditions, however, may influence the differential thermograph and lead to misinterpretations. Inorganic hydrates, for example, where non-equilibrium conditions exist, may give rise to extra peaks as will be described.

The copper sulfate-hydrate system was studied with DTA by Taylor and Klug.[2] These authors obtained a differential thermograph of $CuSO_4 \cdot 5H_2O$, for the temperature range 40 to 160°, essentially the same as that shown by the solid line in Fig. 1. The product formed at 120° after the double peak was identified as $CuSO_4 \cdot 3H_2O$. Since the transition from $CuSO_4 \cdot 5H_2O$ to $CuSO_4 \cdot 3H_2O$ occurred in two stages, as evidenced by the appearance of a double peak, these authors concluded that an intermediate phase existed. Chemical analysis of the product after the first peak of the doublet gave a Cu to H_2O ratio of 1:4. Thus Taylor and Klug reported the existence of $CuSO_4 \cdot 4H_2O$. More recently, Ghosh[3] reported evidence for this compound as a result of thermogravimetric studies. Its existence is mentioned in several authoritative texts.[4-6]

$CuSO_4 \cdot 5H_2O$ as well as $CoCl_2 \cdot 6H_2O$, $MnCl_2 \cdot 4H_2O$, $SrCl_2 \cdot 6H_2O$, $BaBr_2 \cdot 2H_2O$ and $BaCl_2 \cdot 2H_2O$ were studied in this Laboratory with DTA. The differential thermograph of each of these com-

pounds, with the exception of $BaCl_2 \cdot 2H_2O$, showed one more peak than would be expected on the basis of their commonly known hydrates. An intensive effort to find corroboration for new phases using X-ray techniques led, without exception, to negative results. Further observation made it apparent that when these unexpected peaks occur the hydrate does not go directly to water vapor and the next lower hydrate, but that liquid water is formed. The subsequent vaporization of this water gives rise to the extra peak. The occurrence of this phenomenon is determined by the phase relationships of the hydrate and the procedure followed in differential thermal analysis.

Experimental

DTA Apparatus.—The DTA apparatus was a modification of that reported by Whitehead and Breger[7] which makes use of radiation shields in place of conventional insulation so that a vacuum can be attained readily. Pt–Pt 10% Rh thermocouples are used in a ceramic sample holder. The sample well was 5/32 inch in diameter and 5/8 inch deep, having a capacity of 0.1 to 0.2 g. With such small samples, the heat effect is very small and hence considerable amplification is necessary. An amplification factor of the order of 2000 was used most frequently. The high amplification in turn required special techniques to assure exact centering of the differential thermocouple in the furnace. To achieve this, a device was incorporated which allowed the position of the furnace to be varied continuously by small increments. Details are given elsewhere.[1]

The furnace is wound with nichrome wire. Different gases may be introduced by ‹evacuating the furnace and bleeding in the desired gas.

The output from the differential thermocouple is amplified by a Liston–Becker model 14 d.c. breaker amplifier and is then recorded on a Brown recording potentiometer having a scale 0 to 20 millivolts. The rate of temperature rise was programmed on a Brown Potentiometer Pyrometer which also gave a record of the sample temperature. For most work a rate of temperature rise of 10° per minute was used. Commercial C.P. material was used without further purification. The samples for DTA were ground to 100–200 mesh and diluted with an equal weight of calcined alumina. Alumina also served as reference material. Some aspects of the procedure are discussed in greater detail in another paper.[8]

X-Ray Analysis.—All X-ray work was performed on a North American Philips Type 12045 diffraction instrument using a copper target and nickel filter. For identification purposes a Philips 52057A, 57.3 mm. diameter camera was used. For the patterns reported herein a Philips 52056, 114.59 mm. camera was employed. The patterns were checked on a Norelco diffractometer. For work at ele-

(1) Presented in part at the 129th meeting of the American Chemical Society, Dallas, Texas, April, 1956. Further details may be found in a Ph.D. thesis by Hans J. Borchardt, filed in the Library of the University of Wisconsin, June, 1956.

(2) T. I. Taylor and H. P. Klug, *J. Chem. Phys.*, **4**, 601 (1936).

(3) B. Ghosh, *J. Indian Chem. Soc.*, **20**, 120 (1943).

(4) J. E. Ricci, "The Phase Rule and Heterogeneous Equilibrium," D. Van Nostrand Co., New York, N. Y., 1951, p. 140.

(5) S. Bowden, "The Phase Rule and Phase Reactions," The Macmillan Co., New York, N. Y., 1938, p. 71.

(6) N. V. Sidgwick, "The Chemical Elements and Their Compounds," Oxford Press, 1950, p. 155. (Sidgwick erroneously cites Taylor and Klug[2] as a reference for $CuSO_4 \cdot 2H_2O$. These authors make no mention of a dihydrate.)

(7) W. L. Whitehead and I. A. Breger, *Science*, **111**, 279 (1950).

(8) H. J. Borchardt, *J. Chem. Ed.*, **33**, 103 (1956).

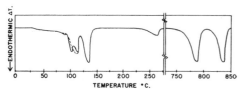

Fig. 1.—Differential thermograph of CuSO$_4$·5H$_2$O.

vated temperatures, two devices were used. One was a Norelco high-temperature camera. Samples for this camera were sealed in quartz capillaries, 0.3 mm. in diameter. The other device was a brass sample holder, 0.5 inch long, 1 inch wide and 1.5 inches deep, open at the top and wound around with nichrome wire so that it could be heated directly. It contained a thermocouple in its interior. This sample holder is mounted on the Norelco diffractometer, the diffractometer shield being removed to accommodate the sample holder. In this manner, diffraction lines of the sample were scanned while the sample was being heated.

Results and Discussion

CuSO$_4$·5H$_2$O.—The differential thermograph of CuSO$_4$·5H$_2$O is shown in Fig. 1. No peaks appeared between 275 and 725°. Above 750° the two-stage decomposition of CuSO$_4$ to CuO with CuSO$_4$·CuO as an intermediate is observed. The identity of CuSO$_4$ and CuO was established by X-ray analysis and reference to standard patterns. A weight–time plot of CuSO$_4$ heated at 720° showed a sharp break at the composition CuSO$_4$·CuO. The X-ray data for the CuSO$_4$·CuO compound are given in Table I.

TABLE I

X-RAY DATA FOR THE COMPOUND CuSO$_4$·CuO

Line no.	1	2	3	4	5[b]	6	7	8
d	6.46	4.91	4.75	4.63	4.02	3.63	3.41	3.29
I/I$_0$[a]	78	6	25	14	4	100	15	8
			band					
Line no.	9	10	11	12	13	14	15[b]	16[b]
d	3.15	2.83	2.78	2.67	2.62	2.54	2.47	2.38
I/I$_0$[a]	16	21	47	8	94	34	10	8
Line no.	17[b]	18	19	20	21	22	23	24
d	2.32	2.25	2.14	2.02	1.97	1.86	1.82	1.76
I/I$_0$[a]	6	52	5	26	10	4	7	23
Line no.	25	26	27	28	29	30	31	32
d	1.70	1.67	1.64	1.62	1.61	1.60	1.58	1.55
I/I$_0$[a]	5	15	4	7	11	14	16	5
					band			
Line no.	33	34	35	36	37	38	39	
d	1.53	1.51	1.48	1.47	1.45	1.41	1.39	
I/I$_0$[a]	5	4	9	15	7	8	17	
		band						

[a] The relative intensities (I/I_0) are the ratios of peak heights on the diffractometer trace. The designated bands appeared as a broad line in the X-ray photograph but were resolved in the diffractometer trace. [b] These lines correspond to intense lines in the CuSO$_4$ or CuO X-ray pattern. They may be due to the presence of these materials as impurities.

The low temperature portion of the pattern lacked reproducibility in early experiments. The first two large peaks (solid line, Fig. 1) appeared on some days as a single peak as shown by the dotted line. The small initial peak, coming before the two solid-line peaks, appeared only when the doublet (solid line) was observed and then only at high heating rates (> 15°/minute). This difference in behavior was traced to the variation in atmospheric humidity, the doublet occurring

only in periods of high relative humidity. The observation was verified by controlling the atmosphere in the furnace. With an initially dry atmosphere, a single peak always occurred. With a furnace atmosphere initially saturated with water vapor, the doublet always appeared.

In order to determine the origin of the extra peak, samples of CuSO$_4$·5H$_2$O were partially dehydrated and X-ray diffraction photographs taken. All lines could be assigned to the lines reported in the literature for CuSO$_4$·5H$_2$O and CuSO$_4$·3H$_2$O. Since the extra peak occurred only at higher vapor pressures, a procedure was followed to give high pressures of water vapor. An intimate mixture containing approximately 55 mole % CuSO$_4$·5H$_2$O and 45 mole % CuSO$_4$·3H$_2$O was sealed in a quartz capillary. On the basis of the size of the capillary and the amount of sample taken, no significant quantity of CuSO$_4$·5H$_2$O dissociated, when the partial pressure of water vapor in the capillary reached 1 atmosphere. This capillary was maintained at 90° for several hours in a high temperature X-ray camera. No new lines appeared. It was further heated to approximately 95° where a blank photograph resulted. Examination of the capillary showed that liquid was present and only a little salt remained. By shifting the position of the capillary a feeble X-ray pattern of the remaining salt was obtained. The pattern corresponded to CuSO$_4$·3H$_2$O.

In order to observe portions of the X-ray pattern continually while CuSO$_4$·5H$_2$O was heated, the previously described sample arrangement on the diffractometer was used. The sample temperature was followed as a function of time at a constant heating rate. Breaks in the time–temperature curve occurred at 92.5°, 102 and 115°. The line of CuSO$_4$·5H$_2$O having a Bragg d value of 3.70 was scanned during heating. No significant changes were observed until the sample had been heated at 92.5° for approximately one minute. At this point the sample shrank and was not available to the X-ray beam. The shrinking and visible wetting of the sample showed clearly that liquid water was forming. Even more convincing was the observation of water vapor steaming from the sample holder for the duration of the 102° break in the heating curve. Since the breaks in the heating curve correspond to the peaks in the differential thermograph, the doublet observed by Taylor and Klug[3] and in this investigation is not due to a tetrahydrate, but simply to the processes

$$CuSO_4·5H_2O(s) \longrightarrow CuSO_4·3H_2O(s) + 2H_2O(l)$$
$$2H_2O(l) \longrightarrow 2H_2O(g) \text{ (from satd. soln.)}.$$

These findings are consistent with the classical vapor pressure–temperature diagram of the CuSO$_4$-hydrate system (Fig. 2). On heating CuSO$_4$·5H$_2$O above its dissociation temperature, trihydrate and water vapor form. This water vapor does not diffuse away from the sample at an appreciable rate. The resulting increase in the local partial pressure of water as the temperature is raised would be given by the dotted line if the system were in equilibrium. In DTA the temperature is probably somewhat higher than indicated since relatively large rates of temperature rise are employed. When the partial pressure of water vapor becomes somewhat greater than 568 mm., the vapor pressure at the quadruple point (A),[9] the remaining $CuSO_4 \cdot 5H_2O$ becomes unstable with respect to the trihydrate and saturated solution. This transition gives rise to the first large peak in the differential thermograph (Fig. 1) and the first break in the heating curve. On further heating, the vapor pressure of the saturated solution increases until it reaches atmospheric pressure, at which point (B) water boils off and the second peak appears, these two peaks comprising the doublet. The next two peaks in Fig. 1 result from the transition of trihydrate to mono (130°), and monohydrate to anhydrous salt (250°), respectively, at a vapor pressure equal to atmospheric pressure.

The failure of a double peak (occasioned by the production of liquid water) to appear when the initial gas is dry, is attributed to the fact that the partial pressure of water vapor in the tube never gets as high as the vapor pressure of the saturated solution and all of the $CuSO_4 \cdot 5H_2O$ is dissociated before a pressure of 568 mm. is reached.

The other evidence for $CuSO_4 \cdot 4H_2O$ is Taylor and Klug's[3] chemical analysis and Ghosh's[4] thermogravimetric work. With regard to the former, one can only conclude that their sample may have consisted of a fortuitous mixture of $CuSO_4 \cdot 3H_2O$ and adhering solution.

Ghosh relied quite heavily on sudden deflections of a simplified thermobalance. His method indicated the presence of $CuSO_4 \cdot 4 \frac{1}{2} H_2O$ and $CuSO_4 \cdot 4 \frac{1}{3} H_2O$ as well as $CuSO_4 \cdot 4H_2O$, neither of which have been observed in any other studies.

Other Hydrates.—The occurrence of extra peaks in the differential thermal analysis of hydrates due to formation of saturated solution is not uncommon. Figure 3 shows the differential thermographs of five other hydrates. The processes giving rise to the peaks are summarized in Table II. Confirmatory X-ray analyses were performed to show when new phases were present and when they were not.

In the course of this work two distinct new lines were found in partially dehydrated $SrCl_2 \cdot 6H_2O$. These were traced to $SrCl_2 \cdot 2H_2O$, being the first two lines exhibited by this compound and occurring at Bragg d values of 5.64 and 4.55. They are the second and third most intense lines, having relative intensities of 79 and 90, respectively. The X-ray data are given in Table III. The lines are consistent with the structure of $SrCl_2 \cdot 2H_2O$ reported by Jensen,[10] corresponding to reflections from the 200 and 011 planes, respectively.

The necessary requirements for the appearance of a liquid phase in the differential thermal analysis of a hydrate are (1) that the hydrate system contains a quadruple point where hydrate, next lower hydrate, saturated solution and water vapor are in equilibrium; (2) that this quadruple point occurs at a water vapor pressure which is less than atmospheric pressure; (3) that the rate of dissociation of the hydrate be rapid (if the sample tem-

(9) E. M. Collins and A. W. C. Menzies, THIS JOURNAL, **40**, 379 (1936).

(10) K. Jensen, *Danske Vidensk. Selsk. Mat. Fys. Medd.*, **20** (No. 5), 22 (1942).

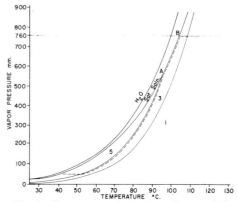

Fig. 2.—Vapor pressure–temperature diagram of the $CuSO_4$–hydrate system. The areas 5, 3 and 1 represent the stable regions of the penta-, the tri- and the monohydrate.

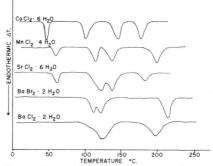

Fig. 3.—Differential thermographs of several hydrated salts. The second peak in each of the top four thermographs is a "false" peak due to vaporization of water.

perature has attained a value greater than the boiling point of saturated solution before appreciable local vapor pressures are established, no liquid can form); and (4) the water vapor which is evolved must be confined to the immediate vicinity of the sample.

Conditions (3) and (4) require that an intermediate rate of temperature rise be employed if a liquid phase is to be observed. With very rapid heating rates the system is too far from equilibrium and condition (3) will not be met. Very low rates of temperature rise will allow time for the water vapor to diffuse away from the sample.

The temperatures and pressures at the quadruple point for the hydrates are summarized in Table IV. All samples comply with the requirement that the pressure at the quadruple point must be below atmospheric. $BaCl_2 \cdot 2H_2O$ is the only material studied where condensation failed to take place, as indicated by the absence of an extra peak in the DTA curve. The relatively small difference between the vapor pressure of $BaCl_2 \cdot 2H_2O$ at the quadruple point and atmospheric pressure requires close adherence to equilibrium conditions if the formation of liquid is to be observed. This condition, obviously, was not met.

Table II
Processes Which Give Rise to the Peaks in Fig. 3

Hydrate	Peak No. and temp.[a]	Process
$CoCl_2 \cdot 6H_2O$	1 (49°)	$CoCl_2 \cdot 6H_2O(s) \rightarrow CoCl_2 \cdot 2H_2O(s) + 4H_2O(l)$
	2 (99°)[b]	$4H_2O(l) \rightarrow 4H_2O(g)$
	3 (137°)	$CoCl_2 \cdot 2H_2O(s) \rightarrow CoCl_2 \cdot H_2O(s) + H_2O(g)$
	4 (175°)	$CoCl_2 \cdot H_2O(s) \rightarrow CoCl_2(s) + H_2O(g)$
$MnCl_2 \cdot 4H_2O$	1 (55°)	$MnCl_2 \cdot 4H_2O(s) \rightarrow MnCl_2 \cdot 2H_2O(s) + 2H_2O(l)$
	2 (102°)[b]	$2H_2O(l) \rightarrow 2H_2O(g)$
	3 (135°)	$MnCl_2 \cdot 2H_2O(s) \rightarrow MnCl_2 \cdot H_2O(s) + H_2O(g)$
	4 (210°)	$MnCl_2 \cdot H_2O(s) \rightarrow MnCl_2(s) + H_2O(g)$
$SrCl_2 \cdot 6H_2O$	1 (66°)	$SrCl_2 \cdot 6H_2O(s) \rightarrow SrCl_2 \cdot 2H_2O(s) + 4H_2O(l)$
	2 (122°)[b]	$4H_2O(l) \rightarrow 4H_2O(g)$
	3 (132°)	$SrCl_2 \cdot 2H_2O(s) \rightarrow SrCl_2 \cdot H_2O(s) + H_2O(g)$
	4 (183°)	$SrCl_2 \cdot H_2O(s) \rightarrow SrCl_2(s) + H_2O(g)$
$BaBr_2 \cdot 2H_2O$	1 (110°)	$BaBr_2 \cdot 2H_2O(s) \rightarrow BaBr_2 \cdot H_2O(s) + H_2O(l)$
	2 (118°)[b]	$H_2O(l) \rightarrow H_2O(g)$
	3 (209°)	$BaBr_2 \cdot H_2O(s) \rightarrow BaBr_2(s) + H_2O(g)$
$BaCl_2 \cdot 2H_2O$	1 (125°)	$BaCl_2 \cdot 2H_2O(s) \rightarrow BaCl_2 \cdot H_2O(s) + H_2O(g)$
	2 (200°)	$BaCl_2 \cdot H_2O(s) \rightarrow BaCl_2(s) + H_2O(g)$

[a] These temperatures bear an uncertainty of about ±5°. [b] Water boiling from saturated solution.

Table III
X-Ray Data for $SrCl_2 \cdot 2H_2O$

Line no.	1[a]	2[a]	3	4[b]	5	6	7	8	9
d	5.64	4.55	3.98	3.28	3.20	2.88	2.80	2.71	2.66
I/I_0	79	90	47		100	24	45	27	71
Line no.	10	11	12	13	14	15	16	17	18
d	2.65	2.61	2.54	2.48	2.44	2.27	2.26	2.19	2.14
I/I_0	34	24	32	34	15	24	32	11	3
Line no.	19	20	21	22	23	24	25	26	
d	2.115	2.103	2.025	1.985	1.973	1.927	1.886	1.857	
I/I_0	27	40	29	8	31	5	19	15	

[a] These lines are not reported in the A.S.T.M. index. [b] Line 4 appears as a separate line in the photograph but did not resolve completely from line 5 in the Geiger counter diffractometer trace at a chart speed of 1/4°/minute using 1° slits. Hence, its relative intensity is not reported. It is a very weak line.

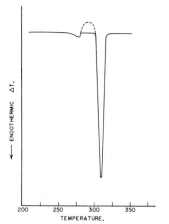

Fig. 4.—Differential thermograph of $NaNO_3$ showing a small peak at about 275° due to a second-order transition, and a large peak at 310° due to fusion. The solid line is obtained with a temperature rise of 15° per minute. The dotted exothermic peak appears when the temperature is raised at the rate of 30° per minute. It is due to increased thermal conductivity caused by fusion.

Small Initial Peaks.—A small initial endothermic peak is observed with $CuSO_4 \cdot 5H_2O$ as well as with

Table IV
Temperature and Pressure at Quadruple Point for Hydrates

Hydrate	$T°C.$	P, mm.	Ref.
$CoCl_2 \cdot 6H_2O$	52.4	48.5	11
$MnCl_2 \cdot 4H_2O$	58.1	63.2	12
$SrCl_2 \cdot 6H_2O$	61.6	89.7	9
$BaBr_2 \cdot 2H_2O$	107.9	664	9
$BaCl_2 \cdot 2H_2O$	101.9	684	9

the other hydrates when high heating rates (> 15°/min.) are used. A similar phenomenon is observed when fusion occurs except that the peak is in the exothermic direction as shown schematically by the dotted line in Fig. 4. These peaks are explained as follows: The formation of liquid in the sample holder is accompanied by a sudden change in the thermal conductivity of the material, inasmuch as it becomes a continuous medium, hence a much better heat transfer agent. This increased heat conductance causes a small surge of heat to the thermocouple junction as the thermal gradient in the sample diminishes. This gives rise to an exothermic peak. In the case of the hydrates, the occurrence of this exothermic peak, while an over-all endothermic reaction is taking place, makes it ap-

(11) "International Critical Tables," Vol. III, McGraw-Hill Book Co., New York, N. Y., 1928, p. 361.

(12) A. von Benrath, Z. anorg. allgem. Chem., **247**, 147 (1941).

pear as if two endothermic peaks are occurring.

This explanation is consistent with the observation that the small peak occurred in the pattern of $CuSO_4 \cdot 5H_2O$ in Fig. 1 only when the doublet appeared, that is, only when liquid formed in the sample and only at high rates of temperature rise.[13] Since the sample wells in this work are very small, thermal gradients are reduced to a minimum. In order to establish an appreciable gradient high heat-

(13) Taylor and Klug[1] observed several small peaks prior to dehydration which they attributed to second-order transitions in $CuSO_4 \cdot 5H_2O$. These were observed at sensitivities much higher than those used in the present work and are apparently not the same as the anomalous peak described above.

ing rates are necessary.

The results of this investigation emphasize precautions which are necessary in the interpretations of some differential thermal analysis measurements.

Acknowledgments.—The X-ray work was conducted with the helpful advice of Professor Sturges Bailey of the Department of Geology, University of Wisconsin.

The authors are grateful for the support of this research by the Atomic Energy Commission Contract AT(11-1)-178 and by a grant for fundamental research by the E. I. du Pont de Nemours Company, Inc.

13

Reprinted from J. Am. Chem. Soc., 79, 41–46 (1957)

The Application of Differential Thermal Analysis to the Study of Reaction Kinetics[1]

By Hans J. Borchardt and Farrington Daniels

Received July 23, 1956

Equations are derived which relate the shape of a differential thermal analysis curve to the kinetics of the reaction giving rise to the curve. For certain reactions, use of these equations allows the order of the reaction, the frequency factor, the activation energy and the heat of reaction to be determined in a single rapid measurement. The equations are applied to the decomposition of benzenediazonium chloride and the reaction between dimethylaniline and ethyl iodide. The results agree very well with data obtained by conventional methods.

Increasing use has been made of differential thermal analysis (DTA) in recent years for studying the processes which a substance undergoes on heating.[2] The differential thermograph is commonly used to determine if a reaction or transition occurs and the temperature at which it takes place. The present article shows that the kinetic parameters for the reaction giving rise to the DTA curve can be accurately determined by an analysis of the shape (i.e., slope, height, area) of the curve. Conditions are assumed in the derivation which can be met more readily by liquids than by solids. The present discussion will therefore concern itself primarily with reactions occurring in solution.

Theory

The apparatus referred to in this derivation is shown in Fig. 1. It consists of two cells mounted in a bath. One cell contains the solution of reactants and the other pure solvent or other inert liquid. The contents of the cells are agitated by the indicated stirrers. The temperature of the bath is raised by a heater (not shown). The temperature of the reactant solution as well as the temperature difference between the contents of the two cells (ΔT) is measured as a function of time. A differential thermocouple (DTC) may be used to measure ΔT. The run is started at a sufficiently low temperature so that the reaction is not occurring at an appreciable rate and it is carried through until the reaction has gone essentially to completion. A curve such as the one in Fig. 2 is obtained.

T_1, T_2 and T_3 are the temperatures of the reactant solution, the liquid in the reference cell and the bath, respectively. $C_{p,r}$ is the total heat capacity

(1) Presented in part at the 129th meeting of the American Chemical Society, Dallas, Texas, April, 1956. Further details may be found in a Ph.D. thesis by Hans J. Borchardt, filed with the Library of the University of Wisconsin, June, 1956.

(2) For an introduction to differential thermal analysis see H. J. Borchardt, J. Chem. Ed., 33, 103 (1956).

of the reactant solution and $C_{p,s}$ the total heat capacity of the liquid in the reference cell.

We first set up the equations of heat balance (eq. 1 and 2). Equation 1 states that the increase in the enthalpy of the reactant solution ($C_{p,r} \, dT_1$) is equal to the heat evolved by the reaction (dH) plus the heat transferred into the cell from the surroundings.

$$C_{p,r} \, dT_1 = dH + K_r (T_3 - T_1) dt \qquad (1)$$

K_r is the heat transfer coefficient of the reactant cell and dt the time interval. Similarly for the liquid in the reference cell we have

$$C_{p,s} \, dT_2 = K_s (T_3 - T_2) dt \qquad (2)$$

where K_s is the heat transfer coefficient of the reference cell.

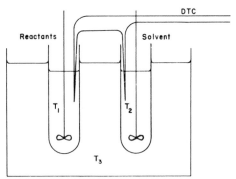

Fig. 1.—DTA apparatus for obtaining kinetic data for reactions occurring in solution.

Thus far two assumptions have been made. The first is that the temperature in the respective cells is uniform. This assumption was made when a single value of the temperature (T_1 and T_2) was assigned to the liquids in the cell. The condition

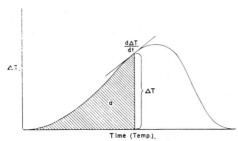

Fig. 2.—DTA curve showing the quantities which are measured in order to evaluate the rate constants for the reaction giving rise to the curve.

of uniform temperature cannot be met exactly by solids, but can be met by liquids which are stirred. It was for this reason that these considerations were limited to reactions in solution. The other assumption is that heat is transferred by conduction alone. This assumption is valid in the temperature range where one would usually be working with solutions. Heat transfer through the thermocouple wires is neglected.

At this point two additional conditions are assumed. One is that $K_r = K_s = K$. The heat transfer coefficients will be equal if identical cells are used and if they are filled to the same level. We therefore drop the subscript and characterize these cell constants by a single symbol K. The other assumption is that $C_{p,r} = C_{p,s} = C_p$. The heat capacities of the two liquids will be nearly the same if the reactant solution is dilute and if pure solvent is used as the reference liquid. If the reaction involves several components, a solution containing all but one of the reactants (so that no reaction can take place) would approximate closely the heat capacity of the reactant solution and serve well as the reference liquid. Because the cells must be filled to the same level, the volumes of the two liquids are the same; hence the heat capacities per unit volume should match. This is fortunate since the heat capacities of solutions and solvents are often nearly the same on a volume basis.

Since the subscripts on C_p and K have been dropped, eq. 2 may be subtracted from eq. 1 giving

$$dH = C_p \, d\Delta T + K \Delta T \, dt \qquad (3)$$

where the conventional symbol ΔT has been substituted for $T_1 - T_2$. In order to obtain an expression for the total heat transferred, eq. 3 is integrated between $t = 0$ and $t = \infty$. Assuming C_p and K to be independent of temperature (time) over the interval where the reaction occurs, we obtain

$$\Delta H = C_p(\Delta T_\infty - \Delta T_0) + K \int_0^\infty \Delta T \, dt \qquad (4)$$

Since ΔT is zero at both $t = 0$ and $t = \infty$ (see Fig. 2), the first term is just zero. The integral is the total area under the curve (A). Therefore

$$\Delta H = KA \qquad (5)$$

This equation which states that the area under the DTA curve is directly proportional to the heat

transferred in the reaction was first derived by Spiel[3] in a somewhat different manner. Since ΔH is the total heat transferred, the heat of reaction per mole is given by KA/n_0 where n_0 is the initial number of moles of reactant.

It is now assumed that the heat evolved in a small time interval is directly proportional to the number of moles reacting during that time.

$$dH \propto -dn; \quad dH = -\frac{KA}{n_0} dn \qquad (6)$$

The constant of proportionality is clearly the heat of reaction per mole. Here the heat of reaction is assumed to be constant over the temperature interval where the reaction occurs. Substituting eq. 6 for dH in eq. 3 and differentiating with respect to time gives

$$-\frac{dn}{dt} = \frac{n_0}{KA} \left[C_p \frac{d\Delta T}{dt} + K\Delta T \right] \qquad (7)$$

This equation gives the actual rate of reaction at any temperature in terms of the slope $(d\Delta T/dt)$ and height (ΔT) of the curve at that temperature.

The number of moles present (n) at any instant is equal of the initial number of moles (n_0) minus the number of moles that have reacted.

$$n = n_0 - \int_0^t -\frac{dn}{dt} dt \qquad (8)$$

Substituting into eq. 8 the expression for the rate of reaction (eq. 7) gives

$$n = n_0 - \frac{n_0}{KA} \left[C_p \int_0^t \frac{d\Delta T}{dt} dt + K \int_0^t \Delta T \, dt \right] \qquad (9)$$

Integrating

$$n = n_0 - \frac{n_0}{KA} [C_p \Delta T + Ka] \qquad (10)$$

The second integral in eq. 9 is the area that has been swept out (a) at time t. The quantity a is shown in Fig. 2.

The expression for the rate constant of a reaction of order x with respect to one component is

$$k = -V^{x-1} \frac{dn/dt}{n^x} \qquad (11)$$

where V is the volume and n the number of moles (rather than concentrations). Substituting eq. 7 and 10 for dn/dt and n, respectively, and rearranging gives

$$k = \left[\frac{KAV}{n_0} \right]^{x-1} \frac{C_p \frac{d\Delta T}{dt} + K\Delta T}{[K(A - a) - C_p \Delta T]^x} \qquad (12)$$

For the case of a first-order reaction $(x = 1)$ eq. 12 simplifies to

$$k = \frac{C_p \frac{d\Delta T}{dt} + K\Delta T}{K(A - a) - C_p \Delta T} \qquad (13)$$

For the case of a reaction with respect to several components of the form $lL + mM + \cdots \rightarrow products$, having the rate expression

$$\frac{dz}{dt} = (L - z)^l \left(M - \frac{m}{l} z \right)^m \qquad (14)$$

(3) S. Spiel, Rept. Inv. No. 3765, Bur. of Mines, U. S. Dept. of Interior, 1944.

the equation becomes

$$k = \frac{\left[\frac{KAV}{L_0}\right]^{l+m-1}\left[C_p \frac{d\Delta T}{dt} + K\Delta T\right]}{[K(A-a) - C_p \Delta T]^l \left[K\left(\frac{M_0}{L_0}A - \frac{m}{l}a\right) - C_p \Delta T\right]^m \cdots}$$

(15)

where z is the number of moles of L reacted in time t and L_0 and M_0 are the initial number of moles of L and M, respectively.

The various quantities are evaluated as follows.

K need only be evaluated once since it is a characteristic of the apparatus. It is best determined by means of eq. 5. A quantity of heat is dissipated in the cell either electrically or by carrying out a reaction where the heat effect is known. The ratio of heat evolved to the area under the resulting curve gives K. The dimensions of K are calories min.$^{-1}$ deg. C.$^{-1}$.

A is the total curve area. The dimensions are min. deg. C.

V/n_0 is the reciprocal of the initial concentration of reactants. If the volume changes considerably with temperature, this fact should be taken into consideration. The choice of units depends upon the dimensions in which the rate constant is desired.

C_p is the total heat capacity of the reactant solution or reference liquid. The heat capacities of most solutions are not well known. It will be shown in the experimental section that little error results if the value for the heat capacity of the solvent is used. The dimensions of C_p are cal. deg. C.$^{-1}$.

$d\Delta T/dt$, ΔT and a are the slope, height and area, respectively, of the curve at the temperature (time) at which k is being evaluated. The dimensions of these quantities are, respectively, deg. C. min.$^{-1}$, deg. C. and deg. C. min. The only term which remains is x (or $l + m$ in the case of eq. 14), the order of the reaction. If a value of x were assumed, k could be calculated at all temperatures over the range where the curve extends. The plot of ln k versus $1/T$ (activation energy plot) would yield a straight line if, and only if, the correct value of x had been assumed. It is this fact and procedure which is used to determine the order of the reaction. One varies x, usually by integers, until the activation energy plot is linear. The value of x thus obtained is the order of the reaction.

The correct plot of ln k versus $1/T$ enables one, of course, to calculate the activation energy and frequency factor. Equation 5 gives the heat of reaction. The activation energy and the heat of reaction enable one to calculate the activation energy for the reverse reaction.

These equations assume that the reaction is carried out at constant pressure. Analogous equations are obtained for constant volume. Here C_v replaces C_p, and ΔE, the change of internal energy, replaces ΔH.

These considerations may also be applied to differential enthalpic analysis (DEA). In this method, the temperature difference between the active and reference material is maintained at zero by supplying heat to either the active or reference materials during a reaction. Thus dH/dt is meas-

ured and plotted directly as a function of time (temperature). This approach has been pursued by Eyraud.[4] In this case the total area is equal to the heat transferred

$$\Delta H = A \qquad (16)$$

If it is assumed that the heat evolved is directly proportional to the number of moles reacted, it follows that

$$-\frac{dn}{dt} = -\frac{n_0}{A}\frac{dH}{dt} \qquad (17)$$

An expression for the amount of reactant present at any instant is obtained in the same manner as before.

$$n = n_0 - \frac{n_0 a}{A} \qquad (18)$$

Thus the rate constant is given by

$$k = \frac{\left(\frac{AV}{n_0}\right)^{z-1}\frac{dH}{dt}}{(A-a)^z} \qquad (19)$$

Equation 19 is not limited to differential enthalpic analysis. It would apply to any procedure where the rate of change of a physical property, proportional to the rate of reaction, is measured as a function of temperature and time under conditions where the temperature is changing. The physical property should be nearly independent of temperature. A different proportionality constant would appear in eq. 17 and 19 if the physical property was other than the heat transferred during a reaction.

Discussion of Assumptions.—The following list summarizes the assumptions made in the derivation of the equations for DTA and the conditions which must practically be met; (1) The rate of reaction is very small at the lowest temperature practically obtainable; (2) the reaction goes essentially to completion before the highest obtainable temperature is reached; (3) the reaction must be accompanied by a measurable heat effect; (4) the temperature in the cells is uniform; (5) heat is transferred to the solutions by conduction alone; (6) $K_r = K_s$; (7) $C_{p,r} = C_{p,s}$; (8) C_p, K and ΔH do not vary over the temperature interval where the reaction occurs; (9) no heat is transferred through the thermocouples; (10) dn is proportional to dH; (11) the kinetics of the reaction can be described by a single rate constant; (12) the activation energy does not vary with temperature.

Assumptions 4, 5, 6 and 7 have been discussed. It will be shown in the Experimental section that inequalities in the cell constants and heat capacities have little effect on the results. Assumptions 8 and 12 are met almost exactly for temperature intervals that one might encounter in normal reaction studies. Heat transfer through the thermocouple is taken account of for all practical purposes, when the cells are calibrated. Assumptions 10 and 11 restrict the method to reactions which meet these conditions experimentally. Thus complex reactions are generally excluded. Some simple reaction which produces a gaseous product may not meet condition 10. If the gas forms a non-ideal solution with the solvent and exceeds its solubility in the

(4) C. Eyraud, Compt. rend., **238**, 1511 (1954).

course of the reaction, an interfering heat effect may occur. This problem may be overcome in some instances by working at constant volume rather than constant pressure. Assumptions 1 and 2 are essentially restrictions on the solvent. The freezing and boiling points of the solution should be such that the DTA curve occurs well within these temperatures. Problems due to vaporization of solvent may be overcome by conducting the experiments at constant volume. With regard to assumption 3, reactions accompanied by a heat transfer of 5 kcal./mole can be quantitatively studied with the simple apparatus described below. Improved instrumentation could probably lower this limit.

Experimental

Apparatus.—The cells consist of two Pyrex tubes, 1.25 inches in diameter and 5 inches long. They have a capacity of about 60 ml. The differential thermocouple (DTC) is a 0.028 inch o.d. copper tube with a coaxial constantan wire, insulated from the copper by fiberglass. This type of thermocouple is available from the Precision Tube Company, North Wales, Pa. under the trade name, Coaxitube. The thermocouples were coated with Kel-F to protect them from being attacked by the reactants. The bath temperature is raised by a conventional heater controlled by a Variac. It should be noted that a linear temperature rise is not required. The output from the thermocouple runs to a Liston–Becker Model 14 d.c. breaker amplifier. From here the signal goes to a Brown recording potentiometer having a scale 0–20 mv. A fine-scale chart paper having 20 divisions/inch is used. The bath temperature (T_b) is measured with a calibrated mercury thermometer which can be read directly to 0.1°.

It was assumed that the temperature difference between the bath and reactant solution was negligible within the limits required for activation energy calculations, hence the bath temperature was used for this calculation.

Calibration.—The heat of dilution of H_2SO_4 was used to determine K. One ml. of $H_2SO_4 \cdot 4.964H_2O$ was added to 35 ml. of water in the reaction cell. This dilution is accompanied by the evolution of 32.21 cal. of heat.[5] A curve having an area of 3.72 sq. inches resulted. This gives K as 8.66 cal./in.². A chart speed of $1/2$ in./min. was used throughout. It was further found that a ΔT of 1° caused a pen displacement on the recorder of 8.95 in. With these two conversion factors, K becomes 38.8 cal./min. °C. The conversion factor for ΔT from inches to degrees was determined by sending a 30 microvolts signal into the amplifier. The observed pen displacement of 6.61 in. together with the known e.m.f. of the copper–constantan couple (40.6 microvolts/°C.) gives the figure of 8.95 in./°C.

Reactions.—The thermal decomposition of benzenediazonium chloride has been studied by Crossley, Kienle and Benbrook[6] who found the reaction to be first order with

an activation energy of 27.2 kcal./mole and a frequency factor of $10^{16.2}$ sec.$^{-1}$. The reaction is

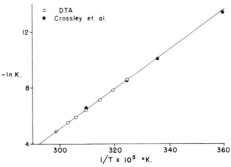

(20)

The products are nitrogen and chlorobenzene or phenol or both, depending upon the conditions.[7] Benzenediazonium chloride was synthesized after the method described by the above authors. Thirty-five ml. of a 0.4 M aqueous solution was inserted into one cell and 35 ml. of distilled water into the other. The initial bath temperature was 2°. The bath temperature was raised at a rate of approximately 1°/min. The reaction is exothermic. The resulting curve is shown in Fig. 3. This figure shows that the curve did not return exactly to $\Delta T = 0$ at the conclusion of the reaction. For the purpose of the calculations, the baseline was taken to be the slanted line joining the curve extremities. In order to obtain data for the activation energy plot, the height, slope and area of the curve were measured at 7 temperatures. The rate constant was calculated at these temperatures from eq. 13. The data are given in Table I.

TABLE I

FIRST-ORDER RATE CONSTANTS FOR THE DECOMPOSITION OF BENZENEDIAZONIUM CHLORIDE

Temp., °C.	k, sec.$^{-1}$	Temp., °C.	k, sec.$^{-1}$
35	1.82×10^{-4}	54	2.74×10^{-3}
40	3.85×10^{-4}	57	4.00×10^{-3}
45	7.90×10^{-4}	61.2	7.38×10^{-3}
50	1.64×10^{-3}		

Figure 4 shows the first-order activation energy plot and a comparison with the data of Crossley and co-workers.[6] It may be noted that the best straight line through Crossley's points has a slightly lower slope than the line through the DTA points. Thus, the activation energies do not agree exactly. By DTA one obtains an activation energy of 28.3 kcal./mole and a frequency factor of $10^{16.3}$ sec.$^{-1}$. From eq. 5 the heat of reaction is found to be -37.0 kcal./mole. This gives $28.3 - (-37.0) = 65.3$ kcal./mole as the activation energy for the reverse reaction. Two other runs were performed to determine the reproducibility of the results. Activation energies of 28.3, 29.1 and 28.5 kcal./mole were obtained.

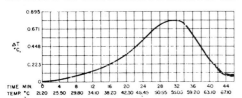

TIME MIN. 0 4 8 12 16 20 24 28 32 36 40 44
TEMP. °C 2120 2550 2980 3410 3820 4230 46.45 5055 5605 5920 6310 6710

Fig. 3.—DTA curve for the decomposition of benzenediazonium chloride. Temperature difference between reactant solution and inert liquid is plotted as a function of temperature and time.

Fig. 4.—Activation energy plot for the decomposition of benzenediazonium chloride. The circles represent the data obtained by differential thermal analysis (DTA) and the solid points the data of Crossley and co-workers.[6]

(5) F. D. Rossini, D. D. Wagman, W. H. Evans, S. Levine and I. Jaffe, "Selected Values of Chemical Thermodynamic Properties," Nat. Bur. Standards Cir. 500, 1952, pp. 42–43.

(6) M. L. Crossley, R. H. Kienle and C. H. Benbrook, THIS JOURNAL, **62**, 1400 (1940).

(7) The formation of both chlorobenzene and phenol in the decomposition of benzenediazonium chloride makes this a complex reaction. The fact that good agreement is obtained suggests that the heats of formation of chlorobenzene and phenol do not differ greatly.

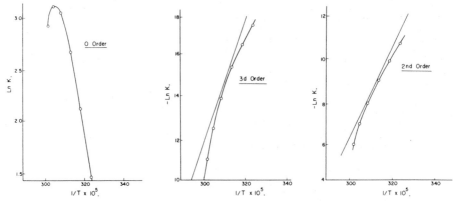

Fig. 5.—Activation energy plots for the benzenediazonium chloride decomposition assuming the order of the reaction to be zero, second, and third.

In order to test the validity of the procedure for determining the order of the reaction, k was recalculated assuming the orders to be zero, second and third, respectively. The plots are shown in Fig. 5. Distinctly non-linear curves result, verifying the procedure suggested. In this connection it may be pointed out that the data points in the first-order plot fit a straight line with an average deviation of only 0.18%.

The reaction of N,N-dimethylaniline with ethyl iodide to give N,N,N-dimethylethylanilinium iodide is discussed by Moelwyn-Hughes.[8] The reaction is

$$\text{C}_6\text{H}_5-\text{N(CH}_3)_2 + \text{C}_2\text{H}_5\text{I} \longrightarrow \left[\text{C}_6\text{H}_5-\text{N(CH}_3)_2\text{C}_2\text{H}_5\right]^+ + \text{I}^-$$

It is a bimolecular reaction, but if dimethylaniline is used as solvent and is accordingly in large excess, the reaction follows pseudo first-order kinetics. The activation energy and frequency factor are 14.0 kcal./mole and $10^{6.12}$ sec.$^{-1}$, respectively.[8] Since these values for the kinetic parameters are considerably different from those of the benzenediazonium chloride decomposition, it seemed an appropriate reaction to study.

Thirty-five ml. of dimethylaniline was inserted into the reactant cell and 40 ml. into the reference cell. When thermal equilibrium was established, 5 ml. of ethyl iodide was added to the reactant cell. The run was started at 25° and terminated at 150°. The resulting curve extended from 45 to 140°, the maximum occurring at 117.5°. The rate of temperature rise was approximately 2.3°/min. The time-interval was 40 minutes. Inasmuch as the reaction is first order,[8] rate constants were calculated at seven temperatures from eq. 13. The values are given in Table II.

TABLE II

FIRST-ORDER RATE CONSTANTS FOR THE REACTION BETWEEN N,N-DIMETHYLANILINE AND ETHYL IODIDE

Temp., °C.	k, sec.$^{-1}$	Temp., °C.	k, sec.$^{-1}$
84.6	3.16×10^{-4}	111	1.96×10^{-3}
91.2	4.51×10^{-4}	117	3.66×10^{-3}
98.0	7.39×10^{-4}	125	5.55×10^{-3}
104.5	1.27×10^{-3}		

The activation energy plot is shown in Fig. 6. The straight line is based on the activation energy of 14.0 kcal./mole and frequency factor of $10^{6.12}$ sec.$^{-1}$, quoted by Moelwyn-Hughes.[8] The circles represent the data obtained by DTA.

(8) E. A. Moelwyn-Hughes, "The Kinetics of Reactions in Solution," Oxford, 1933, p. 42.

Some experimental difficulties were encountered with the present apparatus at elevated temperatures. There was considerable vibration in the recorder pen at temperatures above 70°. This accounts for the presence of more scatter in the points on the activation energy plot. On the basis of the total curve area, the heat of reaction was found to be −20.4 kcal./mole. This gives 34.4 kcal./mole as the activation energy for the reverse reaction.

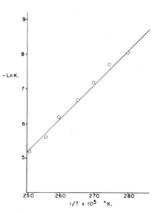

Fig. 6.—Activation energy plot for the reaction between N,N-dimethylaniline and ethyl iodide.

Discussion

It is interesting to note the order of magnitude of the quantities in eq. 13. For the benzenediazonium chloride decomposition, $C_p \mathrm{d}\Delta T/\mathrm{d}t$ varies from 0.634 at 35° to −2.70 at 61.2°. $K\Delta T$ over this range varies from 4.67 to 13.1 going through a maximum of 28.1 at 54°. $K(A - a)$ varies from 486 to 25.5 and $C_p\Delta T$ from 4.22 to 11.83 having a maximum of 25.4 at 54°. It is thus apparent that the quantities $(C_p \, \mathrm{d}\Delta T/\mathrm{d}t)$ and $C_p\Delta T$ are usually an order of magnitude smaller than the quantities to which they are added and subtracted. Neglecting these smaller terms, eq. 13 becomes

$$k = \frac{\Delta T}{A - a} \qquad (21)$$

and eq. 12

$$k = \frac{\left[\dfrac{A V}{n_0}\right]^{z-1} \Delta T}{(A - a)^z} \qquad (22)$$

These equations clearly indicate why this method is so insensitive to errors. The heat capacity essentially drops out. Slopes, which cannot be accurately measured have only a small effect on the final equation. The cell constant also tends to cancel out. The dominating terms in the equation are the height (ΔT) and remaining area of the curve ($A - a$), both of which can be measured accurately. In view of this fact, one would tend to assign more significance to the points calculated at lower temperatures where the remaining area is large and less susceptible to errors and where the smaller terms have the least influence.

It should be noted that eq. 22 is identical to eq. 19, the rate constant expression in differential enthalpic analysis. This demonstrates that the curves obtained by DTA and DEA are to a good degree of approximation the same.

The exceptional feature of this approach to kinetics is the rapidity with which the measurements are performed. The complete procedure including analysis of the data requires about one day, where a conventional approach may require several weeks to obtain the same data.

Acknowledgment.—The support of the U. S. Rubber Company in granting one of the authors (HJB) a fellowship is acknowledged with thanks.

Madison, Wisconsin

14

Reprinted from *Textile Res. J.*, **30**, 624–626 (Aug. 1960)

The Differential Thermal Analysis of Textile and Other High Polymeric Materials

Textile Research Institute
Princeton, New Jersey
May 17, 1960

To the Editor
TEXTILE RESEARCH JOURNAL

Dear Sir:

Recently the thermal behavior of various textile fibers and other high polymers has been studied by differential thermal analysis (DTA). In DTA thermally detectable transitions or reactions that occur as a substance is heated at a constant rate through the temperature range of interest are measured. This is done by determining the temperature difference (ΔT) that obtains between a thermally inert reference substance (calcined aluminum oxide was used in this work) and the substance being examined. ΔT is usually continuously recorded as a function of sample or furnace temperature to yield a curve or thermogram of ΔT in degrees Centigrade plotted against T in degrees Centigrade that is uniquely characteristic for each different sample substance. The DTA curve is characterized by the appearance of peaks due to deflections from the thermally steady state. The technique has been fully described in the literature [14, 17] and was first reported in 1922 [14]. However, the application of DTA to high polymers is of recent date and has been limited largely to a few general surveys [5, 12] and more detailed investigations of the simple transi-

tions occurring at temperatures below 300° C. [4, 9, 10], with the exception of work by Morita [13] and a very recent paper by Anderson and Freeman [1]. Very little is reported on textile fibers [5, 19], and no detailed study of the high-temperature degradation of textile materials by DTA has been reported in the literature at this writing.

Fiber and fabric samples of 75–150 mg. were heated at a rate of 10° C./min. from room temperature to 550° C. The techniques and apparatus employed were based on the work of Gordon and Campbell and others [7, 17] and will be described at a later date. The results obtained have been shown to be quite reproducible.

In Figure 1 the thermograms obtained in air and in nitrogen for Orlon[1] and in nitrogen for pure polyacrylonitrile (PAN) of molecular weight 200,000, in an amorphous form, are shown. In the case of Orlon both curves are dominated by a sharp exothermic reaction with a peak at 308° C. The air curve shows a second, less important, exothermic peak at 328° C., followed by a dip. These latter reactions have been eliminated under a nitrogen atmosphere. There is also a notable lack of any endothermic reaction, which suggests that Orlon does not undergo depolymerization as do cellulose and nylon. In nitrogen, PAN yields a DTA curve essentially identical in configuration with that of Orlon except that the sharp exotherm now occurs at 272° C. instead of 308° C. Since the primary differences between these two materials are those of crystallinity and chain orientation, it is suggested that the difference of 36° C. is related to these structural features. This prominent exothermic reaction, which is nonoxidative in character, is believed to represent the cross-linking of the PAN chains by the elimination of HCN and/or the formation of naphthyridine-like rings along the chains, as first suggested by Houtz [8] in his discussion of "black Orlon" since reactions such as cross-linking or cyclization should in general be of an exothermic character. Pyrolysis studies on PAN have indicated that HCN is liberated above 200° C., and evidence for the formation of fused rings along the PAN chains has been obtained [3]. Madorsky suggests that

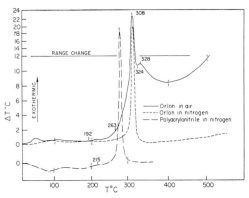

Fig. 1. DTA curves of Orlon and polyacrylonitrile.

[1] Du Pont trademark.

PAN becomes stabilized on heating by cross-linking, with additional stability conferred by the formation of conjugated double bonds in the chains [11].

In Figure 2, curves for drawn Dacron [1] fiber in air and nitrogen are shown together with the curve obtained for undrawn Dacron fiber under nitrogen. In the case of the drawn fiber, the endothermic reaction with a peak at 260° C. in air and 262° C. in nitrogen represents polymer melting; the melting points so obtained compare favorably with literature values [15]. In air, exothermic activity occurs following melting, followed by a sharp rise in the curve at about 440° C. to a prominent peak at 470° C. Under nitrogen, the suppression of oxidative reactions reveals an exothermic reaction at 390° C., followed by an important endothermic reaction with a peak at 447° C. This latter reaction probably represents the depolymerization of polyethylene terephthalate polymer. The DTA curve for undrawn Dacron is quite different from that of drawn Dacron up to 230° C. and quite similar thereafter. Scott [16] has examined powdered amorphous polyethylene terephthalate by DTA up to a temperature of 300° C. and has found a second order transition at 70° C., followed by an exotherm at 140° C. said to be due to polymer crystallization. The determination of glass transition temperatures for nonfibrous polymers by DTA has indicated that the change in specific heat occurring at the glass transition results in a shift in the base line of the DTA curve [10]. Thus the DTA curve for undrawn Dacron fiber in Figure 2 indicates a second order transition at 77° C., followed by a crystallization process at 136° C.

The curve then follows the same pattern as that found for the drawn fiber. The disparity in the depth of comparable peaks and in the area under comparable peaks in the drawn and undrawn fiber is due to the fact that sample size was larger for the undrawn fiber.

In Figure 3, thermograms for nylon 66 fabric are shown. Around 100° C., a weak endotherm appears in both curves due to the loss of sorbed water. In air an exothermic reaction initiating around 185° C. is interrupted by a small endotherm which is polymer melting (m.p. *ca.* 255° C.). In nitrogen the exothermic reactions are completely eliminated, suggesting that those reactions were oxidative in nature. The two endotherms in nitrogen represent sharp melting of the polymer, followed by a depolymerization reaction beginning around 350° C. and completed at or near the peak temperature of 406° C. It should be noted here that different mechanisms

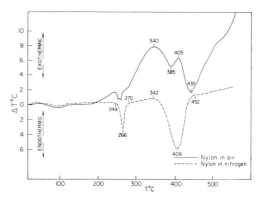

Fig. 3. DTA curves of nylon 66.

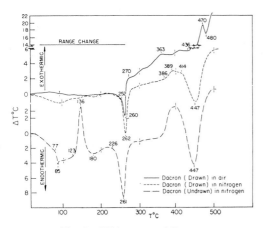

Fig. 2. DTA curves of Dacron.

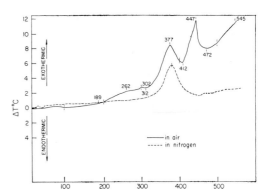

Fig. 4. DTA curves of Neoprene W.

may be operative during the thermal degradation of a polymer in air and in nitrogen.

The DTA curves for Neoprene W are given in Figure 4. The interesting feature in these curves is the exothermic reaction with a peak at 377° C. that persists under a nitrogen atmosphere. Since HCl may be eliminated by heating Neoprene rubber, it seems reasonable to assume that the polymer is being cross-linked at this temperature, as indicated by the nonoxidative exothermic reaction. Roff [15] reports that Neoprene hardens by cross-linking at elevated temperatures. Other textile fibers examined include cotton, nylon 6, cellulose acetate fiber, Arnel,[2] and polypropylene. In addition, other polymers such as butyl rubber, chlorosulfonated polyethylene, and polyethylene have been studied.

From the results obtained it would appear that specific reactions that occur during the thermal degradation of polymers such as rearrangements, cross-linking, and depolymerization may be detected and identified. Other results have also indicated that relatively small changes in polymer composition or the presence of substituents on the polymer backbone can readily be detected by DTA. The indications are that DTA may be a powerful technique for mechanism studies, as well as a means for the characterization of textile materials and evaluation of the thermal stability of textile fibers and fabrics. The study of the degradation reactions at high temperatures in textile fibers represents a new application of DTA. When used in conjunction with pyrolysis studies and thermogravimetric analysis (TGA), it is expected that many of the reactions and reaction mechanisms involved in the thermal decomposition of high polymers will be clarified. The area under a DTA peak is reported to be proportional to the heat of reaction, ΔH, so that quantitative data may also be obtained [2, 6]. Ke [9] has recently determined the crystallinity of isotactic polypropylene from the heats of fusion obtained from DTA data. Thus it is further suggested that crystallization

---

[2] Celanese trademark.

processes, degree of crystallinity, and orientation phenomena in textile fibers may also be studied by DTA.

We are grateful for valuable discussions with W. J. Kauzmann. We are also indebted to the U. S. Naval Supply Research and Development Facility for supporting this work.

Further details of this work will be submitted for publication in this journal in the near future.

Literature Cited

1. Anderson, D. A. and Freeman, E. S., *J. Appl. Polymer Sci.* **1**, 192 (1959).
2. Borchardt, J. J. and Daniels, F., *J. Am. Chem. Soc.* **79**, 41 (1957).
3. Burlant, W. J. and Parsons, J. L., *J. Polymer Sci.* **22**, 249 (1956).
4. Chackraburtty, D. M., *J. Chem. Phys.* **26**, 427 (1957).
5. Costa, D. and Costa, G., *Chimica e industria (Milan)* **33**, 71 (1951).
6. Erikkson, E., *Ann. Agri. Coll. of Sweden* **19**, 127 (1952).
7. Gordon, S. and Campbell, C., *Anal. Chem.* **27**, 1102 (1955).
8. Houtz, R. C., TEXTILE RESEARCH JOURNAL **20**, 786 (1950).
9. Ke, B., *J. Polymer Sci.* **42**, 15 (1960).
10. Keavney, J. J. and Eberlin, E. C., *J. Appl. Polymer Sci.* **3**, 47 (1960).
11. Madorsky, S. L., *J. Research Nat. Bur. Standards* **63A**, 261 (1959).
12. Morita, H. and Rice, H. M., *Anal. Chem.* **27**, 336 (1955).
13. Morita, H., *Anal. Chem.* **28**, 64 (1956).
14. Murphy, C. B., *Anal. Chem.* **30**, 867 (1958).
15. Roff, W. J., "Fibers, Plastics, and Rubbers," New York, Academic Press (1956).
16. Scott, N. D., *Polymer* **1**, 114 (1960).
17. Smothers, W. J. and Chiang, Y., "Differential Thermal Analysis: Theory and Practice," New York, Chemical Publishing Company Inc. (1958).
18. Varma, M. C. P., *J. Appl. Chem.* **8**, 117 (1958).
19. White, T. R., *Nature* **175**, 895 (1955).

ROBERT F. SCHWENKER, JR.
LOUIS R. BECK, JR.

15

Reprinted from *Anal. Chem.*, **32**, 1582–1588 (Nov. 1960)

Differential Thermal Analysis by the Dynamic Gas Technique

ROBERT L. STONE

Robert L. Stone Co., 3314 Westhill Drive, Austin 4, Tex.

▶ The differential thermal analysis apparatus used in this study provides means of controlling pressure in the system, composition of the gas which flows through the test powders during the test, and also the temperature of the system. Five test procedures are prescribed where only one of the variables is dynamic and the other two are held constant. Thermograms are given which illustrate the following types of reactions: oxidation—e.g., manganese oxide, lignite, cotton fiber, coffee, and plastics; thermal decomposition of carbonates (dolomite), hydrates (copper sulfate), and salts (ammonium perchlorate); suppression of reactions by pressure and/or by gas—e.g., lignite, ammonium perchlorate, Teflon, etc.; crystallographic inversions—e.g., ammonium perchlorate, quartz; evaporation of liquids by evacuation; catalytic activity; and melting and freezing of pure substances and mixtures—e.g., paraffin, soap, etc.

T HE PURPOSE of this paper is to illustrate the use of the dynamic gas method of differential thermal analysis (DTA) in studying many types of reactions wherein the variables, temperature, pressure, and gas-phase composition are controlled and are individually or collectively varied during the DTA run. Excellent reviews of the literature on the historical development of the DTA technique include a paper by Murphy (4) and a book by Smothers and Chiang (9).

Some authors, in using DTA results, hand trace the thermograms produced by the recorder to make a smooth-looking curve. Frequently very valuable diagnostic features are lost by doing so. For this reason, the illustrations used in this paper are photos of the original, untouched thermograms because certain very slight inflections are very important.

The two principal uses of the DTA apparatus are as an analytical tool for the identification of mineral or chemical species present in the test sample, and as a means of observing previously undetermined characteristics of a known substance. This latter use frequently leads to patentable subject matter.

The successful use of the method is based on the operator's familiarity with and use of the following: (1) Any substance which is stable under fixed values of pressure, temperature, and composition will always be stable under these fixed values but, according to the second law of thermodynamics, when any one of these variables is changed, the system must change to re-establish equilibrium. When any one of these variables exceeds certain limits, an abrupt redistribution of energy among phases takes place as the system readjusts itself. (2) Any change—reaction—is theoretically observable by DTA so long as it either absorbs or evolves heat, whether the mass be solid, liquid, or gas. Very sensitive instruments are required for weak reactions. (3) Any observable reaction involves either making or breaking of chemical bonds, absorption bonds, or van der Waals' forces. (4) Changes in energy per unit weight (or volume) of a gaseous liquid, or solid system during a reaction can be the result of a single reaction or the sum of several concurrent or rapidly successive reactions. (5) DTA does not spell out what net reaction is taking place when an exothermic or endothermic effect is recorded. It simply records that there is a change in energy content taking place. Knowledge of possible reactions must be used to develop the meaning of the thermogram.

The historical DTA technique and apparatus provided a means of varying

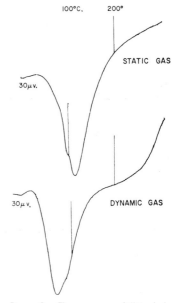

Figure 2. Thermograms of illitic shale showing difference in results between static gas and dynamic gas methods

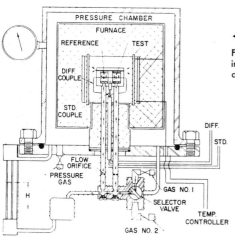

◀

Figure 1. Schematic drawing of pressure chamber and furnace assembly

only one of these variables, namely, temperature—with no control over the other two variables, pressure and gas-phase composition. Rowland (6, 7) developed apparatus for control of composition of the gas around the sample holder at atmospheric (room) pressure. In the present equipment, the gas is forced through the powder during the test. The importance of having the gas flow through the powder rather than around the sample holder is illustrated by the following discussion where a closed type of sample cavity is used.

At room temperature when the powder is placed in the test cavity, the gas is room air. Then, as temperature is raised, the air is replaced by CO_2, water vapor, or whatever vapor is being released by the test material. The gas thus evolved escapes through the cracks in the sample holder and at the same time air is diffusing back into the powder. Thus, at the start of a CO_2-releasing reaction, the interstitial gas is air and at the height of the reaction the gas is CO_2; then, as the reaction

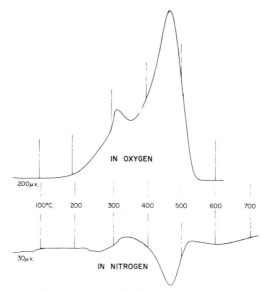

Figure 3. Suppression of oxidation of a lignite by dry nitrogen

Sample, 2.5% powdered lignite +97.5% α-alumina
Both thermograms at atmospheric pressure

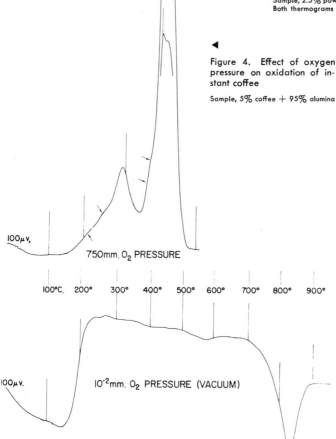

◄

Figure 4. Effect of oxygen pressure on oxidation of instant coffee

Sample, 5% coffee + 95% alumina

slows down, the CO_2 is gradually replaced by air. If the reaction is one whose equilibrium temperature is a function of CO_2 pressure, the shape and temperature of the loop of the thermogram will be affected by the rate of temperature rise, the degree of packing of the sample, the tightness of the sample holder, the grain size, etc. Experience with the dynamic gas method has shown that for crystallites or grains larger than about 2 microns, these factors have very little effect on the thermogram, lending more reliability and reproducibility to the results. Grinding may have pronounced effects on crystals as shown by Bradley et al. (1).

DTA thermograms using a constant temperature and varying gas composition were obtained by Crandall and West (2) in studying the oxidation of cobalt. In this case, the gas flowed around the sample rather than through it.

The apparatus used for this paper provides accurate control over all three variables. The equilibrium line (or surface) can be crossed in three ways: by holding temperature and pressure constant and changing the gas environment; by holding temperature and gas composition constant and varying pressure; or by holding pressure and gas composition constant and varying temperature. The observability of any reaction brought about by any of the three methods of crossing equilibrium condition is a function of the instru-

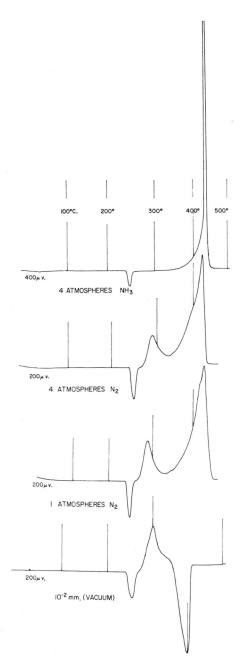

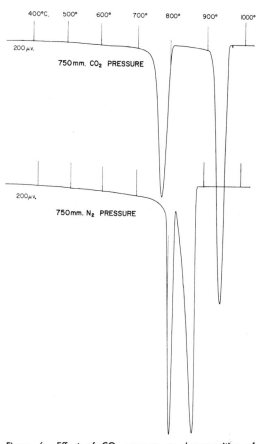

Figure 6. Effect of CO₂ pressure on decomposition of dolomite, CaCO₃.MgCO₃

Sample, 50% NBS No. 88 dolomite +50% alumina
Both thermograms at 1-atm. pressure

◄

Figure 5. Effect of pressure and gas composition on decomposition of NH₄ClO₄

Sample, 50% C.P. NH₄-ClO₄ + 50% alumina

mentation of the apparatus. The present apparatus is so sensitive that certain reactions causing a temperature change of as little as 0.003° C. can be detected without outside interference.

APPARATUS AND PROCEDURES

The basic apparatus as described by

Rase and Stone (5) consists of the instrumentation, and the pressure chamber and furnace assembly. The complete assembly is manufactured by the Robert L. Stone Co. Figure 1 is a schematic of the DTA part of the apparatus which makes control of pressure and atmosphere possible. The furnace and the sample holder (complete with thermocouples) are placed inside a gas-tight metal cylinder. The desired pres-

sure is created inside the cylinder by using compressed N₂ or compressed air, then the desired controlled atmosphere or dynamic gas is streamed via the tubes into and through the powdered samples as indicated by the path of arrows. The gas then escapes into the pressure chamber. The dynamic gas, streaming through the sample, sweeps away any gas being evolved by the decomposition reactions, thereby maintaining a known composition of gas around the particles of test powder at all times.

By using an inert gas (A, N₂, etc.) as the dynamic gas, one can study the effects of pressure alone on the test substance. This is the Le Chatelier-Braun effect. By using an active gas—one that participates in the reactions being observed—one can study the reaction according to the Clausius-Clapeyron and van't Hoff equations.

Theoretically with this apparatus all three of the variables could be varied simultaneously, but, in practice, most tests involve holding two of the variables constant while the third is

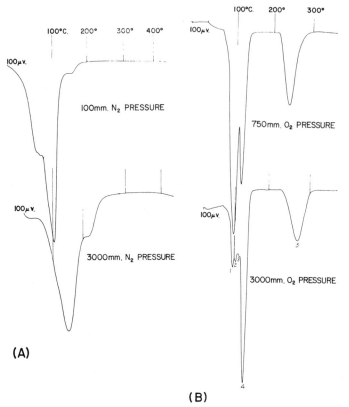

(A)

(B)

Figure 7. Effect of gas pressure

A. Removal of low temperature water of a Ca-montmorillonite, sample undiluted, dried at 28° C. and 50% R. H.
B. Dehydration of CuSO₄.5H₂O. Sample, 1 part CuSO₄.5H₂O + 5 parts alumina

the reactions involved in burning an instant coffee. The slight inflections indicated by the arrows and the relative intensities of the large doublet at 400° C. differ with the brand of coffee. The significance of the features of these thermograms is being investigated. Figure 5 shows the effects of reaction-participating gas (NH₃) on the decomposition of ammonium perchlorate. Figure 6 also shows the effect of the reaction-participating gas—the effect of CO_2 pressure on the decomposition of the $CaCO_3$ portion of dolomite.

The choice of pressure for the test is as important as the gas. Figure 7,A, shows the Le Chatelier-Braun (pressure of inert gas) effect on the removal of interplanar water and cation-hydration water from a Ca-montmorillonitic clay. The strong shoulder in the low pressure run below 100° C. is indicative of vermiculite, and the doublet at 100° C. shows the presence of H^+ and Na^+ as exchange cations. Figure 7,B, shows the effect of pressure on the dehydration steps in $CuSO_4$. $5H_2O$. The high-pressure run (the lower thermogram) shows five endotherms and attempts are now being made to determine by thermogravimetric means whether these represent five possible steps in weight loss. Figure 8 shows the effect of oxygen pressure on the reactions of unsintered Teflon.

Procedure 2 (change gas composition during the run) is rapidly becoming important; it is employed where the effects of a condensable gas—e.g., water vapor—are to be observed. A dry gas is used to a temperature above the condensation temperature of the gas and then the condensable gas is injected. Such runs are classed under Procedure 2 because they involve two dynamic gases during the run. Figure 9 illustrates the application of the procedure on cotton fiber. It shows the difference in reactions under an oxygen atmosphere and in a water vapor atmosphere. Similar results have been reported by Schwenker (8).

Another case where the gas is changed during the run is in studies involving a material that contains organic matter which is best identified in an oxygen

varied. The same specimen may be subjected to several sets of conditions during the test—for example, at temperature T_1, the gas composition may be changed with the pressure constant, and at another temperature, T_2, the pressure may be changed with the gas composition constant. There are innumerable combinations, each step involving a simple procedure.

There are five basic procedures, (Table I), and of these, 1 and 5 are the simplest and most commonly employed:

Procedure 1 is the simplest of the normal DTA variety, and offers a high degree of versatility. The pressure can be anything from a high vacuum to 100 p.s.i.g., and the gas can be any dry gas—N₂, CO₂, O₂, He, etc. However, moist gases having a water vapor content of 3% or less can be used. The low water vapor content of the dynamic gas causes the loops of water-releasing reactions to occur at lower temperatures than in the static gas technique. For example, in Figure 2, the loop representing loss of mechanical

water reaches its maximum at 80° C. with dynamic dry gas as compared to 110° C. when the test is made under static gas conditions (both tests at 1 atmosphere pressure).

The choice of gas depends on the chemical or mineral being investigated and on the information sought. The thermograms of lignite in Figure 3 illustrate the oxidation reactions in O₂ atmosphere and distillation reactions in the N₂ atmosphere. Figure 4 shows the effects of oxygen pressure on

Table I. Basic Procedures

Procedure Number	Temperature	Pressure	Gas Composition
1	Rising, fixed rate	Constant	Constant
2	Rising, fixed rate	Constant	Change at a specific temperature
3[a]	Rising, fixed rate	Change at a specific temperature	Constant
4	Constant	Cycle or change	Constant
5	Constant	Constant	Cycle

[a] Because of difficulty in performing the operations of this procedure, the data are usually obtained by using Procedure 1 at two or more pressures.

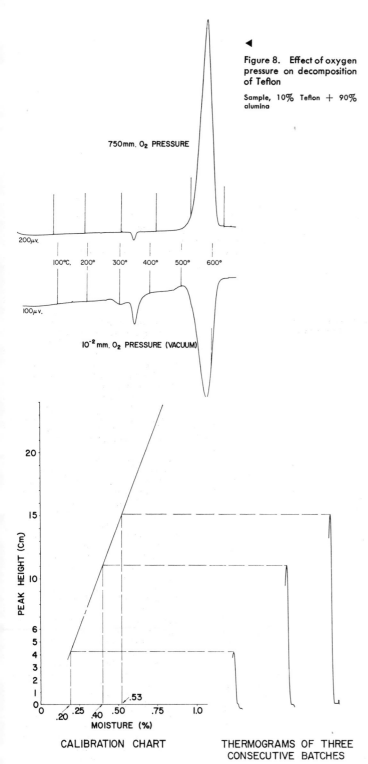

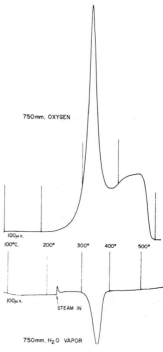

Figure 8. Effect of oxygen pressure on decomposition of Teflon

Sample, 10% Teflon + 90% alumina

750mm. O_2 PRESSURE

10^{-2} mm. O_2 PRESSURE (VACUUM)

Figure 9. Effect of gas composition on behavior of cotton fiber

Sample, 5% medicinal cotton + 95% alumina

CALIBRATION CHART

THERMOGRAMS OF THREE CONSECUTIVE BATCHES

Figure 10. Method of determining moisture content of dry powders by evacuation

Sample, undiluted; thermogram, room temperature; instrument range, 20 μv.

atmosphere. In such cases, an atmosphere of O_2 is maintained up to about 650° C. and is then changed to N_2 or CO_2 to prevent oxidation of the sample holder. Such practice is common in studying clays.

Procedure 4 (change pressure at constant temperatures). In this procedure, the pressure may be either reduced or raised one time for the test, or it may be cycled or repeated as many times as desired. The temperature can be any value within the limits of the apparatus. With this procedure the reaction responsible for the thermal effect is caused by crossing the equilibrium line via pressure rather than temperature change.

Figure 10 is an example of this procedure as a very rapid method for determining moisture content of a nearly dry powdered substance. The procedure is extremely simple because no furnace is involved and a complete test cycle requires only 2 or 3 minutes. The sample holder is charged at room temperature and room pressure. The system is then evacuated at a fixed rate and the loop in the thermogram starts the instant the pressure reaches the vapor pressure of the water at room temperature. For routine tests, a calibration curve (Figure 10) is first

137

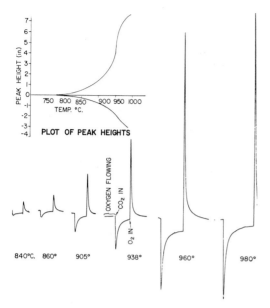

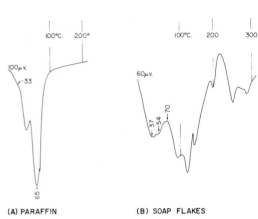

(A) PARAFFIN   (B) SOAP FLAKES

Figure 12. Melting of paraffin and of a commercial soap flakes in dry nitrogen

Temperature, 1 atm.; sample, undiluted; liquid-type sample holder

Figure 11. Oxidation-reduction of a manganese oxide-containing ceramic body by gas cycling with CO_2 and O_2

Pressure, 1 atm.; instrument range, 50 μv. for all peaks

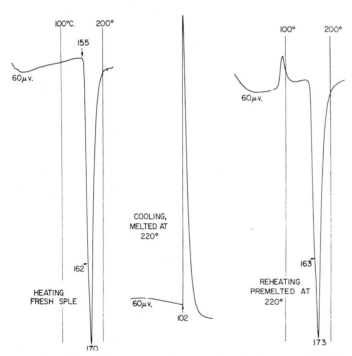

Figure 13. Effect of premelting on freezing-melting thermograms of 4,4'-iso-propylidenediphenol (99.92%)

Samples, undiluted; liquid-type sample holder

obtained by running samples of known moisture contents.

The decomposition of carbonates, hydrates, hydroxides, etc., can be studied by the same technique. With such compounds, pressures greater than 1 atmosphere are generally employed. The decomposition of $CaCO_3$ has been studied by this procedure at pressures up to 1000 p.s.i.a. and a manuscript describing the results is in preparation. If the decomposition reaction is reversible, the compound can be reconstituted by raising the pressure.

Procedure 5 involves an abrupt change of gaseous environment so that the equilibrium line is crossed in the most drastic manner possible. This is in contrast to Procedure 4 where the equilibrium line is crossed slowly. The abrupt change in gas composition causes the test substance to be transferred quickly from an environment in which it is stable to an environment in which it is highly unstable. The result is that the reaction proceeds extremely rapidly.

This technique is so radically different from the usual procedures that very little information has been published. However, the types of reactions that can be studied are well defined. Among them are: oxidation-reduction, including quantitative determination of low percentages of combustibles; heat of adsorption of gases; determination of surface area of fine powders; and catalytic activity.

In this procedure, the gases are cycled—from gas A to gas B to gas A to gas B, etc.—as many times as desired. When the second gas is injected, the loop reaches its maximum height in 3 or 4 seconds. Therefore, if the data are obtained by simple injection of the second gas, the results are obtained very quickly so that many test samples can be measured per hour.

The tests can be made at any temperature and pressure within the limits of design of the apparatus. For example, the test may be at room pressure and temperature with the two gases being N_2 and NH_3; or at 300° C. and 40 p.s.i.a. using N_2 and O_2 as the gases, etc. Catalyst testing as described by Rase and Stone (5) is an example of this technique in which the cycling test was done at 330° C. at 1 atmosphere pressure using dry nitrogen and water vapor. Locke and Rase (3) used a similar procedure.

Figure 11 is an interesting example of the gas cycling technique applied to an oxidation-reduction problem: The problem was to establish the temperature at which the test substance became unaffected by oxygen. A sample of the material was cycle-tested at several temperatures. The heights of the exothermic and endothermic peaks were plotted vs. temperature. The temperature at which the curves cross the temperature axis is the temperature at which the substance could not be reduced by CO_2. The same data can be obtained by weight gain and weight loss methods but at least 1 week is required as compared to less than 2 hours by DTA.

Change of State Data (Inversions, Melting, Vaporization). Crystallographic inversions of powdered materials are observed with powder-type sample holders. The α-quartz to β-quartz inversion is the classical example.

Melting and freezing experiments can be done without dilution of the sample by using a sample holder designed specifically for handling liquids and molten materials. In Figure 12, the thermograms of a commercial paraffin and of commercial soap flakes are given as illustrations. Frequently it is desirable to observe first the melting behavior of a compound and then to study its freezing behavior. The thermograms of 4,4'-isopropylidenediphenol, shown in Figure 13, illustrate such a technique.

GENERAL SUMMARY

The controlled pressure and controlled atmosphere differential thermal analysis apparatus can be used for many types of thermal analyses that cannot be performed with older types of equipment. All three of the variables, temperature, pressure, and gas composition, are controlled. The effect of pressure alone is illustrated with thermograms of montmorillonite and of $CuSO_4.5H_2O$ which show several stages of removal of hydrate water. The effect of pressure of the reaction-participating gas is illustrated with dolomite (CO_2), ammonium perchlorate (NH_3), etc. In the case of the ammonium perchlorate, the 300° C. exotherm is eliminated and the decomposition at 405° C. becomes violent. The oxidation of compounds is illustrated with cotton fiber, lignin, and coffee. Melting and decomposition thermograms are given for Teflon, paraffin, and soap. Special techniques involving cycling of the dynamic gas at constant temperature are illustrated with a manganese oxide-containing ceramic wherein reduction-reoxidation thermograms are obtained at several temperatures and the peak heights are plotted as a function of temperature. A similar technique is used to determine moisture content of nearly dry powder (less than 1% moisture) wherein the test powder is evacuated and the test is carried out at room temperature in about 1 minute. Other techniques are illustrated.

LITERATURE CITED

(1) Bradley, W., Burst, J. F., Graf, D. L., *Am. Mineralogist* **38**, 207–17(1953).
(2) Crandall, W. B., West, R. R., *Am. Ceram. Soc. Bull.* **35**, 66–70 (1956).
(3) Locke, Carl, Rase, Howard, *Ind. Eng. Chem.* **52**, 515 (1960).
(4) Murphy, C. B., ANAL. CHEM. **30**, 867–72 (1958).
(5) Rase, Howard, Stone, R. L., *Ibid.*, **29**, 1273–7 (1957).
(6) Rowland, R. W., *Calif. Dept. Nat. Resources, Div. of Mines, Bull. No. 169*, 151–63 (1955).
(7) Rowland, R. A., Jonas, E. C., *Am. Mineralogist* **34**, 550–8 (1949).
(8) Schwenker, R. F., Textile Research Institute, personal communication.
(9) Smothers, W. J., Chiang, Yao, "Differential Thermal Analysis: Theory and Practice," Chemical Publishing Co., New York, 1958.

RECEIVED for review April 19, 1960. Accepted August 1, 1960. Division of Analytical Chemistry, 137th Meeting, ACS, Cleveland, Ohio, April 1960.

Reprinted from *Anal. Chem.*, **34**(13), 1841–1843 (1962)

Identification of Organic Compounds by Differential Thermal Dynamic Analysis

Jen Chiu

S<small>IR</small>: One of the traditional analytical methods for identifying organic compounds determines the melting points of crystalline derivatives prepared from the sample and a standard reagent (5). This procedure is often tedious, and sometimes misleading because of the complications of side products. Recently we attempted to replace this multistep process by a one-step thermal dynamic technique using a differential thermal analysis apparatus. The sample was heated with a specific reagent at a programmed rate in a selected atmosphere. The thermogram showed the derivative-forming reaction, the physical transitions of the sample or the reagent in excess, and the physical transitions of the intermediates and products in a single run. Only a few milligrams of sample and reagent were required. Preliminary results are very promising. This new approach could lead to a new scheme of organic analysis.

EXPERIMENTAL

Differential thermal analysis measures the thermal effects occurring in the sample by continuously recording the temperature difference between the sample and a reference material as a function of the sample temperature. The principles and general applications can be found in excellent reviews (3, 4, 6).

The apparatus used consisted of four major parts: cell assembly, temperature programmer, amplifier, and recorder. The last three parts and thermocouple junctions were similar to those described by Vassallo and Harden (8). For crude experiments the temperature programmer could be replaced by an ordinary powerstat.

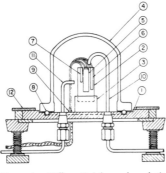

Figure 1. Differential thermal analysis cell assembly

A schematic diagram of the cell assembly is shown in Figure 1. This simple design gives high sensitivity and resolution, good control of atmosphere, quick cooling, fast change of thermocouples, and no cell cleanup. An aluminum block, 2 (0.75 × 1.5 inches), held in a Marinite seat, 3, is used as the heat sink. The heat source is a 0.25 × 1 inch, 30-watt cartridge heater, 5 (Hot Watt Co., Danvers, Mass.), which can be operated up to 500° C. without difficulty. Ordinary melting point capillaries, 6 (1.5 to 2.0 × 25 mm.; KIMAX 34505), containing the reaction mixture and roasted glass beads (100- to 140-mesh, Potters Bros., Carlstadt, N. J.), respectively, are placed in two holes (0.070 × 1 inch) symmetrical in position to the heater. The 28-gage, glass-insulated Chromel-Alumel thermocouples are directly inserted into the reaction mixture and the reference material to measure the sample temperature and the temperature differential. A similar thermocouple, 7, is placed close to the cartridge heater to control the temperature programming. The whole assembly is enclosed by a borosilicate glass bell jar, 4 (3.125 × 3.375 inches), and sealed by a neoprene O-ring, 8, on the bottom aluminum plate, 1, which is supported by a three-leg clamp, 12. Electrical connections of the thermocouples and the heater are made through

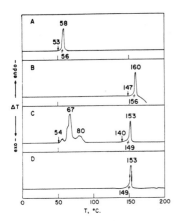

Figure 2. Thermograms showing formation of p-nitrophenylhydrazone of acetone

A. Acetone
B. p-Nitrophenylhydrazine
C. Reaction mixture of acetone and p-nitro-phenylhydrazine
D. Rerun of residue from C

taper pins (AMP, Inc., Harrisburg, Pa.) to the plug board, 9, sealed in the bottom plate. Thus the inside of the bell jar can easily be evacuated through a short piece of $1/4$-inch o.d. copper tubing, 10, attached to the bottom plate by standard joints. One can cool down the aluminum block rapidly by partially evacuating the inside of the bell jar and letting liquid nitrogen flow onto the block through another $1/4$-inch o.d. copper tubing, 11. A nitrogen atmosphere can be obtained from such cooling or by a side path to a nitrogen source. The reference junctions of the sample thermocouple and the programming thermocouple are immersed in a triple-point cell to maintain a constant temperature of 0.01° C. (7).

All the analyses were performed in a static nitrogen atmosphere at a heating rate of 15° C. per minute. Sample size was in the range of 1 to 5 mg. No precise weighing was necessary. Excess reactant was indicated by its boiling or melting endotherm. Reactants were mixed in the capillary below 0° C.

RESULTS AND DISCUSSION

The preparation of derivatives by classical methods has been discussed fully by Shriner, Fuson, and Curtin (5); this information should be a good starting point in scouting for reagents suitable for the new technique. Being quick and dynamic in nature, the present method requires a specific reagent which forms a derivative rapidly with the sample at either room temperature or elevated temperatures. A catalyst may be used, if necessary. The derivative so produced should show a discernible physical transition or a characteristic thermogram. One reactant more volatile than the other should be used in

excess, in order to simplify the thermogram. Usually the reaction is fast when one of the reactants serves as a solvent for the other. We are currently searching for novel reagents to replace the conventional ones.

The following examples illustrate the features of this method.

Figure 2 shows acetone (Merck, reagent grade) forming a hydrazone derivative with p-nitrophenylhydrazine (Eastman Kodak, White Label). Thermogram A indicates the boiling endotherm of acetone with deviation point at 53° C., inflection point at 56° C., and peak maximum at 58° C. The inflection point is evidently closer to the reported boiling point 56.5° C. (2). Similarly thermogram B shows a melting endotherm at 160° C. for p-nitrophenylhydrazine, followed by exothermic decomposition. However, a mixture of p-nitrophenylhydrazine with excess ace-

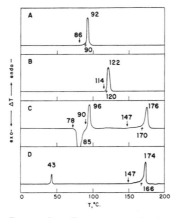

Figure 3. Thermograms showing formation of triethylamine picrate

A. Triethylamine
B. Picric acid
C. Reaction mixture of triethylamine and picric acid
D. Rerun of residue from C

tone shows complex endotherms beginning at 54° C. and peaking at 67° C. This is assumed to be the net result of evaporation of excess acetone, solution of p-nitrophenylhydrazine in acetone, and hydrazone formation. The melting of the hydrazone is clearly shown by the late 153° C. endotherm. An immediate rerun of the residue shows only the melting endotherm of the derivative as seen from thermogram D. The reported melting point of acetone p-nitrophenylhydrazone is 152° C. (5), which falls between the inflection temperature, 149° C., and the peak temperature, 153° C., obtained by differential thermal analysis.

Figure 3 shows the identification of triethylamine (Eastman Kodak, White

Label) by forming a derivative with picric acid (Eastman Kodak, White Label). As shown in thermogram C, the derivative-forming reaction is represented by the 85° C. exotherm. The 96° C. endotherm is comparable to the boiling endotherm of triethylamine as shown in thermogram A. The absence of the 122° melting endotherm of picric acid shown in thermogram B indicates no picric acid remaining. The 176° C. endotherm is ascribed to the melting of the triethylamine picrate formed. As shown in thermogram D, another endotherm appears at 43° C. This endotherm, probably a crystalline transition of the picrate, offers another characterizing feature in addition to the melting point of the derivative. The reported melting point of this derivative is 173° C. (5).

Besides the use of distinct physical transitions of the derivatives, organic compounds can sometimes be characterized by the thermospectrum of a certain reaction. For instance, we found that dextrose can differentiate between primary amines and others. A characteristic thermogram consisting of a series of endothermic and exothermic effects is shown by most primary amines, such as n-propylamine, n-butylamine, isopropylamine, isobutylamine, sec-butylamine, tert-butylamine, ethylenediamine, 2-aminoethanol, cyclohexylamine, benzylamine, and allylamine.

Figure 4 shows the thermograms of dextrose (National Bureau of Standards), n-propylamine (Eastman Kodak, White Label), di-n-propylamine (Eastman Kodak, White Label), and mixtures of dextrose with the two amines. The double-melting endotherm of dextrose

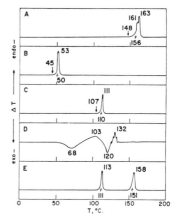

Figure 4. Thermograms showing reaction of dextrose with propylamines

A. Dextrose
B. n-Propylamine
C. Di-n-propylamine
D. Mixture of dextrose and n-propylamine
E. Mixture of dextrose and di-n-propylamine

141

shown in thermogram *A* and the boiling endotherm of *n*-propylamine shown in thermogram *D*. Instead, a complex thermospectrum appears, indicating a stepwise reaction with endothermic and exothermic effects overlapping. On the other hand, thermogram *E* of the mixture of dextrose and di-*n*-propylamine shows distinctively the boiling endotherm of the amine at 113° C. and the melting endotherm of dextrose at 158° C. as compared to thermograms *C* and *A*, respectively. No reaction between these two is apparent.

The use of differential thermal analysis to study chemical reactions is not new (*3, 4, 6*). Bollin and Kerr used a similar technique for pyrosynthesis of minerals from their constituents (*1*). Great potential exists in applying this technique for the identification of an organic compound (including polymers), first by its own thermogram, and then by the thermogram of its reaction with a specific reagent. Compared with the classical method, the thermal dynamic approach is less time-consuming, uses only small amounts of sample and reagent, requires less analytical skill, and provides more characterizing features. In addition to determining the melting point of the principal derivative, the new technique reveals products other than the derivative, and supplies information on the over-all reaction.

LITERATURE CITED

(1) Bollin, E. M., Kerr, P. F., *Am. Mineralogist* **46**, 823 (1961).
(2) "Handbook of Chemistry and Physics," 43rd ed., Chemical Rubber Publishing Co., Cleveland, Ohio, 1961.
(3) Mackenzie, R. C., Mitchell, B. D., *Analyst* **87**, 420 (1962).
(4) Murphy, C. B., ANAL. CHEM. **30**, 867 (1958); **32**, 168R (1960); **34**, 298R (1962).
(5) Shriner, R. L., Fuson, R. C., Curtin, D. Y., "Systematic Identification of Organic Compounds," 4th ed., Chap. 9 and 10, Wiley, New York, 1956.
(6) Smothers, W. J., Chiang, Y., "Differential Thermal Analysis. Theory and Practice," Chemical Publishing Co., New York, 1958.
(7) Stimson, H. F., *J. Res. Natl. Bur. Stds.* **65A**, 139 (1961).
(8) Vassallo, D. A., Harden, J. C., ANAL. CHEM. **34**, 132 (1962).

JEN CHIU
Plastics Department
E. I. du Pont de Nemours & Co., Inc.
Wilmington, Del.

RECEIVED for review October 2, 1962. Accepted October 22, 1962.

Reprinted from *Anal. Chem.*, **34**(1), 132–135 (1962)

Precise Phase Transition Measurements of Organic Materials by Differential Thermal Analysis

D. A. VASSALLO[1] and J. C. HARDEN

Polychemicals Department, E. I. du Pont de Nemours & Co., Inc., Wilmington, Del.

▶ A differential thermal analysis method is presented for the precise, semiautomatic determination of phase transition temperatures. Several DTA methods are compared for estimating transition temperatures. The effects are given of variation in thermocouple position and gage, and in heating rate. A recommended procedure gives a precision of ±0.3° C. for 2- to 5-mg. samples over the temperature range from −150° to +450° C. Liquid or solid samples are tested directly; gases, after condensation into cooled tubes within the apparatus. Conventional melting point capillary tubes are used in the simultaneous determination of melting, boiling, and inversion temperatures of organic materials.

DIFFERENTIAL thermograms of organic materials taken over a wide temperature range have been proposed for use as "fingerprints" of the materials studied, because of the unique nature of the complex scans (6). This use has been limited generally to intralaboratory studies, since such thermograms are strongly dependent on operational variables. The situation is compounded by the lack of generally accepted or readily available standardized equipment. A proliferation of applications and a paucity of definitive work have promoted duplication of effort, so that experience with the equipment at hand remains the primary requirement for meaningful results. This is evident even in the determination of well known phase transition temperatures, T_t, where precision has been usually reported as ±3° to 5° C. (6, 9, 10). Other merits of the differential thermal analysis (DTA) technique are likely to be discounted if this supposedly objective instrumental method is not shown capable of better precision and accuracy.

This relatively poor precision is adequate for characterizing most inorganic systems, where transitions usually are well separated. For studies of organic substances most workers have used the same techniques. Because of the complexity of thermograms of organic compositions, however, greatly improved precision is essential, since overlapping or closely occurring bands are the rule rather than the exception. Precision is affected by instrumentation and by sample dilution. The most common sampling condition for inorganic compounds employs addition of reference material to the sample compartment, to equate heat capacities as much as possible. Often samples are diluted 60 to 90% with reference material, resulting in broad thermal peaks. For this reason transition temperatures have been estimated from different parts of the peak. For example, recommendations have included the first deviation from base line (2), the inflection point (7), or intersection with the base line of the low temperature side of the peak (5), and the peak temperature (8).

Although DTA is a dynamic method, melting points, T_m, and boiling points, T_b, occur at specific temperatures. Therefore, heating rate and sample size may have a marked effect on the observed temperature, since a finite time is required to undergo transition. If the heating rate is high or the sample is large, an error can be expected as in manual methods for measuring melting point. Just as the response of a thermometer affects manual melting point determinations, the size and response

of the temperature measurement device in DTA can be limiting. With the foregoing in mind, several DTA methods for measuring T_t were studied to minimize the effects, if any, of heating rate, sample size, and detector response.

An improved DTA technique has been developed which provides rapid, precise, and accurate T_t measurements based on the use of small undiluted samples and continuous temperature measurement within the sample. A precision of ±0.3° C. is easily obtained over a wide range of heating rates. Liquid or solid samples are conveniently handled in melting point capillary tubes into which the thermocouples are inserted. To extend the application of the technique, a cell was devised to cover the temperature range from −150° to +450° C. This extended range required blanketing the sample compartment with flowing dry inert gas to prevent condensation at low temperature and oxidation or hydrolysis at high temperature.

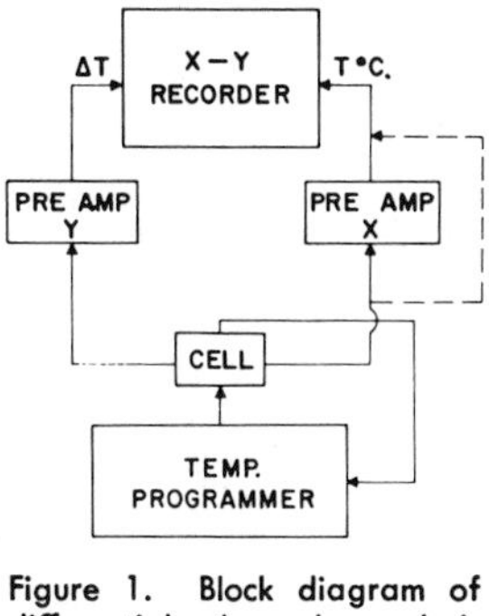

DESIGN OF APPARATUS

With the exception of the temperature programmer (1), the electronic components of the DTA apparatus are commercial products for sensitive differential temperature measurement, accurate temperature measurement, and $X - Y$ potentiometric recording. The block diagram in Figure 1 illustrates the flow of thermocouple signals and the general plan of the apparatus.

The differential thermocouple e.m.f. is preamplified by a Leeds & Northrup low level d.c. preamplifier No. 9835-A and fed to the Y axis of a Moseley Model 4S $X - Y$ potential recorder (15 × 10 inch recording surface). The temperature e.m.f. is similarly fed to the X axis of the recorder with or without preamplification, depending on the sensitivity required.

A preamplifier has been included in the temperature-measuring circuit when accuracy better than ±0.5° C. (the reading error of the recorder at its narrowest range) is desired, but the basic recorder sensitivity has proved sufficient for most purposes. The preamplifier was used to test the various methods without limitation from the

Figure 1. Block diagram of differential thermal analysis components

[1] Present address, Sabine River Works, Orange, Tex.

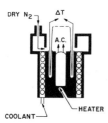

Figure 2. Cross section of heater-cooler differential thermal analysis block

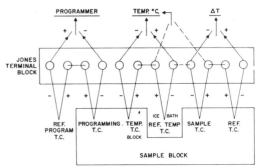

Figure 3. Thermocouple junctions for temperature programming, temperature, and differential temperature measurements

readability of the recordings. When the temperature preamplifier is used, a "bucking potential" from a voltage divider is placed before the amplifier in series with the thermocouple signal, to increase the range of temperature measurement. Both temperature and differential temperature e.m.f. can be attenuated at the recorder. Since the maximum gain of the preamplifier is 200X, the narrowest Y range is 2.5 μv. or 0.04° C. on the 0.5-mv.-per-inch range of the recorder. Normal operational sensitivy is 25 μv. per inch for differential temperature measurement.

The temperature programming thermocouple is in a "feedback" circuit with a proportionally controlled linear temperature programmer (1). Temperature programming from any selected temperature is achieved by a "bucking" reference voltage, usually a thermocouple placed in a bath at the selected temperature.

The cell, shown in cross section in Figure 2, was designed to accommodate small samples contained in 1.5 to 2.0 × 30 mm. melting point capillary tubes (KIMAX 34505). It is contained in a standard 1-quart vacuum flask. The cartridge heater (30 watts 1 × 0.25 inch, Hot Watt, Danvers, Mass.) is in the center of the 1 × 1 inch aluminum block with sample and reference holes placed symmetrically about the heater and a hole to accommodate the temperature - programming thermocouple close to the heater. A double coil of copper tubing serves as a cooling line. One coil is immersed in coolant in the lower part of the vacuum flask, while the second coil surrounds the block. The block is cooled by a flow of air or nitrogen through the lower coil, thence to the coil surrounding the block. The outer sheath is stainless steel of lower thermal conductivity.

All components except the heater-cooler cell are housed in two interlocking Emcor FR-14A cabinets.

The cell and DTA system has several features which provide speed and simplicity of operation with versatility.

Rapid controlled heating and cooling are achieved by placing both the heater and cooler in contact with the block — the heater at the center and the cooler at the periphery. Heating rates of

2.5°, 5°, 10°, 20°, 25°, and 30° C. per minute can be selected from the programmer. It is thus possible to select a heating rate commensurate with the resolution and thermal effect being studied. The cooling may be controlled by adjusting the rate of flow of inert gas or liquid coolant through the cooling coils. This feature gives a temperature range of −150° to +450° C.

Samples are contained in disposable capillary tubes, which are easily filled and packed.

All thermocouple junctions are arranged on a Jones terminal block (Figure 3), which is set into the Marinite cover of the cell. Unusable or broken thermocouples may be quickly replaced. The two methods of connecting the thermocouples are shown in Figure 3. Thermograms recorded with respect to block temperature are obtained by connections shown in the solid line. An ice reference junction was used for all temperature measurements. When recording with respect to sample temperature, the temperature measurement circuit is modified, as shown by the dashed lines, and the preamplifier is grounded.

By use of extension leads, the cell may be placed at a distance from the electronic components. Thus, if a noxious atmosphere or samples which degrade to form noxious gases are used, only the cell need to be moved to a hood.

The method may be applied objectively when little sample is available — e.g., a gas chromatographic fraction or an expensive reagent.

PROCEDURE

After connecting suitable thermocouples — e.g., gold-constantan or copper-constantan for low temperature (−150° to +150° C.), or Chromel-alumel for high temperature (−50° to +450° C.) — and their transmission wires, the circuits are checked for continuity. Since most of our work has involved endothermic phenomena, the differential thermocouples are adjusted to give a positive deflection for a temperature lag in the sample compartment. This results in a scan which resembles

a heat capacity curve. The sample and reference are packed in 1.5- to 2.0-mm. melting point capillary tubes cut to a length of 30 to 40 mm. Between 2 and 5 mg. of sample are used. The heat capacities are adjusted by adding reference substance, 100- to 200-mesh porcelain, to the reference tube. Porcelain is thermally inert over a wide temperature range and is not hygroscopic. Silicone oil also performed well as reference material, but has temperature limitations. The differential thermocouples are centrally positioned in the sample and reference, and the tubes inserted into the heater block.

Other variables are adjusted. The proper blanketing atmosphere is selected, with the proper heating rate. starting temperature, and sensitivities on the recorder. The temperature programmer is started and the recorder pen actuated. If the variables have been selected properly, the instrument requires no attention until the run is completed. (A limit switch can be included to prevent heater burnout.) In scouting runs, a faster heating rate is used and the large "peaks" are attenuated during the run. Proper variables can then be estimated from this scan.

While the apparatus is cooling, another sample can be prepared and positioned in the capillary.

The thermocouples are cleaned by burning off organic residues. An unusable thermocouple can be replaced in a matter of seconds at the terminal block connection.

RESULTS AND DISCUSSION

Comparison of Methods of Measuring Transition Temperature. To compare various methods for measuring T_t, the above apparatus was used with constant cell geometry. Chromel - Alumel thermocouple outputs were taken from different sources and the temperature was estimated from different portions of the DTA peak. The two major conditions were temperature measurement in the reference material (equivalent to block

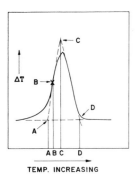

Figure 4. Transition temperature estimation methods

temperature) and temperature measurement in the active sample. The temperatures estimated for melting point, T_m, or boiling point, T_b, were selected from the most often recommended portions of the DTA peak, as illustrated in Figure 4. Point A is the intersection of the extrapolated straight-line portion of the low temperature side of the peak with the base line, and point B is the inflection point of the low temperature side. This inflection point has been obtained electronically by derivative techniques (3). Point C is extrapolated peak temperature and point D the extrapolated return to base line. Table I gives the sample or reference temperatures at which the various points occurred during melting of benzoic acid and boiling of toluene at a heating rate of 10° C. per minute, using 40-gage Chromel-Alumel thermocouples.

With the exception of point D in Table I, T_t estimates using sample temperature are generally closer to the true values. The closest estimates of T_t were achieved by use of sample temperature at point C or reference temperature at point A. Similar data resulted when thermocouple gage and heating rate were varied. The difference between the use of sample temperature at point A or point C was usually less than 1° C., with the exception of the case in which 40-gage thermocouples were used at high heating rates to estimate boiling point.

T_b estimated for toluene at points A, B, and D are higher than at point C. This is probably due to superheating which is detected by the very responsive 40-gage thermocouples. Even at a 40° C. per minute heating rate, 28-gage thermocouples show only an abrupt stop in temperature rise. Some of these effects are shown in Figure 5, where the character of boiling endotherms under various conditions is illustrated. The relatively poor results obtained for boiling points compared to melting points by measuring reference temperature could be related to superheating also.

The effect of heating rate and thermocouple gage is shown in Table II for the two best methods (A and C). Point A, representing the best measure using reference temperature, compares with the worst measure using sample temperature, point A, but contrasts poorly with the best measure from sample temperature, point C. The precision, σ, of any measurement at 10° C. per minute with 28-gage thermocouples was 0.5° C. (m.p.), 0.7° C.

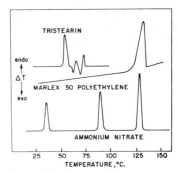

Figure 5. Character of boiling endotherms using sample and reference temperature measurement
R_h 10° C.

(b.p.) for point A reference and 0.3° C. (m.p.), 0.3° C. (b.p.) for point C sample. When various heating rates and thermocouple gages are included as in Table II, the precision using sample temperature remains constant ($\sigma = 0.3$ m.p.; 0.3 b.p.), whereas precision is decreased for reference temperature measurement ($\sigma = 0.7$ m.p.; 1.9 b.p.). Since outstanding precision and accuracy were indicated for both melting and boiling points by measuring sample temperature at point C, peak temperature, with wide variations in heating rate and thermocouple size, this method was selected for other studies.

Variation of T_t with Heating Rate. Since there was little difference in accuracy between 40- and 28-gage thermocouples, the 28-gage thermocouple was selected as a standard because of its ruggedness, although the peak heights developed were about 15% less. Sensitivity was not a problem, for even at 5° C. per minute R_h, a peak height of 0.8° C. (50 μv.) was obtained on 4 mg. of benzoic acid ($\Delta H_f = 33.9$ cal. per gram), while a differential of 0.05° C. was easily dis-

Table I. Comparison of Various Methods for Transition Temperature Measurement[a]

Benzoic Acid,[b] Melting Point, °C.

Temperature Measured	A	B	C	D
Reference	121.5	125.0	128.0	131.0
Sample	120.3	121.0	121.7	131.9

Toluene[c] Boiling Point, °C.

Method	A	B	C	D
Reference	112.5	114.2	115.9	116.2
Sample	114.6	113.2	111.2	118.8

[a] See Figure 4.
[b] NBS, m.p. = 121.8°.
[c] Merck, b.p. = 1.0° boiling range including 110.6° C.

Table II. Effect of Varying Heating Rate and Thermocouple Size on T_t Estimation

		Benzoic Acid, M.P., °C.		Toluene, B.P., °C.	
T.C. Gage	R_h, °C./Min.	A_{ref}	C_{sample}	A_{ref}	C_{sample}
28	10	120.8	121.8	109.1	111.1
40	10	121.5	121.7	112.5	111.3
28	40	120.6	121.8	109.5	111.1
40	40	122.0	121.9	116.6	111.3
σ		0.7	0.27	1.9	0.27

Figure 6. DTA scans of representative materials

tinguished with this system. Table III shows the variation of T_m measurement over a wider range of heating rates. For comparison, both reference temperature (point A) and sample temperature (point C) were studied. Again, sample temperature is superior—T_m is constant to ±0.2° C. even with the partially crystalline Marlex 50 polyethylene. T_m is constant when sample temperature is measured, whereas the T_m through reference measurement increases with heating rate.

Representative thermograms of ammonium nitrate, Marlex 50 polyethylene, and tristearin are shown in Figure 6. These materials represent three types of morphology. Ammonium nitrate undergoes three crystalline transitions between room temperature and its melting point at 170° C. Marlex 50 polyethylene has a broadened melting endotherm, normal to partially crystalline high polymers. Further examples of this behavior can be seen in the work of Ke (4). Tristearin is monotropic—the lowest melting crystalline form of this polymorphic material is obtained by crystallization from the melt. As tristearin is heated, the various polymorphic forms melt and sequential recrystallization and melting occur.

Table III. Variation of Melting Point with Heating Rate

(T_f, °C., and method of measurement)

	Benzoic Acid		Marlex 50	
R_h	A_{ref}	C_{sample}	A_{ref}	C_{sample}
5	121.5	121.8	136.0	134.2
10	121.6	121.7	138.0	134.4
15	122.1	121.9	138.7	134.2
25	124.0	121.9	139.5	134.2
40	125.5	121.9		134.2
80		121.8		134.4

Table IV. DTA Transition Temperatures of Various Compounds

	Melting Point, $T°$ C.		Boiling Point, $T°$ C.	
Compound	Found	Reported	Found	Reported
n-Butane	−135.0	−135.5	−0.5	−0.55
n-Pentane	−129.5	−129.7	36.2	36.0
n-Hexane	−94.5	−95.3	69.0	68.8
n-Heptane	−90.3	−90.6	98.2	98.4
n-Octane	−57.0	−56.8	125.5	125.6
n-Nonane	...	...	150.2	150.7
n-Decane	...	...	173.0	174.0
n-Dodecane	...	...	215.5	216.0
Benzoic acid	121.8	121.8	...	...
Water	0.0	0.0	100.0	100.0
Toluene	...	...	111.1	110.6
Benzene	5.2	5.5	80.5	80.1
Acetic acid	16.5	16.6	118.4	118.1
Alathon 10 polyethylene resin	110.5	110.5[a]	...	...
Marlex 50 polyethylene resin	134.2	134.5[a]	...	...
Teflon TFE fluorocarbon resin	327.5	327.0[a]	...	...
Teflon FEP fluorocarbon resin	272.0	272.5[a]	...	...
Delrin acetal resin	170.5	171.0[a]	...	...

[a] Melting points taken with Kofler hot stage microscope, or by x-ray techniques.

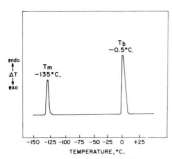

Figure 7. Low temperature DTA scan of n-butane

Crystallization of tristearin from solution gives only the highest melting form. A low temperature, simultaneous determination of T_m and T_b for a normally gaseous material is shown in Figure 7. A small amount of n-butane was condensed into a capillary tube and run from −150° to +10° C.

The T_f's for hydrocarbons are particularly sharp.

The transition temperatures of several other materials are shown in Table IV, using a 28-gage thermocouple and a 15° C. per minute heating rate. T_f measurements below 100° C. were taken with a calibrated gold-constantan thermocouple, those above 0° C. with a Chromel-Alumel thermocouple, and the 0° to 100° C. range was used as a check. Average error calculated as the average deviation of the measured values from reported values was 0.3 for T_m and 0.45 for T_b. This average error thus approaches the standard deviation of ten determinations of benzoic acid T_m ($\sigma = 0.2°$ C.). It is probable that even more precise data can be obtained by DTA with strict control of variables such as rigid standardization of thermocouples and greater amplification of the temperature signal to reduce systematic errors.

LITERATURE CITED

(1) Dal Nogare, S., Harden, J. C., ANAL. CHEM. **31**, 1859 (1959).
(2) Frederickson, A. F., *Am. Mineralogist* **39**, 1023 (1954).
(3) Gordon, S., Campbell, C., ANAL. CHEM. **31**, 1188 (1959).
(4) Ke, B., in "Organic Analysis," Vol. IV, p. 361, Interscience, New York-London, 1960.
(5) Keavney, J. J., Eberlin, E. C., *J. Appl. Polymer Sci.* **3**, 47 (1960).
(6) Morita, H., Rice, H. M., ANAL. CHEM. **27**, 336 (1955).
(7) Partridge, E. P., Hicks, V., Smith, G. W., *J. Am. Chem. Soc.* **63**, 454 (1941).
(8) Smyth, H. T., *J. Am. Ceram. Soc.* **34**, 221 (1951).
(9) Stross, F. H., Abrams, S. T., *J. Am. Chem. Soc.* **73**, 2825 (1951).
(10) Varma, M. C. P., *J. Appl. Chem.* **8**, 117 (1958).

RECEIVED for review June 30. 1961. Accepted October 9, 1961. Delaware Science Symposium, Wilmington, Del., February 15, 1961.

18

Reprinted from *Anal. Chem.*, **34**(9), 1101–1105 (1962)

Differential Thermal Analysis of Organic Samples

Effects of Geometry and Operating Variables

EDWARD M. BARRALL II[1] and L. B. ROGERS[2]

Department of Chemistry and Laboratory for Nuclear Science, Massachusetts Institute of Technology, Cambridge, Mass.

▶ Quantitative analyses of 1 to 10 mg. of organic materials have been demonstrated with salicylic acid. Under carefully controlled conditions, measurements of the endothermal depth had approximately half the error of an area measurement. The standard deviation for a depth measurement was 0.07 mg. of acid or 5% relative, whichever was larger; that for area was 0.15 mg. or 7%. A study of variables has shown that equations developed for pure samples can be extended to diluted samples. In addition, the specific heat of melting of salicylic acid, 28.2 ± 0.5 cal. per gram, has been determined using silver nitrate to calibrate the apparatus.

DIFFERENTIAL THERMAL ANALYSIS (DTA) has been most widely used by geologists and mineralogists for studying clays (*15*), coals (*5*), and organic material of geological origin (*6*). Its use by chemists to characterize pure substances has been relatively slight but is increasing rapidly (*19*). Despite the almost universal reporting of melting points in the organic literature, phase changes of organic compounds have received the least attention (*7-10, 16-18, 22*). The technique should be more widely used now that the effects of reaction kinetics on the curve shape (*1, 8, 4, 11, 12*) and the direct calculation of heats of transformation (*2, 3, 8, 22*) from the curve have provided a

sound basis for interpretation of data.

The writers and Vassallo (*21*) independently concluded that the usefulness of DTA would be considerably greater if the accuracy, precision, and sensitivity were improved, if the mass of the apparatus were reduced so that cooling could be effected rapidly to permit more analyses to be made, if factors affecting flatness of the baseline were controlled, and if the required amount of sample

[1] Present address, California Research Corp., 576 Standard Ave., Richmond, Calif.
[2] Present address, Chemistry Department, Purdue University, Lafayette, Ind.

were decreased to a few milligrams. To explore these factors, Vassallo and Harden (*22*) worked with the pure sample in a melting-point capillary, whereas the writers elected to dilute the sample with inert material (*18, 20*). Both approaches have been used in the past, and each has advantages of its own which will be discussed later.

Ordinarily, two thermocouples are employed to measure the differential temperature between the sample and the inert reference, and a third thermocouple is used to record the temperature of the block. Effects associated with the heating rate, the amount of active material, total amount of sample and diluent, and locations of the thermocouples are reported below. The present study has been confined to the melting of salicylic acid, a reversible phase transformation in the sense that no decomposition occurs. Some of the effects are a function of the kinetics and order of the phase change and hence might be quite different for another substance.

EXPERIMENTAL

Salicylic acid, obtained from Eastman Kodak Co., was recrystallised from ethyl alcohol and had a Fisher-Johns melting point of 158.1° to 160.0° C.

Mallinckrodt silver nitrate was used for the calorimetric calibrations. This material had a melting point of 212° C. block and a specific heat of 16.2 cal. (*13*).

The inert support and reference material was a purified carborundum from Union Carbide and Carbon Chemical Co. labeled "abrasive, 500 mesh." Before use, it was screened between 400-, 500-, and 600-mesh sieves. The 500-mesh fraction was chosen for inert support, washed with acetone and chloroform (to remove organic impurities), and washed successively with concentrated nitric acid, hydrochloric acid, and sodium hydroxide (to remove metals and metal oxides). After washing with distilled water, it was ignited overnight at 800° C. After cooling, the material was stored under nitrogen to avoid contamination by the laboratory atmosphere.

Twenty-gram samples were prepared by mixing crystalline salicylic acid with carborundum powder moistened with a few drops of ethyl alcohol in a mortar. The mixture was ground lightly until the solvent had evaporated. Final removal of the ethyl alcohol was carried out in a vacuum oven operated at 80° C. and 4 mm. pressure. The exact salicylic acid content of a large sample was determined by weight loss either after extraction with ethyl alcohol or after ignition. When typical samples were split into two 10-gram portions, results by the two methods agreed within 0.05%.

The most useful concentration range for most organic materials of carborundum is 0.1 to 10%. In this range the melt clings to the particles of carborundum and does not flow. As a

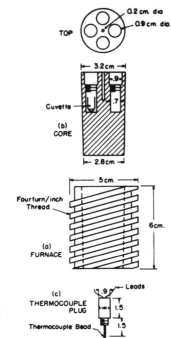

Figure 1. Differential thermal analyses block

a. Furnace block, Alcoa aluminum rod stock
b. Cuvette carriage core
c. Thermocouple and mounting

result, an analysis can be repeated many times on the same sample with a high degree of reproducibility.

Samples for the calorimetric calibration were prepared by mixing silver nitrate with carborundum in a mortar using a few drops of water to moisten the mixture. Water was removed by heating the preparation in a drying oven at 100° C. for 2 hours. The silver nitrate content of the sample was determined by a potassium thiocyanate titration of an aqueous extract from a 2-gram sample.

Apparatus. A Leeds & Northrup d.c. amplifier No. 9835A, a Houston Instrument Co. HR92-4 X-Y recorder with 1 mv. per inch response on both axes, a Model 40 F & M linear temperature programmer, and a sample block shown in Figure 1 were assembled.

The sample block incorporates several useful features. The frustum-shaped well, first used by Vassallo (*21*), permits rapid removal of hot cores and rapid cooling of the apparatus to room temperature by means of a stream of compressed air. The removable core permits a stationary location of the furnace and heating wires without limiting the apparatus as a whole to one size of sample chamber or cell geometry. The

relatively small mass of the block permits rapid equilibration at any temperature and ensures close following of the temperature program. The furnace was heated by a noninductively wound Chromel wire insulated with glass fiber. An iron-constantan thermocouple located in the center of the block was used to govern the programmed temperature unit. The heating was on-off rather than proportional so some cycling must have occurred. However, this was usually negligible.

Samples were placed in round-bottomed borosilicate cuvettes prepared from 4 mm. i.d. tubing. Total height was about 8 mm. Samples were packed by firmly tapping the cuvette five times on the desk-top. If very gentle tapping is employed, up to 10 times as many taps may be required.

A Rubicon potentiometer, Model 2780 B, was used for calibrating the thermocouples and also for providing bucking voltages fed from an auxiliary Rubicon precision potentiometer. This permitted the temperature scales to be expanded greatly. Appropriate voltage dividers were also used. The temperature detectors were made from No. 27 gage copper-constantan duplex thermocouple wire and were set in the aluminum plugs with a ceramic cement. The thermocouples used to determine the absolute temperature of the system employed an ice bath as the cold junction and were calibrated using a platinum resistance thermometer in the range 0° to 300° C. At 160° C., 200 μv. (sensitivity setting of 200) corresponded to 4.1° C. and 130 mm. The maximum differential temperature, T_{max}, could be measured to ±0.1° C. absolute or ±0.03° C. relative to another temperature on the same thermograph for a rapid reaction. When the location of T_{max} is sensitive to heating rate, the uncertainty may be as great as ±0.5° C.

RESULTS

Location of Thermocouple for Determination of System Temperature. This detector may be located in any one of three places with respect to the sample undergoing thermal change: 1. directly in the middle of the block; 2. in the reference material; and 3. in the sample. As will be shown below, most of the discrepancies observed on going from one heating rate to another can be explained on the basis of a lag time in the passage of heat through the metal block and the material in the sample and reference cells. When the temperature of the block is programmed linearly with time and the thermocouple measuring that temperature is located in the block, the temperature of the sample follows a different curve (Figure 2). In the transition region, the temperature of the sample changes more slowly than the block and gives the greatest lag at the temperature at which the melting rate is largest. Following the transition, the sample cell changes

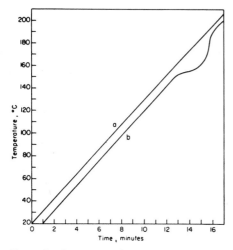

Figure 2. Comparison of temperatures in block and sample cups, rate of 11° C./min.

a. Metal block
b. Cup holding 0.0934-gram sample of 6.87% salicylic acid on carborundum

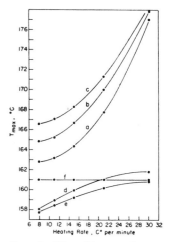

Figure 3. Effect of location of system thermocouple and heating rate on temperature at T_{max} using successive runs on different sample weights of 8.3% salicylic acid on carborundum

Case 1		Case 2	
a.	0.0952 gram	*d.*	0.0952 gram
b.	0.1125 gram	*e.*	0.0822 gram
c.	0.1555 gram	Case 3	
		f.	0.0952 gram

temperature more rapidly until it again lags only slightly behind the block. The existence of a lag of the sample behind the block, even in the absence of a transition, is evidenced in Figure 2. This causes the progressive shift, shown in Table I, of the endothermal minimum to higher temperatures with faster rates of heating. Any interpretation of the maximum differential temperature obviously requires careful standardization of procedure and calibration of the temperature lag (*22*).

When the system thermocouple is in the reference cell (case 2), the reference and the sample both lag behind the programmed block temperature. Provided the thermal conductivities of the sample and reference are similar, the lag in both wells will be nearly the same except in the transition region where the isothermal sample lag again appears.

In case 3, the temperature axis for the system is the actual temperature of the sample which, in the region of a transition, is a nonlinear function of the programmed temperature of the block. In contrast to cases 1 and 2, a change in the heating rate from 5.9° to 15.0° C. per minute produced no detectable shift in curve *f*, Figure 3, or in the endothermal minimum. Furthermore, the endothermal dip was more nearly symmetrical and the recorded melting range was narrower than in cases 1 and 2.

Results similar to the above would be expected for other melting points with-

Table I. Temperature Lag at Different Heating Rates for a 0.0795-Gram Sample of 6.89% Salicylic Acid on Carborundum

Heating Rate, ° C./Minute	ΔT at T_{max}, ° C.
5.6	0.050
7.9	0.221
11.0	0.506
15.0	1.031

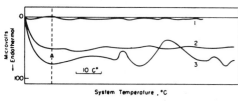

Figure 4. System temperature vs. ΔT for carborundum in both cells

Curve 1. Symmetrically located differential detectors
Curve 2. Unsymmetrically located differential detectors in uniformly packed samples
Curve 3. Unsymmetrical detectors in a nonuniformly packed sample

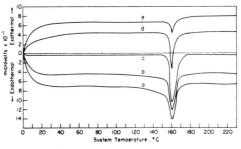

Figure 5. Effect of unbalance in heat capacities between reference and a 0.0954-gram sample of 2.25% salicylic acid using heating rate of 11° C. per minute

Curve	Reference Weight, Grams
a	0.0450
b	0.0745
c	0.0958
d	0.1225
e	0.1650

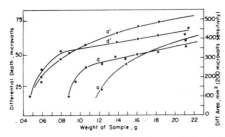

Figure 6. Effect of sample weight on differential area and differential depth using heating rate of 7.9° C. per minute and system temperature detector in sample

2.26% salicylic acid, curves *a* (area) and *d* (depth)
6.87% salicylic acid, curves *a'* (area) and *d'* (depth)

out decomposition, but irreversible systems would behave quite differently. They will naturally show a shift of any thermal signal with change in heating rate. Kissinger (*11, 12*) and Borchardt and Daniels (*4*) have demonstrated this.

If DTA data are to be compared with melting points in the literature, it is important to note that the temperature at maximum melting, T_{max}, should not correspond exactly to the melting point, T_{mp} (*22*). However, T_{max} may be used instead because, under constant operating conditions, it is more sensitive to impurity content than the visually determined melting point.

Asymmetrical Location of Differential Thermocouples. If the thermocouples are not located in identical positions within the sample and reference cells, irregularities will be observed. However, even when they

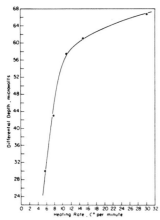

Figure 7. Effect of heating rate on differential depth using system detector in 0.0966-gram sample of 8.60% salicylic acid

are located symmetrically within the cells and pure carborundum is used in both of those cells, packing of the samples may lead to differences in geometry which produce differences shown by the curves in Figure 4. (The actual tracings had only slightly more short-term noise in them than the curves in the figure.) Curve 1 shows the type of blank obtained with careful packing such that the thermocouples are symmetrically located not only with respect to the walls of the block but also with respect to the top and bottom of the material in the sample and reference cells. However, if one detector is displaced approximately 2 mm. toward the side, top, or bottom of the sample, curve 2 is obtained. The displaced thermocouple heats more rapidly than the symmetrically located detector and gives rise to the portion of the curve up to line *A*. After the temperature difference has been established, it stays constant for the remainder of the run. Curve 3 shows that loose packing leads not only to a displacement but also to maxima and minima, presumably due to shift of the support particles as they expand during the heating process. For work of high sensitivity and precision, the exact symmetrical location of the detectors in a well packed sample is obviously necessary.

Thermal Asymmetry due to Unequal Loading of Material. Progressively greater unbalance in loading was effected by holding the weight and concentration of the diluted sample constant and changing the weight of the reference. As shown in Figure 5, differential depth and area varied in a systematic manner, but no translation of the minimum temperature occurred because the system thermocouple was in the sample (case 3). A general loss of sensitivity accompanied greater unbalance of the cell contents. In addition, the change in location of the base

line on the differential axis reflects the fact that although both samples lag behind the furnace temperature, the cell with the lowest heat capacity lags the least. Figure 5 shows that the ability to detect a phase change is greatest when the lags are equalized so that the differential background signals are zero.

Effect of Sample Size on Differential Area Using Equal Cuvette Loadings and Constant Sample Concentrations. Vold's equations (*23*) predict a linear relationship between ΔH and peak area for an undiluted sample system. Kissinger (*12*) noted a departure from this behavior using diluted samples. Therefore, a test of Vold's relationship was made using a constant sample composition, equal weight loading in sample and reference cells, and maintenance of detector symmetry in the centers of the sample and reference.

Figure 6 shows that a marked departure from linearity in area—ΔH occurs below a critical sample size which in turn depends on the percentage of active material in the sample. The depth of the differential minimum varies in much the same way. One would expect the critical sample size to be a function also of the diameter of the sample cup and size (mass) of the thermocouple junction. The critical size should be different, but easily determinable, for any other apparatus.

Effect of Heating Rate on Peak Altitude and Area. A sample will absorb thermal energy and melt as fast as heat is supplied (within limits). Therefore, the transition time will be shorter and the differential mini-

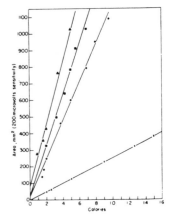

Figure 8. Effect of heating rate on differential area for silver nitrate on carborundum

● 7.9° C. per minute
■ 11.0° C. per minute
▲ 15.0° C. per minute
X extrapolated 0° C. per minute

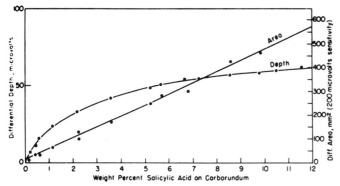

Figure 9. Effect of concentration of active material on differential altitude and area using heating rate of 7.9° C. per minute and a 0.0885-gram sample

mum depth lower, the faster heat is supplied. Figure 7 shows the increase observed for differential depth with heating rate. The ratio of differential depth to differential width varied according to Vold's equation (23).

If DTA could be carried out under nearly equilibrium condition, the differential area, which is proportional to the heat of melting, should be independent of heating rate. On the other hand, Figure 8 shows that, in practice, the area increases with heating rate and the increment depends on the number of calories involved. The true area due to the heat of melting may be determined by extrapolating the results at different heating rates to zero rate and calculating the extra area attributable to temperature lag between sample and reference (see Figure 8). Vold (23) has done this for pure samples using a heating curve in conjunction with a differential thermogram. By following the above precautions to remove lag errors and instrumental variations, the specific heat of melting of salicylic acid was calculated, on the basis of 10 thermograms, to be 28.2 ± 0.5 cal. per gram. The temperature of maximum melting was 161.0° C. The specific heat of melting of ammonium nitrate on the basis of three determinations was 15.0 ± 2 cal. compared to a reported value of 16.2 cal. (15).

Concentration of Active Material. When the total sample size and heating rate were constant, the differential depth varied in a systematic, but nonlinear, manner with the concentration of active material in the sample. The area changed linearly. Figure 9 shows data for the differential depth and area using samples of known salicylic acid content. In the 0.1 to 2.25% salicylic acid range, the differential amplifier was operated at 100-μv. sensitivity (100 μv. = 130-mm. chart); in the range 2.25 to 10%, it was

set at 200-μv. sensitivity (100 μv. = 65-mm. chart). When the salicylic acid contents of samples of unknown concentration were determined both by extraction and by DTA (using Figure 9), the results in Table II were obtained. The agreement with extraction data for these few runs was within one standard deviation in each case. From the points used to establish Figure 9, one should be able to determine up to 10 mg. of salicylic acid from the area with a standard deviation of 0.15 mg. and from the depth of the minimum, with a standard deviation of 0.07 mg. The data in Table II had standard deviations of 0.15 mg. and 0.07 mg. for area and depth, respectively.

DISCUSSION

Use of a diluent provides a larger sample for easier handling although some sensitivity is undoubtedly lost. It is also true that the active material may undergo a change while being mixed with the diluent (14, 18) and hence give unexpected results. On the other hand, Vassallo (21) recommends for polymeric samples that effects of history be minimized by melting the sample before making an analytical determination. Premelting also ensures good contact with the thermocouples and thereby decreases some of the random variations.

However, the transformation obviously must be reversible. If it is irreversible, the data would be misleading just as in the case of prior reaction with diluent. Hence, one must select the approach which seems to be most suitable for a particular situation.

Thermodynamically irreversible reactions such as pyrolytic dehydrations of polyhydroxy compounds should be amenable to a similar treatment provided the DTA chamber has been carefully sealed. Studies of this type are now in progress.

CONCLUSIONS

The above results show that precision DTA using the dilution technique overcomes some of the difficulties encountered in working with a pure sample alone—i.e., mismatched heat capacities, local cooling effects, and major differences in particle size. To use diluted samples, a complete instrumental calibration and evaluation are necessary. Relationships derived for pure samples may be applied to diluted samples only after a detailed study of variations introduced by the diluting agent. In addition, diluted samples introduce a consideration of critical total sample size, which must be evaluated for each instrument before quantitative work can be done. For restricted systems, if all variables other than weight per cent of active material are held constant, differential depth may be used instead of areas in quantitative work.

Precision calorimetry on diluted samples has been demonstrated by the calorie-area calibration for silver nitrate, and by the weight per cent-minimum depth calibration for salicylic acid.

ACKNOWLEDGMENT

We are indebted to Jen Chiu for many constructive criticisms of the manuscript and D. A. Vassallo for plans of his apparatus and for several helpful discussions.

LITERATURE CITED

(1) Baumgartner, P., Duhaut, P., *Bull. Soc. Chim. France* 1960, 1187.
(2) Blumberg, A., *J. Phys. Chem.* 63, 1129 (1959).

Table II. Percentages of Salicylic Acid on Carborundum Related to Differential Minimum Depth and Area Compared with Extraction Weight Loss of 10-Gram Samples

Extraction Weight Loss	Depth		Area	
	Indiv.	Av.	Indiv.	Av.
0.42	0.31; 0.47	0.39	0.45; 0.39	0.42
1.51	1.49; 1.59	1.54	1.55; 1.45	1.50
3.58	3.59; 3.67	3.63	3.50; 3.62	3.56
7.65	7.86; 7.73	7.8	7.63; 7.72	7.7
9.20	9.28; 9.23	9.3	9.16; 9.23	9.2

19

Reprinted from *J. Chem. Educ.*, **40**(2), A87–A116 (1963)

VI. Differential Thermal Analysis

Saul Gordon, *Department of Chemistry, Fairleigh Dickinson University*

When subjected to sufficiently high temperatures, virtually all substances will undergo physical and chemical changes ranging from simple changes in state to complete decomposition that can be studied by the various methods of thermoanalysis. During the course of these physicochemical reactions or phenomena, there will be concomitant sequential or overlapping changes in the enthalpy that lend themselves to a direct measurement by the technique of differential thermal analysis (DTA). As a thermoanalytical method, DTA may be used to characterize and experimentally evaluate a system by continuously measuring the changes in thermal properties due to physical and chemical reactions as they occur at elevated temperatures as a function of increasing temperature. The experimental approach is basically a rather simple modification of the more widely known method of thermal analysis used to evaluate phase diagrams, e.g., melting point versus composition for alloys or salt mixtures. Instead of using cooling curves for the single or multi-component samples, the system under investigation by DTA is heated to elevated temperatures at a predetermined rate of temperature rise. In addition, the temperature of prime interest is not the sample temperature per se, but the difference in temperature between the sample and a thermally inert reference compound which is heated concurrently with the sample under exactly the same conditions. This measurement is made by means of a pair of series-connected (electrically bucking) thermocouples, one of which is placed in the sample and the other in the reference material, as shown schematically in Figure 1. Both substances are contained in the same furnace block or otherwise heated environment so that they will experience a common heating cycle, generally a linear heating rate of 2–20°C per minute. If neither the sample nor the reference com-

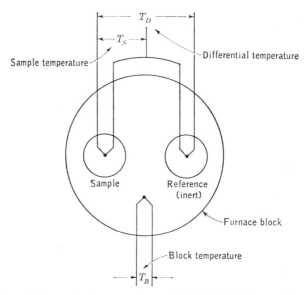

Figure 1. Schematic diagram for thermocouple assembly in typical DTA apparatus.

[*Editors' Note:* Figures 3, 9, 10, and 15 have been omitted because the original halftones were unavailable.]

pound undergoes a thermal reaction over the temperature range involved, a continuous recording of the difference between the temperature of the sample and the reference or furnace temperature should result in a base-line which is parallel to the temperature axis. However, if the sample experiences an endothermal reaction, such as a first order crystalline transformation or even melting, these isothermal phenomena will cause the sample temperature to remain nominally constant while the temperature of the reference material continues to rise at the programed rate. Upon completion of the heat absorbing reaction the temperature of the sample, which has now lagged behind the reference temperature, will rapidly increase until the two substances are again at the same temperature. Further heating will cause both temperatures to rise at the same rate. A plot of the resulting difference in temperature as a function of temperature will produce a deflection from and subsequent return to the baseline, in the form of an endothermal band which is conventionally plotted in a downward direction. A similar effect is obtained if the reaction is non-isothermal, as in the case of virtually all physical and chemical interactions, the specific nature of which will determine the shape of the respective endothermal band. Conversely, if the sample undergoes an exothermal reaction, the heat evolved will cause the temperature of the sample to increase more rapidly than that of the reference compound so that the sample temperature

now exceeds that of the temperature-programed reference compound. When the reaction responsible for this evolution of heat has ceased, the sample will cool down to the temperature of the reference sample, from which point they will both continue to increase in temperature at the programed heating rate. A continuous recording of this difference in temperature as a function of sample or reference temperature will result in an exothermal band, conventionally plotted in an upward direction. The combination of endo- and exothermal bands obtained in DTA thermograms for any given sample represent "thermal spectra" that are very characteristic of any given system in that they indicate the temperatures and temperature ranges for fundamental changes in state, chemical composition and inter- or intra-molecular reactivity of the substance(s), as illustrated in Figure 2. These properties are, in turn, directly related to the thermochemistry and chemical kinetics of the physical or chemical atomic and molecular bonding which are inherent in properties and phenomena such as first order phase transitions, second order transitions and changes in state, such as fusion, boiling, and sublimation. This would also include adsorption and desorption, desolvation, decomposition, oxidation or reduction, solid state reaction, and all other reactions for which the corresponding enthalpic changes will produce measurable changes in the temperature of the sample relative to the reference substance. These are summarized in

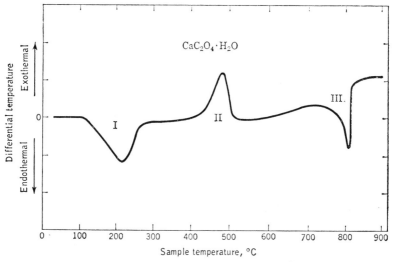

Figure 2. DTA thermogram for $CaC_2O_4 \cdot H_2O$ in an O_2 atmosphere; 8°C. per minute heating rate
I, endothermal dehydration; II, exothermal oxidative degradation; III, endothermal decomposition.

153

Table 1. Since non-isothermal reactions will have rates of reaction that vary as a function of the temperature, in accordance with the Arrhenius type of relationship $(K = Ae - E^*/RT)$, the thermal band shape will be determined largely by the activation energy, E^*.

Other experimental factors such as the size, packing, geometry, and physicochemical history of the sample, electromechanical design of the instrumentation,

Table I. Physicochemical Phenomena and Reactions Amenable to Thermoanalytical Study by Differential Thermal Analysis

| | Enthalpic Change | |
	Endo-thermal	Exo-thermal
Physical		
Crystalline transition (1st order)	×	×
2nd order transition[a]	×	
Fusion	×	
Vaporization	×	
Sublimation	×	
Adsorption		×
Desorption	×	
Absorption	×	
Chemical		
Chemisorption		×
Desolvation	×	
Dehydration	×	
Decomposition	×	×
Oxidative degradation		×
Oxidation in gaseous atmosphere		×
Reduction in gaseous atmosphere	×	×
Redox reaction	×	×
Solid state reaction	×	×

[a]Often characterized by base-line shift due to change in specific heat.

gaseous environment, heating rate and the temperature (sample, reference or furnace) chosen for the abscissa will all affect the shape of the thermal bands obtained. The area under the band will be a function of the sample size, specific enthalpic change and specific heat, density, thermal conductivity and thermal diffusivity of the sample. These are experimental parameters that lend themselves to quantitative as well as qualitative applications and evaluations.

Although the technique of DTA has been applied very effectively and exten-sively by mineralogists, ceramicists, and metallurgists for many decades, it has only been during the past ten years that chemists have begun to utilize the method for diverse applications. To a great extent this has been the result of the recent development and commercial availability of the necessary temperature programing equipment and the high gain, low noise level dc amplifiers and sensitive dc recorders that are now so widely employed in the chemical laboratory. Additional impetus has come from the availability of thermobalances for thermogravimetric analyses that almost invariably provide the investigator with experimental data that can be evaluated adequately only with the complementary information derived from other thermoanalytical techniques, particularly DTA and more recently, effluence analysis. To this should be added the valuable information that may be derived from derivative thermoanalytical techniques. This has led several investigators to advocate the inclusion of DTA and other thermoanalytical experimentation in the undergraduate physical chemistry curriculum, for which easily constructed apparatus and specific experiments have been recommended.

During the past few years complete instrumentation packages for DTA studies have been developed and are now available from several domestic and foreign manufacturers. In virtually all of the apparatus designs there will be found the following modules or components: (*a*) sample and reference holder, (*b*) thermocouple assembly for measuring the sample and reference temperature and/or the block or furnace temperature, (*c*) furnace or other type of heating device, (*d*) suitable controls for regulating the pressure (vacuum and elevated pressures), and the composition of the gaseous environment in which the sample is being heated, either statically or with a controlled dynamically changing flow of gas through the sample, (*e*) temperature controller for obtaining fixed or variable heating rates, as well as constant temperatures for isothermal experiments, (*f*) stable, high-gain, low-noise dc amplifier for amplifying the temperature differences measured by the thermocouples, and (*g*) a recorder for plotting the difference in temperature as a function of sample, reference or furnace temperature, or of time.

In selecting an instrument for DTA studies, prime consideration must be given to experimental objectives and parameters, such as state and chemical comical and chemical reactions anticipated, the use of static versus dynamic gaseous atmospheres and pressure or vacuum requirements, need for effluence analysis of reaction products, upper and lower furnace temperature limits, variability of heating rates, and compatability of the sample and its reaction products with the materials of construction used in the apparatus.

The Thermoanalyzer, shown in Figure 3, is a differential thermal analysis apparatus manufactured by the American Instrument Co., Silver Spring, Maryland ($4800), that utilizes a high temperature pressure-vacuum electric furnace based on a design developed and reported in the literature by W. Lodding and L. Hammell. A metal block with holes for the sample and reference is mounted on two single bore ceramic thermocouple insulators through which pass swaged thermocouples that protrude through the block into the packed sample and reference. These thermocouples, with shielded or bare junctions, are used for measuring the temperature differentials and either the sample or reference temperature. A third thermocouple is available for insertion

into the block for measuring block or furnace temperature. The thermocouple insulator assemblies are connected to the base in such a way that a dynamic flow of gas can be passed through the sample and reference during a DTA experiment at various pressures or under vacuum, by adjusting the flowmeter needle valves, as indicated in Figure 4. This provides the means by which a uniform gaseous atmosphere can be maintained throughout the sample and also serves as the carrier for removing volatile reaction products and transporting them to external collecting devices or analytical instruments for concurrent or subsequent analysis. A kit consisting of another block and quartz or metal sample cups is available for studies of fusible samples that would otherwise melt and flow into the annular space between the thermocouple and the single-bore insulator. The cups are inserted into holes in this block and are seated on the thermocouples. A controlled flow of gas may be passed through this assembly to surround the cup with the required inert or reactive gaseous environment and carry off the volatile reaction products for

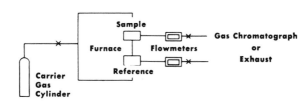

Figure 4.   Flow Control System for Aminco Thermoanalyzer.

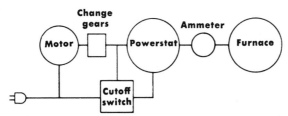

Figure 5.   Furnace power programmer system in Aminco Thermoanalyzer.

155

venting or analysis. Twin 1000°C resistance wound furnaces with removable radiation shields are provided for sequential analyses, thereby allowing the hot furnace to cool down for subsequent use during the following run. The required rate of temperature increase is obtained by linearly increasing the input voltage to the furnace with a motor-driven variable autotransformer, Figure 5. A wide choice of heating rates is obtained by the use of interchangeable gears, the standard values being 2, 4, 8, and 16°C per minute. Provision is made for the automatic cutoff or "hold" of furnace power at preset power levels with an adjustable limit switch. Amplification of the temperature differentials is provided by a stable dc amplifier with fixed levels of amplification ranging from 1 to 1000 in step ratios of 1, 2, 5, and 10. Although a recorder assembly is not provided with the basic instrument, the amplifier output can be conveniently connected to virtually any type of dc *X-Y* or strip chart laboratory recorders. Preprinted graph paper is available with a 0–1230°C chromel-alumel thermocouple *x*-axis suitable for use with *X-Y* recorders that have a sensitivity of 5 mv per inch or 50 mv full-scale, such as the Moseley or Houston recorders. The Thermoanalyzer is a thermoanalytical complement to the Aminco Thermograv, an automatic recording vacuum thermobalance for thermogravimetric analysis.

E. I. DuPont de Nemours

The DuPont 900 Differential Thermal Analyzer ($7500), Figure 6, available from the Instrument Products Division in Wilmington, Delaware, is a compact apparatus using disposable capillary melting point tubes of various diameters to contain

Figure 6. DuPont 900 Differential Thermal Analyzer.

micro or macrosize samples. These tubes are packed with sample and reference substances and inserted into a metal block that is heated by a replaceable resistance wire cartridge heater centrally located in the block, Figure 7. Simple plug-in thermocouples are in turn inserted into the contents of the two tubes. Additional

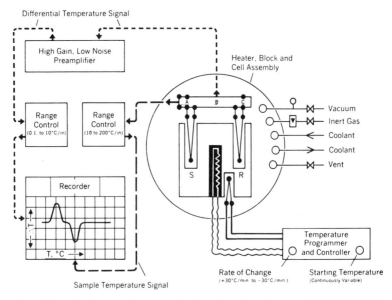

Figure 7. Schematic diagram showing construction of and circuitry for DuPont 900, Differential Thermal Analyzer.

block thermocouples are provided for temperature programing of the system and for recording the temperature difference as a function of block temperature. Provision is made for liquid nitrogen cooling of the analytical block assembly for runs involving temperatures below ambient. The sample cell assembly is enclosed by a small bell jar which provides a compact size for this assembly and allows rapid determinations to be made with a series of samples either under vacuum or in an atmosphere of controlled composition Operational amplifiers and solid state components are employed in the temperature programing circuitry that may be set for linear heating and cooling rates over the continuously variable range of 0 to 30°C per minute, throughout the −100 to +500°C temperature range of the general purpose cell assembly. It may also be preset for automatic high temperature cut-off or for cycling between any selected high and low temperature limits. A high gain, low noise solid-state dc amplifier is used in conjunction with a built-in *X-Y* recorder to continuously plot the thermograms at various sensitivities ranging from 4 to 400 μv per inch, corresponding to 0.1 to 10°C per inch with chromel-alumel thermocouples, for the temperature differential axis; and 0.4 to 8 mv per inch, or 10 to 200°C per inch for the temperature axis. The ink-pen plot may be made on preprinted millivolt or temperature scaled paper. An optional base-line compensation kit is available for providing electronic compensation for the base-line drift normally associated with improperly matched sample and reference substances.

Eberbach Corp.

The first commercially available DTA apparatus seems to have been the portable unit which is manufactured by Eberbach Corp., Ann Arbor, Michigan ($580), Figure 8. This unit was originally designed by R. A. Nelson and constructed in the Bureau of Plant Industry and Soils for widespread use by geologists in the U. S. Geological Survey in their field evaluations of various minerals and soils. The samples and reference substance are packed into three crucible cells, each of which is fitted with thermocouple assemblies for measuring the temperature differences by means of a pair of series-connected thermocouples which are connected to a sensitive galvanometer, and a third thermocouple for measuring the furnace temperature by connecting it to the built-in pyrometer. Each of the three stations is used in sequence by placing the furnace over the crucible and connecting it to a 115 v ac or dc source. The heating rate, which is not programed, is just the nonlinear rate at which the furnace will heat

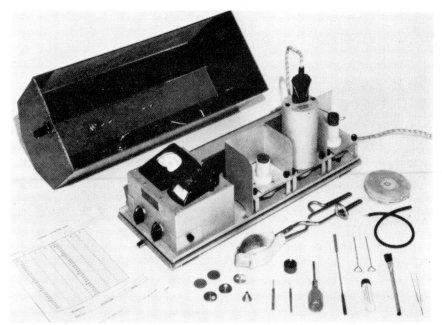

Figure 8.   Eberbach Corp. Portable DTA apparatus.

up when either full line voltage or a rheostat-reduced power level is applied. During the course of this heating cycle the operator reads the two meters at arbitrarily chosen furnace temperature intervals, e.g., every 20 degrees, and manually records the data for subsequent plotting. At the end of the run the furnace is allowed to cool down to a suitable temperature, which may be well above ambient, and is placed over the second crucible for the next run. The same procedure is repeated for the third crucible. By calibrating the instrument with samples having known enthalpic changes and restricting the measurements to temperature ranges in which significant thermal bands appear if the particular mineral or soil component is present, very useful qualitative and semiquantitative data can be obtained.

Harrop Precision Furnace Co.

The Harrop DTA apparatus, Figure 9, is manufactured in Columbus, Ohio. It is essentially a custom-built unit comprised of various component options combined into one assembly in accordance with the needs of the purchaser. Major modules are the control panel and basic wiring cabinet for the primary instrument power with a magnetic amplifier and current limiter for a 1 kva saturable reactor through which the furnace is controlled, and space as well as input connectors for the temperature programer, high gain dc amplifier, and X-Y recorder, Figure 10. A manifold and associated flowmeters, valves, and connectors are provided for experiments in controlled gaseous atmospheres at ambient or elevated pressures or under vacuum. The temperature programing system is a Leeds & Northrup 3-mode current adjusting control system with strip chart recorder using a motor-driven cam and interchangeable gears for obtaining a wide choice of heating as well as cooling rates, Figure 10. By using a zero-centered 10 mv dc recorder with the L & N indicating preamplifier, differential temperature ranges equivalent to ± 25 to 1000 μv may be obtained. A modified Houston X-Y recorder is offered for plotting differential temperature as a function of temperature. Platinum or nichrome wire wound furnaces are available for temperature ranges up to 1400 or 1100°C, respectively. The recording and control systems are designed for optional use with a thermobalance for complementary thermogravimetric studies.

Robert L. Stone Co.

DTA apparatus available from the Robert L. Stone Co. of Austin, Texas, consists of various combinations of several basic modules covering a wide range of prices and capabilities. Basic to all models is a furnace platform to which is added (*a*) any one of several interchangeable sample holders for solids or liquids; single or multiple samples; sample holders and specimen cups fabricated from various metals and alloys for use in a variety of atmospheres and with corrosive samples; and for simultaneous DTA and effluence gas analysis (EGA); (*b*) a furnace for either 1100 or 1400°C operation; and (*c*) a pressure top for extending the pressure range from atmospheric to vacuum or elevated pressures up to 100 psi. Additional accessories provide for dynamic cycling of the gaseous atmosphere either through or around the sample, and, with the vapor generator and preheater coils, it is feasible to use a condensable substance such as steam for the dynamic gas. The schematic diagram in Figure 11 shows the construction and operation of the furnace platform assembly with dynamic gas flow system, thermocouple arrangements, furnace, and pressure chamber. Several recorder-controller assembly models are available with temperature programer and strip chart recorders for both differential temperature and reference temperature, the latter with a built-in device for marking the thermograms with pips at 100°C intervals. The types of programers available range from the simpler modified West JSB-1 motor-driven cam actuated pyrometer with a nominal 10°C per minute heating rate, to the more sophisticated Leeds & Northrup with CAT that provides twelve exact heating rates from 0.4 to 25°C per minute and the automatic temperature limit switch functions of rise-off, rise-hold, and rise-cool. The high gain dc amplifier has an adjustable gain ranging from 5 to 4000 μv, or with the standard Pt-Pt, 10% Rh thermocouple, a full-scale differential temperature of 0.5 to 350°C. With the two-pen recorder and an additional amplifier, two samples may be run concurrently, or the second recording channel may be used for simultaneous DTA-EGA with the appropriate sample holder assembly, a hot-wire gas thermal conductivity cell and a power supply assembly. The power supply is also available in combination with a recorder to complete the operational assembly shown in Figure 12. A new thermobalance is available for thermogravimetric evaluations that may be used with either a separate control system or the DTA recorder-controller assembly.

Technical Equipment Corp.

The Deltatherm ($5875), manufactured by Technical Equipment Corp. in Denver, Colorado, is a console mounted DTA

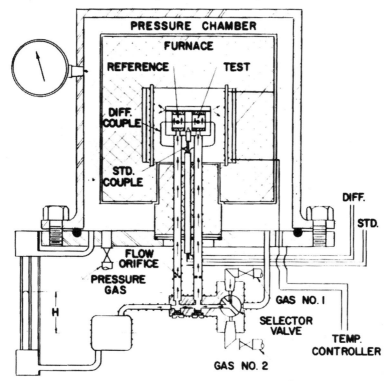

Figure 11.   Schematic drawing of the pressure chamber and furnace platform assembly in the R. L. Stone DTA apparatus.

Figure 12.   R. L. Stone Co. DTA apparatus consisting of the DTA assembly, effluence gas analysis (EGA) detector and power supply, and recorder-programmer assembly.

apparatus, Figure 13. This instrument incorporates a temperature programing system and four-channel dry stylus electrosensitive paper recorder, with twin furnace positions and interchangeable multi-sample plug-in blocks having gold plated thermocouple connecting plugs and printed circuits for the interconnecting thermocouples, Figure 14. A series of metal and ceramic blocks are available for pressure, vacuum, and dynamic gas atmosphere studies with nine wells to accommodate four sample-inert reference pairs and one for the control thermocouple. The basic unit is equipped for a fixed heating rate of 10°C per minute but may be replaced with an optional rate-control unit for more closely controlled heating rates of 2 to 20°C per minute that are obtained with interchangeable gears. A group of six plug-in stable dc amplifiers with printed circuitry and transistorized solid state choppers are

Figure 14. Multiple sample holder with printed circuitry and plug-in module construction used in Deltatherm.

Figure 13. Technical Equipment Corp. Deltatherm.

used for the four sample-reference differential temperature channels, the temperature indicator that also provides a series of solid calibration lines across the chart at 50°C intervals, and in the rate control temperature programer, Figure 15. The sensitivity of 25 μv per inch (0.6°C/in.) may be used over the full 12 in. width of the recorder with overlap of all four traces. Any of four fixed levels of step attenuation may be chosen to minimize or eliminate this overlap. The reference well temperature is constantly displayed for visual panel monitoring on a printed tape potentiometer. A multi-purpose meter is provided for selector-switch read-out of line voltage, furnace current, or recorder chart position for each of the four DTA traces. The use of two furnaces, and an optional third furnace that can be operated externally to the apparatus, allows for virtually continuous operation of the instrument. A companion thermobalance has been developed for complementary thermogravimetric studies.

Summary

Although the major contributions involving DTA studies have been made by investigators who have constructed or assembled their own apparatus, many of which are described in the following references, it is obvious that they did so because well designed instrument packages were not commercially available. This situation has now been greatly alleviated by several companies, as previously described. Nevertheless, those with limited equipment budgets should find many useful suggestions for instrumentation and experiments in the references cited below. This survey of DTA instruments has been restricted to instruments that are domestically manufactured because their foreign counterparts are not readily available at this time.

References

ARSENEAU, D. F., "DTA Apparatus," J. CHEM. ED., **35**, 130 (1958).

BORCHARDT, H. J., "DTA. An experiment for the Physical Chemistry Laboratory," J. CHEM. ED., **33**, 103 (1956).

CAMPBELL, C., GORDON, S., AND SMITH, C. L., "Derivative Thermoanalytical Techniques. Instrumentation and Applications to Thermogravimetry and DTA.," *Anal. Chem.*, **31**, 1188 (1959).

CHIU, JEN, "Identification of Organic Compounds by DTA," *Anal. Chem.*, **34**, 1841 (1962).

GARN, P. D. AND KESSLER, J. E., "Effluence Analysis as an Aid to Thermal Analysis," *Anal. Chem.*, **33**, 1247 (1961).

GORDON, S., "Thermoanalysis," McGraw-Hill, "Encyclopedia of Science & Technology," **13**, 556–9 (1960).

GORDON S. AND CAMPBELL, C., "Thermoanalytical Techniques. Their Importance in the Instrumental Methods Laboratory," Abstracts of Papers, 131st meeting, A.C.S., April 1957.

GORDON, S. AND CAMPBELL, C., "Automatic and Recording Balances," *Anal. Chem.*, Review Issue, **32**, 271R (1960).

GORDON, S. AND CAMPBELL, C., "Differential Thermal Analysis," in "Handbook of Analytical Chemistry," McGraw-Hill Book Co., New York, **1963**, Sec. 8 Thermoanalytical Techniques.

HENDRICKS, S. B., GOLDICH, S. S., AND NELSON, R. A., "A Portable DTA Unit for Bauxite Exploration," *Econ. Geol.*, **41**, 64 (1946).

KE, B., "Application of DTA to High Polymers," in Mitchell, J. Jr., *et al.*, "Organic Analysis," Vol. 4, Interscience Publishers, New York, **1960**, 361–93.

LEWIN, S. Z. "Thermobalances," J. CHEM. ED., **39**, A575 (1962).

LODDING, W. AND HAMMELL, L., "High Temperature Pressure-Vacuum Furnace," *Rev. Sci. Instru.*, **30**, 885 (1959).

LODDING, W. AND HAMMELL, L., "DTA of Hydroxides in Reducing Atmosphere," *Anal. Chem.*, **32**, 657 (1960).

MACKENZIE, R. C. (ed.), "The Differential Thermal Investigation of Clays," Mineralogical Society, London, **1957**.

MACKENZIE, R. C., "Scifax DTA Data Index (punched cards)," Cleaver-Hume Press, Ltd., London, **1962**.

MACKENZIE, R. C. AND MITCHELL, B. D., "DTA. A Review," *The Analyst*, **87**, 420 (1962).

MURPHY, C. B., "DTA. Review of Fundamental Developments," *Anal. Chem., Annual Reviews*, **30**, 867 (1958); **32**, 168R (1960); **34**, 298 (1962).

SMOTHERS, W. J. AND CHIANG, Y., "DTA. Theory and Practice," Chemical Publishing Co., **1958**.

STONE, R. L., "DTA by the Dynamic Gas Technique," *Anal. Chem.*, **32**, 1582.

WENDLANDT, W. W., "An Inexpensive DTA Apparatus," J.CHEM.ED., **37**, 94.

WENDLANDT, W. W., "Reaction Kinetics by DTA. A Physical Chemistry Experiment," J. CHEM. ED., **38**, 571 (1961).

161

20

Reprinted from *Anal. Chem.*, **36**(11), 2162–2166 (1964)

Determination of Specific Heat and Heat of Fusion by Differential Thermal Analysis

Study of Theory and Operating Parameters

D. J. DAVID

Mobay Chemical Co., New Martinsville, W. Va.

▶ Utilization of differential thermal analysis under nearly equilibrium conditions permits the determination of heat of fusion and specific heat on a variety of inorganic and organic compounds from a single calibration of an easily handled material like tin. Variables such as sample size, heating rate, and sample state were studied for their effect upon the heat of fusion. These variables did not exhibit effects upon the results within the limits of error of the determination. The theory and equations underlying the specific heat determination are discussed and a practical method is presented which is applicable to a wide range of materials. The standard deviation at the 95% confidence level for the heat of fusion and specific heat was found to be 1.5 cal./gram and 0.02 cal./gram/°C., respectively.

DIFFERENTIAL THERMAL ANALYSIS has been applied previously to a variety of both inorganic and organic materials. These applications were initially concerned with minerals and soils (*8, 23*), and subsequently with pure inorganic compounds (*7, 15, 21*). More recently, less well defined organic materials and compounds have been studied. The investigators have reported both qualitative and quantitative results (*1, 2, 6, 11, 17, 22, 30, 31*).

As a result of the gradual development of a variety of differential thermal analyzers, a re-evaluation of qualitative and quantitative variables has resulted in some duplication of effort (*27*).

The variation of peak temperature has been reported to be dependent upon sample size, size of the cylindrical holder, and heating rate (*12, 24, 28*).

The effects of diluent techniques have been covered in various papers. Particle size and packing were found to be important factors that affected the thermograms obtained (*19*). In addition, the formation of complexes with the inert diluent has also been reported (*4, 18*).

However, with the development of more sophisticated and sensitive instrumentation, the effect of sample size, diluent, size of sample holder, heating rate, and difference in heat capacities between reference and sample, and other parameters can be better evaluated.

Base line deviation, especially at the beginning of a run, is a common occurrence in differential thermal analysis. This is due to an imbalance in heat capacities between the sample and reference thermocouples and is affected by symmetry, sample loading, inert loading, and packing (*3*).

Many of the detrimental and non-uniform parameters that accompany the great variety of techniques common to each specific analyzer, may be turned to advantage when the proper technique is employed as in the present application.

The aspects of differential scanning calorimetry (DSC) and applications to quantitative measurements of transition energies have been reported recently by Watson, *et al.* and O'Neill (*20, 29*). DSC measures the transition energy

directly (*29*) while conventional DTA measures ΔT *vs.* sample temperature. Thus, DTA must be calibrated before it can be utilized for quantitative transition energy measurements.

The present paper will show that when DTA is carried out under nearly equilibrium conditions, a single calibration can be performed which is applicable to the determination of the specific heat and heat of fusion of both organic and inorganic compounds.

THEORY

The major factors affecting base line deviation (equilibrium conditions) are mismatched heat capacities, improper heat transfer, symmetry, packing, particle size (sample and diluent), dilution effects, inertness of diluent, and sample concentration.

When a sensitive system is employed in which the sample size is small (1 to 10 mg.), the sample does not have to be diluted, and the system contains fixed thermocouples allowing reproducible results; the disadvantage of base line deviation can be an asset by allowing the determination of the specific heat of the sample.

Assuming the above conditions exist, the deviation in base line would be greater for those materials with a high specific heat value. To obtain a mathematical expression for C_p, we must consider two factors: the effects of the system upon the differential thermocouple; and the effects of the system plus sample upon the differential thermocouple.

In considering heat effects upon the differential couple, the equations of heat balance employed by Borchardt and Daniels are used (5).

$$C_{p,s}dT_2 = dH + K_s(T_3 - T_2)dt \quad (1)$$

$$C_{p,r}dT_1 = K_r(T_3 - T_1)dt \quad (2)$$

T_1, T_2, and T_3 are temperatures of the reference thermocouple, sample thermocouple, and air bath (furnace thermocouple), respectively (see Figure 1). $C_{p,r}$ is the total heat capacity of the reference thermocouple (including cup) and $C_{p,s}$ is the total heat capacity of the measuring thermocouple including cup. The absolute values which will be obtained are dependent upon the effects of thermocouple symmetry, size, and shape of sample containers and furnace, when these variables can be reduced to constants as in a fixed system.

Equation 1 shows that any increase in enthalpy of the sample side of the differential thermocouple is due to the total enthalpic effects of sample plus the heat transferred to the thermocouple by the surroundings. Equation 2 describes the enthalpic effects on the reference side except that the dH term is necessarily not present. K_s is the heat transfer coefficient for the sample (measuring) thermocouple and K_r is the heat transfer coefficient for the reference side.

The assumptions governing the validity of these equations to the present system are: the differential thermocouples are fixed; a small sample size is employed; the sample does not have to be diluted; the sample holder is capable of maintaining even heat distribution consistent with the demands of the temperature programmer; and linear heating rates from one run to the next can be realized.

There are two situations which we must consider—the absence of a sample and the presence of a sample. Let us consider the first case where a sample is not present in the container. In this instance there can be no heat evolution or absorption except that due to slightly different heat capacities of the thermocouples and empty containers caused by symmetry and size considerations.

Then, any change in base line from the horizontal due to differences in heat capacity may be expressed as:

$$C_{p,s}dT_2 - C_{p,r}dT_1 =$$
$$K_s(T_3 - T_2)dt -$$
$$K_r(T_3 - T_1)dt \quad (3)$$

where $C_{p,s}dT_2 - C_{p,r}dT_1 = C_pdT$ and C_p = heat capacity of the system. K_r and K_s, the heat transfer coefficients from T_3 to T_2 and T_3 to T_1, should be equal even under dynamic conditions providing linear reproducible programming rates exist, thermal gradients within the holder are absent, and the total

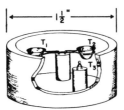

Figure 1.   Sample holder

T_1 = Reference temperature
T_2 = Sample temperature
T_3 = Air bath temperature

heat capacity of the air bath is more than sufficient to provide the thermal energy required to maintain nearly equilibrium conditions within the system.

Rearranging Equation 3 gives

$$C_pdT = K(T_1 - T_2)dt \quad (4)$$

where $K = K_r = K_s \neq 0$

This is true providing the previous assumptions are true and K_r and K_s temperature independent over the temperature range of interest (dT). This is indeed true as will be shown later.

Thus Equation 4 becomes

$$C_{p(w/o \; sample)} =$$
$$\frac{K(T_1 - T_2)dt}{dT} = cal./°C. \quad (5)$$

where

K	= heat transfer coefficient of system in cal./mm.².
C_p	= heat capacity of system in cal./°C.
dT	= temperature range of interest
$(T_1 - T_2)dt$	= ΔTdt; and over a time interval, say t_o
$\int_t^{t_o} \Delta Tdt$	= area encompassed by base line curve and constructed horizontally, measured over temperature range of interest.

The K value does not incorporate a time-temperature function since the chart speed of the recorder and the response of the DC amplifier and recorder were considered fixed.

We now consider the second case where a sample has been placed in the proper container and the enthalpic effects will be registered on T_2 which is one side of the differential couple.

We now choose a time (temperature) interval where $dH = 0$—i.e., the heat of transformation is 0 because the sample is not undergoing a chemical or physical transformation. Thus, the equation

describing the heat capacity due to the presence of a sample is identical to Equation 5. By evaluating Equation 5 without sample and subsequently with sample and subtracting, the heat capacity of the sample is

$$\tilde{C}_p = C_{p(w/o \; sample)} - C_{p(w/sample)} =$$
$$(\text{System effects}) - (\text{System} +$$
$$\text{sample effects}) \quad (6)$$

$$= \frac{K\Delta Tdt}{dT}\bigg|_{(w/o \; sample)} - \frac{K\Delta Tdt}{dT}\bigg|_{(w/sample)}$$

$$= \frac{K}{dT}(\Delta Tdt)_{\text{area due to sample}} \quad (7)$$

But there are thermal effects within the sample. Through the sample itself a thermal lag occurs and the rate at which heat is received by the measuring side of the differential couple becomes highly dependent upon sample size, heating rate, and the thermal diffusivity (α) of the sample. The lag experienced is precisely why the specific heat can be measured since it is characteristic of each material.

Therefore, we must take into account, thermal effects within the sample and consider the diffusion effects by means of the following equation (10, 16, 32).

$$\frac{dT}{dt} = \frac{k}{\rho C_p}\frac{d^2T}{dx^2} \quad (8)$$

where

$\frac{dT}{dt}$	= change of temperature of sample with respect to time
k	= thermal conductivity of sample
ρ	= density of sample
C_p	= specific heat of sample
$\frac{d^2T}{dX^2}$	= rate of change of the temperature through the sample

The average heat capacity of the sample can now be defined by solving for C_p in Equation 8 and adding to Equation 7.

$$\tilde{C}_p = K\Delta T\frac{dt}{dT} + \frac{k}{\rho}\left(\frac{d^2T}{dx^2}\right)\frac{dt}{dT} \quad (9)$$

$$= (\text{sample effects}) + (\text{sample thermal diffusivity effects})$$

$$= \left(\rho + \frac{k}{\Delta TK}\frac{d^2T}{dx^2}\right)\Delta Tdt\frac{K}{\rho dT} \quad (10)$$

Allowing for sample size Equation 10 becomes

$$\tilde{C}_p = \frac{K'K(\Delta T \; dt)_{\text{area due to sample}}}{\rho g dT} \quad (11)$$

which describes the average specific heat of the sample.

Where

$$K' = \left(\rho + \frac{k}{K\Delta T}\frac{d^2T}{dX^2} \right) = 1.68 \text{ (experimentally determined)}$$

g = sample size in grams
$\bar{C}_p$ = average specific heat of sample in cal./gram/°C.

The other terms in Equation 11 have been defined previously. The term $k/\rho C_p$ in Equation 8 is α which is a measure of the thermal inertia or diffusivity of the sample. Although the individual terms in Equation 10 comprising K' cannot be evaluated, K' itself can be.

All determinations of specific heat should be performed using the same heating rate—that is, the rate at which heat is received by the differential couple is a constant. In cases where $dH \neq 0$, the total heat of transformation (solving for dH) is the area generated by the transformation (cal./gram) plus the heat due to the change in specific heat of the sample. The specific heat of the sample is generally not included because of suitable base line construction.

EXPERIMENTAL

Apparatus. The apparatus consists of a controlled pressure and controlled atmosphere differential thermal analyzer Model 12BC2, recorder-controller assembly Model JAC, and furnace platform assembly Model GS-2 manufactured by the Robert L. Stone Co. This instrumentation has been described by Stone (25, 26).

The apparatus is equipped with an L & N Speedomax G dual pen recorder and two ARA Model 4A-DC amplifiers. The highest sensitivity of the amplifiers is 0.5 μvolt/inch. Normal operating sensitivity is from 1 μvolt to 50 μvolt/inch depending upon the type of sample and the purpose of the analysis.

The sample holder and recorder system permit the recording of two thermograms simultaneously. The recorder is linear in time only and has a chart speed of 0.1 inch/minute. A separate recorder in conjunction with an L & N Series 60 linear temperature recorder-programmer is an integral part of the instrument.

The sample holder assembly, Model SH-M4CL, consists of two separate Chromel-Alumel differential thermocouples and a Chromel-Alumel furnace couple which is utilized for programming as well as recording the temperature of the furnace. A quick cool furnace, Model F-1-F, which is positioned over the sample holder by two guide pins was utilized throughout the experiments.

The differential couples are circular in shape and hold aluminum pans in which the samples are placed (see Figure 1). For demonstration purposes, only one of the differential couples has been shown. A top which fits over the Inconel sample holder

in Figure 1 is not shown. In addition, a furnace cover was placed over the furnace to render the system insensitive to room air currents.

Thermocouple and Sample Parameters. Since the thermocouples are relatively fixed in this type of sample holder, once the system has been balanced to provide an acceptable base line, thermal asymmetry will not be an important factor because the base line will remain reproducible.

The initial base line may be adjusted by decreasing or enlarging the size of the loop as well as by movement in a horizontal or vertical direction. A base line is then run using semispherical aluminum pans which vary in weight from 1.75 to 1.80 mg. Evaluation of this base line permits subsequent adjustment of the differential couples to provide a more horizontal base line.

The sample size required in this type of system is generally from 0.1 to 10.0 mg. depending upon whether or not the thermograms are to be used for qualitative or quantitative purposes. Obviously, the nature of the transition is an important factor since first order transitions will require generally less sample and a lower sensitivity than the determination of second order transitions.

A decided advantage of this technique is that the sample pans may be discarded after each run and replaced with new ones. Also, when polymers are run the carbonization of these materials does not contaminate the thermocouples since the sample is not in actual physical contact with the thermocouples. This system provides excellent base line stability from one run to the next.

Calibration. The calibrations were performed utilizing a linear programming rate of 10° C./minute and helium as the dynamic gas to prevent any oxidative effects. The instrument was calibrated, and the K values were determined by utilizing the heat of fusion of tin, 14.0 cal./gram (9). The material under study was placed in the sample container and no reference material was used. K was determined for each differential thermocouple by means of the following equations which describe the response of the system.

$$\frac{\text{Area}}{\text{gram}} = \frac{\text{Area of sample transition} \times \text{range setting of interest (μvolt)}}{\text{Sample wt. (grams)}} \quad (12)$$

$$\frac{\Delta H_f \text{ of standard (cal./gram)}}{\text{Area/gram}} =$$

$$\frac{\text{Cal.}}{\text{Area}} = \frac{\text{Cal.}}{\text{mm.}^2} = K \quad (13)$$

The K values determined were 2.80 $\times 10^{-6}$ cal./mm.² for differential couple 1 and 2.65 $\times 10^{-6}$ cal./mm.² for differential couple 2. Various range settings of the d.c. amplifiers provided identical K values from Equations 12 and 13.

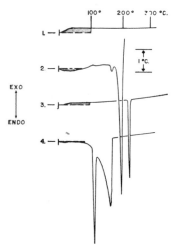

Figure 2. Thermograms showing areas used for C_p determinations

(Sensitivity, 50 μvolt/inch)
1. Thermogram of base line with empty pans
2. Thermogram of 7.55 mg. of dicyandiamide
3. Thermogram of 10.0 mg. of tin
4. Thermogram of 7.66 mg. of benzoic acid

Determination of Specific Heat. To test the validity of the preceding equations governing specific heat, a base line, in duplicate, was run with empty sample pans. The area (b) encompassed between the base line and a line extended horizontally from ambient to the desired upper temperature was measured. (See Figure 2.)

The (a) area, as shown in Figure 2, was measured on a series of samples to the desired upper temperature. When a transition occurred below 100° C., only the area below this transition was usable and the same area measurement to this transition temperature must be carried out on a blank run.

Those areas above the constructed horizontal were considered positive and those areas below it negative. All measured areas due to sample effects were subtracted from the area of the blank base line. This area was then treated according to Equation 11.

From an initial run of benzoic acid, the net correction resulted in a multiplication factor of 1.68 (K'). The equation utilized for all subsequent specific heat determinations reduced to,

$$\bar{C}_p = \frac{KX(\text{Area under base line w/o sample} - \text{area under base line w/sample}) \times 1.68}{g \times \rho \times dT}$$

$$(14)$$

The terms in Equation 14 have been defined previously.

The specific heat of a series of samples was determined. These results are shown in Table I.

Determination of Heat of Fusion. The variation of heating rate, sample

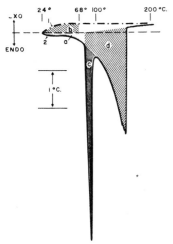

Figure 3. Thermogram of naphthalene showing areas used for ΔH_f and ΔH_r and C_p determinations

(Sensitivity, 50 μvolt/inch)
1. Thermogram of base line with empty pans
2. Thermogram of 8.34 mg. of naphthalene

Areas:

$$b - (-a) = \text{Area due to } C_p$$
$$c = \text{Area due to } \Delta H_f$$
$$d = \text{Area due to } \Delta H_v$$

size, and sample shape was evaluated since these factors have been known to have a major effect upon the results obtained (3, 12, 28).

The heats of fusion were determined on a series of compounds at a program rate of 10° C./minute. The heats of fusion of several compounds were then determined at a program rate of 4° C./minute. This rate was thought to be significantly different from the original to provide a basis for determining any effect upon area.

The sample size was varied from 2.0 to 11.0 mg. in order to determine this effect upon heat of fusion. Benzoic acid was utilized for this evaluation.

The effects of sample state upon benzoic acid were also determined. Benzoic acid is normally obtained as fine crystals. This compound was fused and a portion was utilized for the heat of fusion determination.

In a separate experiment, tin, which was in the physical state of small granules, was flattened into a thin sheet by pressure. The heat of fusion was then determined on an appropriate size sample of the tin sheet.

All heats of fusion are based on the calibration values reported previously. The results of these experiments are listed in Table II. The areas utilized for this measurement are shown in Figure 3, a thermogram of naphthalene. The heats of fusion and vaporization were calculated from the following equation:

$$\Delta H_f \text{ or } \Delta H_r = \frac{KX \text{ area in mm.}^2 \times \text{range setting of interest } (\mu\text{volt})}{g} \quad (15)$$

Table I. Results of Determination of Specific Heat

Sample	Cal. 15°/gram/°C. C_p, determined	Cal. 15°/gram/°C. C_p, known[a]
p-Dinitrobenzene[b]	0.25	0.259 at 119° C.
p-Dinitrobenzene[b]	0.23	0.259 at 119° C.
2,4-Dinitrotoluene[b]	0.34	0.350 at 100° C.
2,4-Dinitrotoluene[b]	0.35	0.350 at 100° C.
Silver nitrate	0.05	0.146 at 50° C.
Silver nitrate	0.09	0.146 at 50° C.
Benzoic acid	0.29	0.287 at 20° C.
Benzoic acid	0.22	0.287 at 20° C.
Tin	0.04	0.054 at 20° C.
Tin	0.04	0.054 at 20° C.
Oxalic acid	0.39	0.338 at −200° to +50° C.
Nickelous nitrate	0.47	0.473 at 80° C.
Dicyandiamide	0.40	0.456 at 0° to 204° C.
Naphthalene	0.40	0.402 at 87 5° C.
Polyethylene	0.56	0.55
Polystyrene	0.35	0.32–0.35

[a] All known C_p values were obtained from References 9, 13, and 14.
[b] Eastman White Label, all other reagents are c.p. grade or better. Polyethylene and polystyrene were received through courtesy of Monsanto Chemical Co.

Table II. Parameters Studied and Results of ΔH_f Determinations

Sample	Sample size, mg.	Heating rate	Sample state	Cal. 15°/gram determined ΔH_f	Cal. 15°/gram known[a] ΔH_f
Tin	10.16	10° C./min.	Normal	13.8	14.0
Tin	10.70	10° C./min.	Flattened sheet	14.2	14.0
Benzoic acid	4.80	10° C./min.	Fused	34.6	33.9
Benzoic acid	6.16	10° C./min.	Normal	34.1	33.9
Benzoic acid	5.73	10° C./min.	Normal	33.6	33.9
Tin	2.45	10° C./min.	Normal	13.6	14.0
Indium	2.42	10° C./min.	Normal	6.9	6.8
Silver nitrate	10.00	10° C./min.	Normal	15.5	17.7
2,4-Dinitrotoluene	5.94	10° C./min.	Normal	28.2	26.4
Naphthalene	6.34	10° C./min.	Normal	36.0	35.6
Benzoic acid	6.67	4° C./min.	Normal	33.9	33.9
2,4-Dinitrotoluene[b]	8.80	4° C./min.	Normal	27.9	26.4
Tin	5.31	4° C./min.	Normal	13.3	14.0

[a] All ΔH_f values were obtained from reference 9.
[b] Eastman White Label, all other reagents are c.p. grade or better.

DISCUSSION

The portion of the thermograms from ambient to 100° C., in Figure 2, show that the sample couple can lag the reference couple appreciably, depending upon the specific heat of the sample. Thermal diffusivity of the sample and heat transfer of the system are also important factors. The rate of transfer per unit time was considered a fixed value (constant), because the speed of response of the pen is fixed and an exact chart speed of 1 inch/10 minutes was used throughout.

Once this lag has been overcome by heat input to the system, the sample curve follows the base line curve closely and may render specific heat determinations in extended temperature regions inaccurate. However, when the specific heat value of a material changes abruptly due to a morphological transformation—e.g., glass transition—this change is readily detected and a reliable value of the change is possible even at elevated temperatures.

Equation 14 is valid only for a number of specific heat determinations in which the instrumental conditions are held constant. During any series of specific heat determinations, the thermocouples must not be moved and the same sets of semispherical sample and reference pans must be used. A change of sample pans, even though weight differences of the pans amounted to only 0.1 mg., affected the value of the result. Asymmetry of the pans which may vary slightly on forming may also be an important factor. Any adjustments of the recorder for sensitivity or dampening also affects this measurement. The furnace should also be covered to prevent drafts which will result in thermocouple drift and identical heating rates must be used.

Additional error may be introduced by the inability to measure accurately some of the small areas experienced. All areas were measured in triplicate with a plane planimeter and the results were averaged. Greater error was experienced for the inorganic samples examined. This may be due in part to the lower specific heat values of silver nitrate and tin in particular, or perhaps differences in the thermal conductivity of these materials as compared to organics.

Many known specific heat values are available only at certain temperatures, while all values reported in this paper are average values. Whenever possible the temperature region used for an area measurement was ambient to 100° C., which made $dT = 76°$ C. If a smaller dT must be used, due to transitions below 100° C., the area measured will be smaller and greater error will be introduced.

The specific heat was determined on 10 separate samples of naphthalene. The standard deviation at the 95% confidence level was found to be ±0.02 cal./gram/°C.

Sample size, heating rate, and sample state did not affect the accuracy of the heat of fusion measurement for this system. These results substantiate Vold's equations (28) and show that nearly equilibrium conditions exist for this system. The depth of the differential temperature and the width of the transition peak vary with heating rate; but the total area does not change within the error of the measurement itself for identical size samples.

These results also show that K is a constant and temperature independent over the temperature ranges investigated and that the physical shape of the samples does not affect the results within the precision of the determination. The fact that accurate ΔH_f values were obtained for a variety of substances over extended temperature ranges shows that any one material could have been used for calibration. Tin was chosen as the calibration standard because of the comparative ease in handling this material and also for determining whether or not the calibration value could subsequently be applied to other materials which manifest transitions at widely divergent temperatures.

The theoretical discussions assume $K_r = K_s = K$. Although it has been shown that K is invariant, in practice it is difficult to achieve the condition that $K_r = K_s$; however, it is possible, but

time consuming, to do so. This does not invalidate the theoretical treatment since K_r can be made equal to K_s by thermocouple adjustment (synthetically), or by quantitative measurement from known materials, such as tin or benzoic acid and then adjusting one by a suitable factor. However, the experimental treatment takes into account slight differences between K_r and K_s by considering the system effects and system plus sample effects, separately.

The calibration was repeated weekly to provide accurate results and establish any indeterminate instrument changes. Slight variations in K were experienced over a 7-day period. The calibration factor, K, from an easily handled material like tin, was applicable for all heats of fusion of those compounds investigated.

In addition, a value of 77.2 cal./gram for the heat of vaporization of naphthalene (ΔH_v) was obtained. This compares favorably with the literature value of 75.5 cal./gram. Figure 3 shows the area utilized for this measurement.

The standard deviation at the 95% confidence level for ΔH_f and ΔH_v of naphthalene were found to be ± 1.5 cal./gram and ±2.0 cal./gram, respectively.

CONCLUSIONS

The application of this type of system permits the determination of ΔH_f and $\bar{C}_p$ on a variety of inorganic and organic compounds from a single calibration of an easily handled material—e.g., tin.

Although every effort must be extended to maintain a constant effect of parameters that cannot be numerically evaluated, variables such as sample size (within limits), heating rate, and sample state did not have deleterious effects upon the results.

ACKNOWLEDGMENT

The author thanks D. H. Chadwick, R. L. Moore, and Irvin Van Horn, of Mobay Chemical Co., for their assistance in this work.

LITERATURE CITED

(1) Anderson, David A., Freeman, Eli S., ANAL. CHEM. **31**, 1697 (1959).
(2) Anderson, David A., Hugh, C., *Ibid.*, **32**, 1593 (1960).
(3) Barrall, E. M., II, Rogers, L. P., *Ibid.*, **34**, 1101 (1962).
(4) *Ibid.*, **34**, 1106 (1962).
(5) Borchardt. H. J., Daniels, F. J., *J. Am. Chem. Soc.* **79**, 41 (1957).
(6) Fauth, M. I., ANAL. CHEM. **32**, 655 (1960).
(7) Gordon, Saul, Campbell, Clement, *Ibid.*, **27**, 1102 (1955).
(8) Gruver, R. M., *J. Am. Ceram. Soc.* **34**, 353 (1951).
(9) Hodgman, C. D., "Handbook of Chemistry and Physics," Chemical Rubber Publishing Co., Cleveland, Ohio, 1955.
(10) Ingard, Uno, Kraushaar, W. L., "Introduction to Mechanics, Matter, and Waves," Addison-Wesley Publishing Co., Inc., Reading, Mass., 1960.
(11) Ke, Bacon, *J. Polymer Sci.* **L**, 79–86 (1961).
(12) Kissinger, H. E., *J. Res. Natl. Bur. Std.* **57**, 217 (1956).
(13) Lange, N. A., "Handbook of Chemistry," Handbook Publishers, Inc., Sandusky, Ohio, 1952.
(14) Manufacturing Chemists' Association, Inc., "Technical Data on Plastics," Washington, D. C., 1957.
(15) Markowitz, M. M., Boryta, D. A., ANAL. CHEM. **32**, 1588 (1960).
(16) Mickley, H. S., Sherwood, T. K., Reed, C. E., "Applied Mathematics in Chemical Engineering," McGraw-Hill, New York, 1957.
(17) Morita, Hirokazu, *Ibid.*, **28**, 64 (1956).
(18) Morita, Hirokazu, Rice, H. M., *Ibid.*, **27**, 336 (1955).
(19) Norton, F. H., *J. Am. Ceram. Soc.* **22**, 54 (1939).
(20) O'Neill, M. J., ANAL. CHEM. **36**, 1238 (1964).
(21) Reisman, Arnold, *Ibid.*, **32**, 1566 (1960).
(22) Rudin, A., Schreiber, H. P., Waldman, M. H., *Ind. Eng. Chem.* **53**, 137 (1961).
(23) Smothers, W. J., Chiang, Yao, "Differential Thermal Analysis," Chemical Publishing Co., New York, 1958.
(24) Speil, S., Berkelhamer, L. H., Pask, J. A., Davies, B., U. S. Bur. Mines, Tech. Paper **664** (1945).
(25) Stone, R. L., ANAL. CHEM. **32**, 1582 (1960).
(26) Stone, R. L., Robert L. Stone Co., Austin, Texas, Bulletin **DTA-103**.
(27) Vassallo, D. A., Harden, J. C., ANAL. CHEM. **34**, 132 (1962).
(28) Vold, M. J., *Ibid.*, **21**, 683 (1949).
(29) Watson, E. S., O'Neill, M. J., Justin, J., Brenner, N., *Ibid.*, **36**, 1233 (1964).
(30) Wendlandt, W. W., *Ibid.*, **32**, 848 (1960).
(31) Wendlandt, W. W., Horton, R. G., *Ibid.*, **34**, 1098 (1962).
(32) Wylie, C. R., Jr., "Advanced Engineering Mathematics," McGraw-Hill, New York, 1960.

RECEIVED for review May 18, 1964. Accepted August 3, 1964.

21

Reprinted from J. Phys. Chem., **68**(10); 2810–2814 (1964)

Heats of Transition for Nematic Mesophases[1]

by Edward M. Barrall, II, Roger S. Porter, and Julian F. Johnson

California Research Corporation, Richmond, California (*Received February 17, 1964*)

Heats of transition have been measured for three pure compounds which exhibit a mesophase of the nematic type. The compounds are p-azoxyanisole, anisaldazine, and N-p-methoxybenzylidene-p-phenylazoaniline. Their liquid crystal or mesophase range is separated by first-order transitions from both a solid crystalline phase and an isotropic or true liquid state. Heats and temperatures for the two transitions of each purified compound were measured with an extensively calibrated custom-built differential thermograph. The nature of nematic mesophase transitions is discussed along with limited thermal data previously available on these compounds.

Nematic mesophases or liquid crystals represent a phase which is distinguishable from both a solid crystalline phase at lower temperatures and an isotropic or normal liquid phase at higher temperatures by first-order transitions.[2] Thermal data are rare for liquid crystals.[3] Definitive data are required for an insight into the field of order and flow of liquid crystals.

In this study, heats of transition for solid–nematic and nematic–isotropic states have been measured for three pure compounds which exhibit liquid crystal phases of the nematic type using differential thermal analysis (d.t.a.). The three compounds studied are p-azoxyanisole, anisaldazine, and N-p-methoxybenzylidene-p-phenylazoaniline. These compounds are hereafter referred to as PAA, AAD, and MBPA. The compound MBPA has been referred to as p-anisal-p-aminoazobenzene.[4] The source, structure, methods of purification, and compositional analyses for these three compounds have been previously described.[4] Published heat of transition data for PAA are evaluated in the light of these results. New heats of transition are reported herein for the other two compounds. The only previous information available is that the heat for the nematic–isotropic transition for MBPA is small, as estimated from double refraction near the transition.[5]

Calorimetric measurements for transition heats and temperatures of transition were made with an advanced-type differential thermograph. The design and calibration of this custom thermograph have been described recently.[6a,b] To achieve greatest precision and accuracy, different d.t.a. block designs and sample preparation procedures were used to measure transition temperature and heat absorption, respectively.

Experimental

Sample Preparation. Samples for the measurement of transition temperatures were prepared by diluting PAA, AAD, and MBPA with 500-mesh carborundum and grinding gently with a few drops of benzene. The carborundum had been purified as described previously.[7] The concentration of the liquid crystal compound was determined by extracting weighed aliquots of the dry carborundum mixture with chloroform and benzene and reweighing. Concentrations are shown in Table I. Samples for the determination of heats of transformation were prepared by precisely weighing about 10 mg. of the pure compound onto a 1-cm. square of aluminum foil. Foils were folded carefully to form small leakproof packets. Ammonium chloride, ammonium bromide, and NBS benzoic acid, prepared in the same way, were used to calibrate the apparatus

(1) Part IV of a series on order and flow of liquid crystals.

(2) P. L. Jain, J. C. Lee, and R. D. Spence, *J. Chem. Phys.*, **23**, 878 (1955).

(3) G. H. Brown and W. G. Shaw, *Chem. Rev.*, **57**, 1049 (1957).

(4) R. S. Porter and J. F. Johnson, *J. Appl. Phys.*, **34**, 51 (1963).

(5) V. N. Tsvetkov, *Acta Physicochim.* (USSR), **19**, 86 (1944).

(6) (a) E. M. Barrall, II, J. F. Gernert, R. S. Porter, and J. F. Johnson, *Anal. Chem.*, **35**, 1837 (1963); (b) E. M. Barrall, II, R. S. Porter, and J. F. Johnson, *ibid.*, **36**, 2172 (1964).

(7) E. M. Barrall, II, and L. B. Rogers. *ibid.*, **34**, 1101 (1962).

Table I : Thermographic Data on Three Liquid Crystal Compounds

Com-pound	Phase transition						Compound concn. in d.t.a. tests	
	Solid–nematic			Nematic–isotropic			Wt. %	Compound wt., g.
	$T_b{}^a$	T_m	T_e	T_b	T_m	T_e		
PAA	113.50	117.60	125.02	133.30	133.85	136.72	10.25	0.0146
AAD	161.96	168.90	175.71	177.28	180.46	185.53	8.30	0.0143
MBPA	142.07	147.20	153.30	177.48	179.07	181.80	8.49	0.0188

^{a} Temperatures indicate the beginning of the d.t.a. endotherm, T_b, the endothermal minimum, T_m, and the temperature, T_e, at which the recorder returned to the previously established base line for 4.7°/min. heating rate. Each temperature is the average of nine separate thermographic runs.

for calorimetry.[6b] The ammonium salts were Baker Analyzed reagent grade chemicals.

Procedure. All thermograms were recorded with the differential temperature signal as a function of sample temperature on an x–y recorder as described previously.[6,7] The sample temperature and the sample half of the differential temperature were measured with the same thermocouple.

Transition temperatures were measured with the thermocouples located in contact with, and in the center of, the 0.1-g. carborundum-diluted samples. The samples were contained in glass tubes. The block (block A) design has been described previously.[6b] Duplicate runs at four heating rates (4.7, 5.8, 11.4, and 12.7°/min.) showed no shift in the vertex of the endothermal minimum with heating rate.

Calibration. The temperature axis was calibrated with the melting points of NBS benzoic acid (m.p. 121.8°), salicylic acid (m.p. 158.3°), and potassium thiocyanate (m.p. 177.0°) diluted in the same manner as the samples in carborundum. The temperature at the minimum of the endotherms corresponded closely to the reported melting points.

The absolute temperature errors on the various portions of the thermographic endotherms were: beginning (T_b), ±0.09°, minimum (T_m), ±0.05°, and end (T_e), ±0.09°. The reason for this variation is the judgment which must be exercised in determining the beginning and end of the endotherm from peak shape. The location of the vertex is less dependent upon such judgments.

Runs at different heating rates were made on the same samples of PAA and AAD, since relocation of the samples during the melting process was not a problem with these materials. Sublimation of the MBPA melt necessitated the use of an identical, freshly weighed sample for each run in block A.

Calorimetry using block A suffers from large errors when samples of differing thermal conductivities and

physical states are studied.[6b] Calorimetric measurements were therefore carried out in the block of different design (block B). Sample and calibration runs were carried out at the same heating rates as in block A. Block B produces peaks which are too broad for precise transition temperature measurements. The peak areas were determined by the automatic integration method.[8] Integration errors were limited to less than ±1.5%.

Ammonium bromide and chloride and benzoic acid were used in the calorimetric calibration since the transition temperatures bracket the temperature range of interest for the compounds studied. The temperatures and heats of solid–solid transition of ammonium chloride (183.1°, 1073 cal./mole) and ammonium bromide (137.2°, 882 cal./mole) have been determined by Arell.[9] The thermal data for the fusion of benzoic acid (121.8°, 33.9 cal./g.) have been tabulated by Rossini, *et al.*[10] The data obtained for these solid–liquid and solid–solid phase changes are found to fit the same calibration curve. This is consistent with the successful removal of extraneous sample variables by using block B.

The precision of the calorimetric studies was determined by running each sample and calibration three times and determining the standard deviation of the set.

Results

Transition Temperatures. Figure 1 shows representative differential thermograms for *p*-azoxyanisole (PAA), anisaldazine (AAD), and N-*p*-methoxybenzyli-

(8) K. W. Gardiner, R. F. Klaver, F. Baumann, and J. F. Johnson, "Gas Chromatography," Academic Press, New York, N. Y., 1962, p. 349.

(9) A. Arell, *Ann. Acad. Sci. Fennicae, Ser. A VI.* **57**, 42 (1960).

(10) F. D. Rossini, D. D. Wagman, W. H. Evans, S. Levine, and I. Jaffe, "Selected Values of Chemical Thermodynamic Properties," National Bureau of Standards Circular 500, U. S. Government Printing Office, Washington, D. C., 1952.

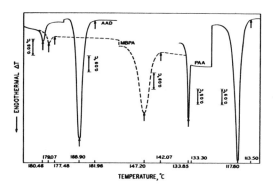

Figure 1. Thermographic traces of three liquid crystal compounds, block A: anisaldazine, AAD; N-p-methoxy-benzylidene-p-phenylazoaniline, MBPA; p-azoxyanisole, PAA.

dene-p-phenylazoaniline (MBPA) at a heating rate of 4.7°/min. Table I gives the temperatures of the beginning, minimum, and end of the phase transition endotherms. These three temperatures define the thermographic characteristics for each mesophase transition.

Figure 2 shows thermograms for PAA at four heating rates. The location of the peaks on the temperature axis is not affected by heating rate. This is because of the choice of geometry for the d.t.a. system. In Fig. 2 the heat absorbed by the solid–nematic transition was 0.420 cal., and 0.0102 cal. for the nematic–isotropic transition of PAA. These results indicate that each transition may be considered a single event without a distinct pretransition caloric effect as has been reported.[11]

There is generally good agreement between the solid–nematic transition temperatures obtained from visual and d.t.a. measurements.[4] The visually observed nematic–isotropic transition is somewhat higher, 1–5°, than the vertex value but generally lower than the endotherm conclusion temperature as seen by d.t.a. This difference was seen by Martin and Müller.[11] The difference increases with decreasing heat of transition and, therefore, may be due in part to pretransition effects which are discussed below.

Transition Heats. Table II lists the results of the calorimetric studies carried out by d.t.a. on three liquid crystal compounds. These new calorimetric values for PAA agree excellently with the combined data obtained by Martin and Müller.[11] These values differ considerably, however, from an earlier extensive study of the nematic–isotropic transition for PAA.[12,13] Existing calorimetric values for PAA may be compared in Table

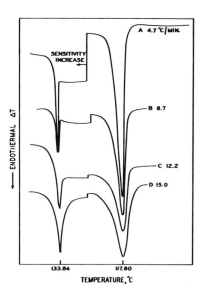

Figure 2. Thermographic trace of p-azoxyanisole at four heating rates, block A. A 0.0150-g. sample of PAA was thermographed at the indicated heating rates. The sensitivity on the ΔT axis for the 117.6° endotherm was: A, 0.09°/in.; B, 0.18°/in.; C, 0.36°/in.; D, 0.72°/in. The sensitivities for the 133.84° endotherm were: A, 0.05°/in.; B–D, 0.09°/in.

III. There have been no previous reports for heats of transitions for the other two compounds.

Table II: Heats of Transition for Three Liquid Crystal Compounds

Liquid crystal compound	Heat absorbed phase transition	
	Solid–nematic, cal./g.	Nematic–isotropic, cal./g.
PAA	28.1 ± 0.9	0.68 ± 0.02
AAD	26.5 ± 0.5	0.59 ± 0.02
MBPA	25.9 ± 0.6	0.41 ± 0.02

High resolution nuclear magnetic resonance (n.m.r.) spectra obtained on these three compounds provide insight into nematic mesophase transitions.[14] In the high temperature isotropic state, spectra indicate normal liquid behavior. In the nematic state,

(11) H. Martin and F. H. Müller, *Kolloid-Z.*, **187**, 107 (1963).

(12) K. Kreutzer, *Ann. Physik*, (5) **33**, 192 (1938).

(13) K. Kreutzer and W. Kast, *Naturwiss.*, **25**, 233 (1937).

(14) R. S. Porter, unpublished results.

Table III: Thermal Data for Transitions of *p*-Azoxyanisole

Worker	Method	Solid–nematic		Nematic–isotropic	
		Temp., °C.	Heat, cal./g.	Temp., °C.	Heat, cal./g.
This work	D.t.a.	117.6	28.8	133.9	0.68
Schenck and Schneider and Buhner[a]	Ice calorimeter	...	29.8	132	0.68
Martin and Müller[b]	D.t.a.	117	28.2	(128) 132	0.69
Dekock[a]	Depression nematic–isotropic temperature	...	...	132	0.68
Hulett[a]	Clausius-Clapeyron pressure dependence	...	...	132	0.71
Kreutzer[c]	Ice calorimeter	...	...	131	1.79

[a] See R. Schenck, "Kristalline Flüssigkeiten," Leipzig, 1905, pp. 84–89. [b] See ref. 11. [c] See ref. 12 and 13.

n.m.r. spectra are much more complex and suggest that all aromatic protons are unique. This means that commonly flexible groups, such as methyl ethers, are free to rotate in the isotropic or true liquid state. In contrast, bonds in the nematic state must be fixed in nonrotating positions. These results are in accord with conclusions that molecules in nematic mesophases are packed such that they have freedom of rotation about one axis only, which is ordinarily the long axis.[15] The heat for nematic–isotropic transition thus involved energy for both molecular separation and internal rotation.

Discussion

Changes in physical properties have been observed at temperatures near first- as well as second-order transitions. Pretransition effects may be expected to be abnormally large in systems capable of forming liquid crystals.[16] By the theory of Frenkel, the largest pretransitions, *viz.*, heterophase fluctuations, are to be expected in cases where two phases differ but slightly from one another and where the heat of transition is small.[17] These features facilitate the formation of nuclei of one phase in another. Indeed, at a few degrees below nematic–isotropic transitions, a number of physical properties have been found to change rapidly with temperature, indicative of pretransition. These include density, specific heat, viscosity, dielectric constant, optical transparencies, and flow and magnetic birefringence.[16] There is also commonly a real discontinuity in these physical properties at nematic–isotropic transitions which is characteristic of first-order transitions.[4,18]

Precise density and magnetic and flow birefringence measurements also indicate pretransitions on the liquid side of the nematic–isotropic transformations.[4] Small aggregates have been found in the isotropic state of the nematic-forming compounds studied here.[5] The aggregates or molecular swarms contain tens to several hundred molecules and exist up to 5° above the nematic–isotropic transition. The fact that they are of such small dimensions is in good agreement with the apparent absence of aggregates in light scattering measurements on the isotropic state.[5]

By the theory of Frenkel, the extent of pretransition phenomena should be related to the heat of transition. Data on the three compounds studied here agree with this concept. The temperature range for pretransitions for the series increases with decreasing heat for the nematic–isotropic transition. This is revealed by viscosity data,[4,19] density data,[4] and by magnetic and flow birefringence measurements.[5]

Experimental evidence of several types clearly indicates pretransitions in liquid crystals. These effects appear adequately interpreted by Frenkel's heterophase fluctuation theory. Therefore, it seems unnecessary to separate the caloric effects due to so-called first-order and pretransition effects as has been done previously.[11]

Acknowledgment. The authors express appreciation to Messrs. D. Trujillo and A. R. Bruzzone for help with the experimental work.

Discussion

A. A. ANTONIOU (National Research Council, Ottawa). I wish to point out that in the system porous glass–water, also,

(15) R. Williams, *J. Chem. Phys.*, **39**, 384 (1963).

(16) W. A. Hoyer and A. W. Nolle, *ibid.*, **24**, 803 (1956).

(17) J. Frenkel, *Zh. Eksperim. i Teor. Fiz.*, **9**, 952 (1939), and "Kinetic Theory of Liquids," Clarendon Press, Oxford, 1946.

(18) E. Bauer and J. Bernamont, *J. Phys. Radium*, **7**, 19 (1936).

(19) R. S. Porter and J. F. Johnson, *J. Phys. Chem.*, **66**, 1826 (1962).

the phase transition for which becomes apparent as a change of the thermal coefficient of expansion, the value of this coefficient passes through a maximum.

R. S. PORTER. Effects both pre- and post- to the first-order transitions can likely be observed for a variety of systems and perhaps are only a matter of degree for many systems. Observations of these effects depend on the sensitivity of the measurement method, e.g., heat capacity and specific volume, and on measurements at temperatures close to the transition. The inherent tendency for the test system to show such effects is, of course, important.

J. H. DE BOER (The Hague, Netherlands). Since entropy changes may—or rather should—give additional, or even primary information about these phase transitions, may I ask whether such entropy changes have been studied?

R. S. PORTER. Such a measurement is definitely desirable, particularly for liquid crystals, as the higher temperature nematic–isotropic transitions involve relatively large volume changes yet small latent heats, whereas just the reverse is true for lower solid–nematic transitions. Complete evaluation, of course, requires heat capacity data on the several phases which are not, as yet, available.

22

Reprinted from *Anal. Chem.*, **36**(7), 1233–1238 (1964)

A Differential Scanning Calorimeter for Quantitative Differential Thermal Analysis

E. S. WATSON, M. J. O'NEILL, JOSHUA JUSTIN, and NATHANIEL BRENNER

The Perkin-Elmer Corporation, Norwalk, Conn.

▶ An instrument for differential thermal analysis has been developed which directly measures the transition energy of the sample analyzed. The instrument performs thermal analyses of milligram level samples at high speeds (scan rates up to 80° C. per minute) and in a temperature range of 173° to 773° K. (−100° to +500° C.). Analytical data are recorded in a fashion graphically similar to that of traditional DTA, but peak amplitude directly represents millicalories per second of transition energy and peak area directly represents total transition energy in millicalories. Direct temperature marking is also displayed. Atmosphere control and vacuum operation are also provided. Samples may be run in closed-cup or open-cup configurations and are conveniently handled in powder or sheet form. Quantitative performance is independent of specific heat of sample, sample geometry, and temperature scanning rate. Qualitative information is equivalent or superior to conventional DTA in terms of speed, sensitivity, resolving power, and operational convenience. However, the principal innovation of the unit is the capability for direct, convenient, and precise quantitative measurement of transition energy.

THE EXAMINATION of the rate and temperature at which materials undergo physical and chemical transitions as they are heated and cooled and the energy changes involved has been the subject of investigations for almost a century.

Conventional differential thermal analysis instruments subject a sample and an inert reference material to a controlled heating program and measure the differential temperature between sample and reference material. The appearance of an increase or decrease in the sample temperature with respect to the reference temperature is attributable to the energy-emitting (exothermic) or energy-absorbing (endothermic) transitions occurring in the sample. Systems of this type were developed by Roberts-Austen (9) and Saladin (10) over fifty years ago and have been improved over the years for mineralogical and inorganic chemical applications and, more recently, in the organic chemical field (13, 15). The desirability of direct calorimetric information rather than indirect thermometric data has been recognized by advanced workers in DTA for many years. Sykes (1935–36)

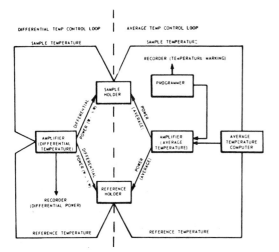

Figure 1. Block diagram, Perkin-Elmer differential scanning calorimeter

EXOTHERM

AREA = CALORIES →

ENDOTHERM

AMPLITUDE = CAL/SEC. →

Figure 2. Readout display

(14), Kumanin (1947) (8), and Eyraud (1954) (4) built and operated equipment in which this objective was indeed accomplished by various elegant means.

A new system for differential thermal analysis has been developed which offers direct calorimetric measurement of the energies of transition observable in the analysis and therefore improves the utility of the DTA technique for quantitative analysis. The system, called a Differential Scanning Calorimeter (DSC), differs from the conventional thermal analyzers in one fundamental respect. The conventional units measure and record the temperature difference between sample and reference channels. The DSC system measures the differential energy required to keep both sample and reference channels at the same temperature throughout the analysis. When an endothermic transition occurs, the energy absorbed by the sample is replenished by increased energy input to the sample to maintain the temperature balance. Because this energy input is precisely equivalent in magnitude to the energy absorbed in the transition, a recording of this balancing energy yields a direct calorimetric measurement of the energy of transition.

Figure 1 shows a schematic representation of the DSC system.

The schematic system may be most easily comprehended if we divide the system into two separate control loops, one for average temperature control, the second for differential temperature control. In the average temperature loop, a programmer provides an electrical output signal which is proportional to the desired temperature of the sample and reference holders. The programmer temperature information is also relayed

to the strip chart recorder and appears as the abscissa scale marking. The programmer signal, which reaches the average temperature amplifier, is compared with signals received from platinum resistance thermometers permanently embedded in the sample holder and reference holder via an average temperature computer. If the temperature called for by the programmer is greater than the average temperature of the sample and reference holders, more power will be fed to the heaters, which, like the thermometers, are embedded in the holders. If the average temperature demanded by the programmer is lower than the average of the two holders, the power to the heaters will be decreased.

In the differential temperature control loop, the major distinction between the DSC-1 and traditional DTA devices is most marked. Signals representing the sample and reference temperatures, measured by the platinum thermometers, are fed to the differential temperature amplifier via a comparator circuit which determines whether the reference or the sample signal (temperature) is greater. The differential temperature amplifier output will then adjust the differential power increment fed to the reference and sample heaters in the direction and magnitude necessary to correct any temperature difference between them. A signal proportional to the differential power is also transmitted to the pen of the galvanometer recorder. The direction of the pen excursion will depend upon the direction of excess power input (sample or reference heater).

The instrumental system may be broken down into the following subgroups for analysis:

A pair of platinum resistance thermometers, the associated circuitry, a double-pole, double-throw mechanical chopper, and two input transformers. Also, a program potentiometer and associated circuitry. This group of components measures sample and reference temperatures and generates a ΔT error signal at 60 c.p.s. A summing network provides the average temperature T_{AV}, which is compared with the set-point temperature T_P generated by the program potentiometer, to generate a second error signal at 60 c.p.s.

A ΔT signal amplifier, an output transformer and a half-wave balanced output circuit. The output of this amplifier is connected differentially to the two heater elements in the sample holders. Every half-cycle, therefore, the differential temperature loop is closed, and ΔT is nulled. An a.c. voltage proportional to differential power is generated and fed to. . .

. . .the readout system, which includes a demodulator, range and zero controls, and a closed-loop d.c. amplifier. This system drives a 5-ma. Texas Instruments Recti/Riter Recorder. Outputs suitable for a Perkin-Elmer Model 194 Integrator and a 10-mv. recorder are also provided.

A T_{AV} signal amplifier, which supplies power to the sample and reference heaters, to close the loop around T_{AV}. This amplifier and the ΔT amplifier are operated on a time-sharing basis, each connected to the heaters for half of the time. A secondary circuit controls a pilot light which allows the operator to monitor the operation of this loop. (See discussion below.)

A scaling circuit which provides line-synchronized pulse trains at eight pulse repetition frequencies, one of which is selected by the Speed Selector switch and fed to. . .

. . .the motor drive circuit. This is a flip-flop circuit, driving the field windings of a stepping motor in the temperature programmer. In addition, there is a pulse generator, triggered by signals from the programmer, which drives a ballistic marker pen mounted on the Texas Instruments Recti/Riter Recorder. This action is inhibited by a signal from the T_{AV} amplifier when that loop loses control. This will

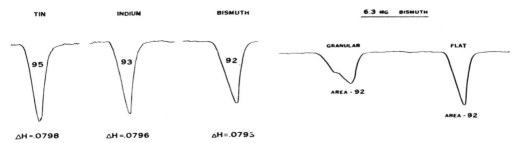

Figure 3. Energies of fusion

Figure 4. Effect of sample geometry on peak shape and area

happen if the operator attempts to set the temperature below, or only slightly above the ambient temperature, or if the operator attempts to program the temperature down too fast. In either case, the heater power will be turned off, and the true sample temperature will be somewhere above the indicated (dial) temperature. The pilot light monitors this situation visually; the absence of a temperature scale on the chart provides a *post facto* record.

The calorimeter part of the instrument; namely, the ΔT loop and ΔW readout, operates continuously, entirely unaffected by the situation in the T_{AV} loop.

The thermal conductivity effluent monitor cell and its associated circuitry, plus a proportional temperature controller for the detector block.

The power supply which provides plus and minus voltages for all parts of the instrument.

The programmer unit which includes the stepping motor, gear train, clutch, counting dial, commutator, and program potentiometer.

The graphic data obtained from the calorimeter superficially resemble those obtained from a traditional DTA. Figure 2 illustrates the display as seen on a galvanometer recorder equipped with an auxiliary pen for temperature marking. As in DTA, the abscissa represents temperature and each mark equals one degree C. or K. Larger marks are inscribed each 5 and 10 degrees. The second pen draws the thermogram itself and, as in DTA, a flat baseline indicates ranges in which

no transition occurs, while excursions (peaks) above and below the baseline represent exothermic and endothermic transitions, respectively.

The distinction between calorimetric and traditional DTA lies in the fact that the amplitude of the pen from the baseline position is directly measureable as a rate of energy output or input (millicalories per second) and the area under a peak equals total transition energy (calories).

The validity of this distinction is illustrated in Figure 3. Here three samples of pure metal were weighed out in quantities calculated to yield, upon melting, approximately equal energies of fusion. The heat capacities and thermal conductivities of two of the metals (Sn and In) are very similar, but that of the third (Bi) is very much different. (The thermal conductivities of the materials also vary considerably: Sn, 0.150 cal. cm.$^{-1}$ sec.$^{-1}$ deg.$^{-1}$; In, 0.059 cal. cm.$^{-1}$ sec.$^{-1}$ deg.$^{-1}$; Bi, 0.020 cal. cm.$^{-1}$ sec.$^{-1}$ deg.$^{-1}$) Figure 3 shows that the areas of the

fusion peaks are essentially equal ($A_{Sn} = 95$, $A_{In} = 93$, $A_{Bi} = 92$) and that the measurement made is of the fundamental energy value undistorted by other sample properties.

In conventional DTA, the thermal conductivity of the sample will markedly influence the area obtained per unit of energy input or output. In practice, this problem is handled by diluting the sample with a large volume of the reference material so that the resultant conductivity is essentially equal to that of the pure reference (*1*). In this way, dissimilar compounds may be quantitatively compared. Naturally, the dilution takes its toll in terms of sensitivity, and runs the risk of possible interaction between diluent and sample.

The freedom from effects of sample geometry is illustrated in Figure 4. Here a sample of bismuth in the form of large lumps was melted and the heat of fusion recorded. Because of poor heat transfer, an irregular peak shape results.

The second record is of the same sample run in a thin layer form. A

[Editor's Note: Figure 6 has been omitted because the original halftone was unavailable.]

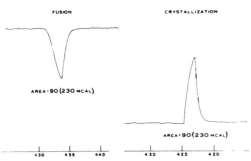

Figure 5. Fusion and crystallization transitions of indium

174

Table I. Sample Weight and Program Rate Dependence of the Differential Scanning Calorimeter

Weight dependence		Program dependence	
Weight of indium, mg.	ΔHf^a	Program speed, °C./min.	ΔHf^a
1.36	7.05 ± 0.096	2.5	6.82 ± 0.008
5.34	6.80 ± 0.088	5.0	6.81 ± 0.006
11.70	6.79 ± 0.042	10.0	6.82 ± 0.004
18.46	6.76 ± 0.075	20.0	6.78 ± 0.013

a Based on ΔHf of tin standard = 14.5 calories/gram. Actual ΔHf of indium = 6.8 calories/gram.

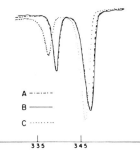

Figure 7. Chain rotation and fusion transitions

well shaped peak results, which, while esthetically superior, is quantitatively identical to the first run since both areas are equivalent.

A similar example is shown in Figure 5 where the fast exothermic transition of crystallization of the indium yields a peak identical in area to that of the much slower endotherm of fusion of the same sample.

The temperature scanning rate of the calorimeter may be varied from as low as 0.6° to as high as 80° per minute. Output sensitivity may also be varied from levels corresponding to 0.002 calory per second half scale recorder reading to as low as 0.032 calory per second for a half scale deflection on the galvanometer recorder. Changes in scan rate produce no change in total area under a peak, though the peak shape (height-to-width ratio) will be altered. [In traditional DTA devices, scan rate changes may be accompanied by area, as well as peak shape changes, further complicating quantification of data (11).]

Changes in sensitivity range (signal attenuation) result in changes in area and peak amplitude in precise ratio to the ranges selected. The results of such changes are shown in Table I, along with results of changes in sample weight. (In the weight dependence determinations, all runs were made at a scanning rate of 10° C. per minute. In the program dependence determinations, sample weight in each case was 5.3 mg. All figures represent the mean

of three single determinations.) In al cases, the true energy of transition is accurately derived.

The calorimeter operates in the range 173° to 773° K. (−100° to +500° C.). Cooling below ambient temperature is accomplished by substituting an enclosure cover containing a liquid nitrogen well and Dewar jacket for the standard enclosure cover. A small volume (200 cc.) of liquid nitrogen, poured into the well, will cool the sample enclosure to 173° K. in about 5 minutes.

The analyzer section of the calorimeter includes the operating head which is seen in exposed view in Figure 6. Samples in the usual range of 0.1 to 10 mg. are placed in small aluminum or gold pans and are set in either of the sample wells shown in Figure 6. Samples may be in powder or sheet form and may be run either encapsulated or open to the ambient atmosphere. Encapsulation of samples is accomplished with a sample pan sealer accessory designed for this purpose. A standard sample pan containing sample and covered with a metal disk is placed into the sealer. Depression of the sealer handle causes a folding inward of the sample pan rim, creating a tight enclosure resembling a flat-bottomed metal aspirin. Samples may be visually observed during a run. In this case, the sample well cover (usually metal) is replaced by a mica disk and an inverted glass beaker is used in place of the stainless steel enclosure cover. In contrast

to most traditional DTA, no actual reference material is required. An empty sample pan is usually placed in the reference well, however.

Quantitative performance is illustrated by the measurement of the transition energies associated with the chain rotation and fusion of dotriacontane. These transitions have been previously investigated by Hoffman and Decker (5) and Ke (6).

Figure 7 shows three superimposed thermograms of dotriacontane. Two of the runs (A and B) were made from a sample shown to be 98% pure by gas chromatographic assay; the third (C) was an 80% pure material obtained from a second supplier. The transition temperatures of the impure material are shifted to lower temperatures and the band width of the chain rotation peaks broadened.

Table II shows the energy measurements obtained from these samples. Note that the fusion energy is essentially identical in both pure and impure materials, but that the chain rotation energy is considerably lower in the impure material.

The actual analyses were performed using an external standard method. A 12.1-mg. sample of pure indium served as the standard and the heat of fusion of indium was used to calibrate the area response of the calorimeter. Area measurements were made with a planimeter.

While the fundamental contribution of the calorimeter to thermal analysis lies in its direct quantitative capability,

Table II. Comparison of Transition Energies of Pure and Impure Dotriacontane

	Pure (A)			Pure (B)			Impure (C)	
Run	Chain Rotation	Fusion	Run	Chain Rotation	Fusion	Run	Chain Rotation	Fusion
1	13.6 cal./g.	37.9 cal./g.	1	13.4 cal./g.	37.8 cal./g.	1	12.3 cal./g.	37.4 cal./g.
2	13.3 cal./g.	38.1 cal./g.	2	13.6 cal./g.	37.4 cal./g.	2	11.5 cal./g.	38.1 cal./g.
3	13.7 cal./g.	37.8 cal./g.				3	11.0 cal./g.	36.6 cal./g.
4	14.0 cal./g.	38.1 cal./g.						
Mean	13.7 cal./g.	38.0 cal./g.	Mean	13.5 cal./g.	37.6 cal./g.	Mean	11.6 cal./g.	37.4 cal./g.

175

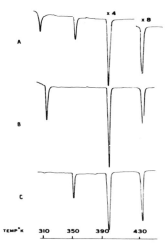

Figure 8. Ammonium nitrate transitions

the qualitative performance of the device is equivalent to that of the best conventional DTA equipment.

Barshad (2) suggested the use of NH₄NO₃ as a calibration standard for the temperature scale of conventional DTA units. The material undergoes four transitions between room temperature and decomposition, which have been carefully reported (3). It has been commonly used to illustrate the qualitative performance of thermal analyzers.

Figure 8 shows a series of heating curve analyses of ammonium nitrate from room temperature to 180° C. (553° K.). Four transition points have been identified in Figure 8a—at 315° K. (32° C.), 353° K. (80° C.), 398° K. (125° C.), and 443° K. (170° C.). These transitions have been ascribed in the literature (3) to the following transitions:

315° K. (32° C.) β rhombic to α rhombic
353° K. (80° C.) α rhombic to rhombohedral
398° K. (125° C.)   rhombohedral to cubic
443° K. (170° C.)   cubic to liquid (fusion)

The clear sharp peaks obtained using only 3 mg. of sample make the identification of the transition temperature unequivocal. Figures 8b and 8c illustrate the effect of prior cooling conditions on the transitions observed upon subsequent heating of the samples. Figure 8b shows the disappearance of the peak at 353° K. indicating that the sample went directly from the β rhombic to rhombohedral configuration. This sample had been previously melted and rapidly cooled with liquid nitrogen. Figure 8c shows the result when the molten sample was cooled to just below room temperature. Here the β to α transition is missing, indicating that the sample did not revert to the β state.

Quantitatively, the determination of transition energies were made by planimetric measurement of peak areas and comparison with a 6.47-mg. indium standard (44 millicalories). Values obtained for the NH₄NO₃ transitions were:

I   Cubic to liquid (fusion)   19.0 cal./g.
II   Rhombohedral to cubic   12.4 cal./g.
III α Rhombic to rhombohedral   4.0 cal./g.
IV β Rhombic to α rhombic   4.0 cal./g.

The literature value for transition I is 19.1 calories per gram (7), and for transition II, 12.3 calories per gram (12), as determined by conventional calorimetric procedures. Transitions III and IV are subject to wide variation with respect to sample history and cannot therefore, be compared with standard values.

It is interesting to note here the difference between the quantitative capabilities of the DSC and conventional DTA systems with respect to this problem. Figure 9 shows a comparison, with expanded ordinate and with no signal attenuation between peaks, of the two high temperature transitions of NH₄NO₃ (rhombohedral to cubic, cubic to liquid state). The energies of transition have been previously shown to be approximately 12 and 19 calories per gram, respectively

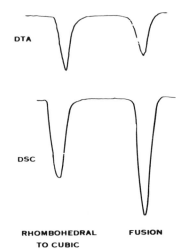

RHOMBOHEDRAL   FUSION TO CUBIC

Figure 9. Comparison of conventional DTA and DSC records of NH₄NO₃ transitions

The DSC presentation obviously reflects this energy ratio, while the DTA areas, obtained on a well constructed system, are essentially equivalent.

Glass transitions are common phenomena in high polymer behavior. These transitions are changes in heat capacity resulting from relaxation of the chain segments in those portions of the polymer structure which are amorphous. It has been previously stated that when using the scanning calorimeter, sample heat capacity does not affect the measurement of transition energy. However, changes in heat capacity, such as those occuring in glass transitions, are observable with the calorimetric system. A change in heat capacity of a sample during a temperature scan will result in a change in the power necessary to keep the sample temperature tracking the temperature of the reference pan. This change in power will be reflected as a shift in the position of the baseline to a new level. While this display is graphically analogous to that shown on a conventional DTA system, the DSC baseline displacement may be interpreted quantitatively. In common with thermometric DTA, the DSC must be run at high sensitivity with high stability in order to observe the very small changes inherent in the glass transition. Figure 10 illustrates such a transition in a polycarbonate resin. The high scanning rate (40° per minute), large sample size (20 mg.), and high sensitivity (4 millicalories per second for full scale deflection) all help to magnify the observed transition. The direction of the transition is from lower heat capacity at a lower temperature to higher capacity at higher temperatures. In principle, the actual

20mg,   RANGE 2   40 °/MIN

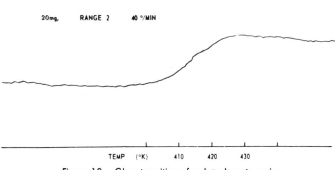

TEMP   (°K)   410   420   430

Figure 10. Glass transition of polycarbonate resin

values of the heat capacities may be determined with the calorimeter, but this has not been done in this instance.

LITERATURE CITED

(1) Barrall, E. M., Rogers, L. B., ANAL. CHEM. **34,** 1101 (1962).
(2) Barshad, I., *Am. Mineralogist* **37,** 667 (1952).
(3) Early, R. G., Lowry, T. M., *J. Chem. Soc.* **115,** 1393 (1919).
(4) Eyraud, C., *et al., Compt. Rend.* **240,** 862 (1955).
(5) Hoffman, J. D., Decker, B. F., *J. Phys. Chem.* **57,** 520 (1953).
(6) Ke, Bacon, *J. Polymer Sci.* **62,** 15 (1960).
(7) Keenan, A. G., *J. Phys. Chem.* **60,** 1356 (1956).
(8) Kumanin, K. G., *Zh. Prikl. Khim.* **20,** 1242 (1947).
(9) Roberts-Austen, W. C., *Proc. Inst. Mech. Engrs.* **1,** 35 (1899).
(10) Saladin, E., *Iron and Steel Metallurgy and Metallography* **7,** 237 (1904).
(11) Smothers, W. J., Chiang, Y., "Differential Thermal Analysis," pp. 120–1, Chemical Publishing Co., New York City, 1958.
(12) Steiner, L. E., Johnston, J., *J. Phys. Chem.* **32,** 912 (1928).
(13) Stone, R. L., ANAL. CHEM. **32,** 1582 (1960).
(14) Sykes, C., *Proc. Roy. Soc.* **148,** 422 (1935).
(15) Vassallo, D. A., Harden, J. C., ANAL. CHEM. **34,** 132 (1962).

RECEIVED for review July 5, 1963. Accepted December 11, 1963. Pittsburgh Conference on Analytical Chemistry and Applied Spectroscopy, Pittsburgh, Pa., 1963.

23

Reprinted from *Anal. Chem.*, **36**(3), 602–605 (1964)

Micro and Semimicro Differential Thermal Analysis (μDTA)

CHARLES MAZIÈRES

Ecole Nationale Supérieure de Chimie, Université de Paris, France

▶ Through a critical study of the various factors involved in differential thermal analysis, apparatus has been designed which permits examination—in controlled atmosphere and between −180° and +1200° C.—of samples weighing from 1 to 200 μg. (10⁻⁶ to 2 × 10⁻⁴ gram) or, alternatively, from 0.1 to 10 mg. (10⁻⁴ to 10⁻² gram). Thermal effects as small as 10⁻⁶ cal. can be detected. Examples are given showing uses of the apparatus.

C HEMISTS, mineralogists, and solid-state physicists have shown increasing interest in differential thermal analysis (DTA) (7). The advantages and disadvantages of the method, as well as the experimental solutions to various problems, have been the object of numerous articles (4).

The present author has given particular attention (5) to the following points: (1) the existence of a thermal gradient between the surface and the core of the sample investigated, resulting in a variable degree of completion of the physicochemical phenomenon giving rise to the thermal effect. This, in turn, results in a broadening of the peak, even in the ideal case of an isothermal transformation. (2) The thermocouples generally used as sensors only indicate their own temperature, which may be notably different from the temperature of the sample, especially if the thermocouple is chemically insulated.

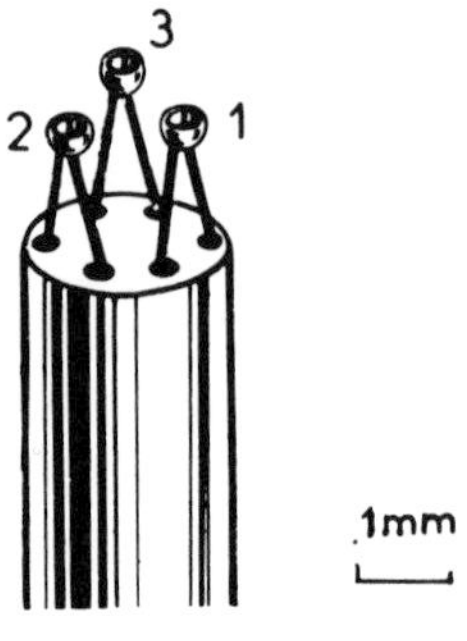

The approach described herein consists essentially in using a sample as

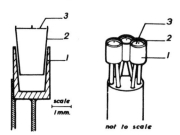

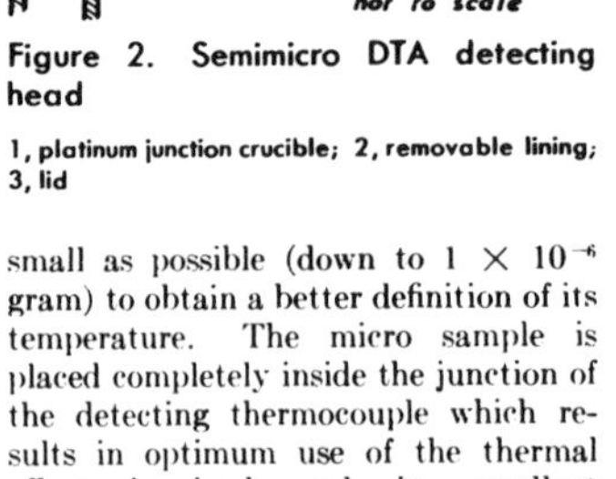

Figure 2. Semimicro DTA detecting head

1, platinum junction crucible; 2, removable lining; 3, lid

small as possible (down to 1 × 10⁻⁶ gram) to obtain a better definition of its temperature. The micro sample is placed completely inside the junction of the detecting thermocouple which results in optimum use of the thermal effect involved and in excellent identification of the temperature of the sample and of that of the thermocouple.

Because of the small mass of the samples involved, a very high sensitivity is essential; this is obtained by foregoing the conventional metal or ceramic block often used to homogenize the temperature.

The thermocouple sample cup was first suggested by Herold and Planje (2); the samples they investigated probably weighed about 1 gram (no data are given); the thermocouples were embedded in a refractory block. Though this design did not solve all the difficulties mentioned above, enclosing the sample within the thermocouple was an improvement on conventional design; as pointed out by one of the referees of the present paper, the effect of sample shape and/or sample thermal diffusity on the area of the peak or on its shape is thus minimized.

On the other hand, Wittels (9) showed that by using the vacuum apparatus of Whitehead and Breger (8), it was possible to analyze samples of calcite as small as 300 μg. However, the thermal effect involved in the decomposition to CaO is quite large; once more, the sensitivity was probably

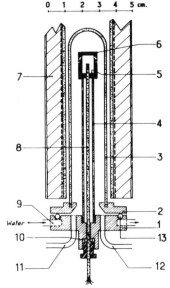

Figure 3. High temperature design

1, 2, water cooled base; 3, refractory sheath; 4, refractory support; 5, 6, platinum base and hood; 7, furnace; 8, refractory six-duct sheath; 10, 13, O-rings; 11, 12, gas inlet/outlet

Figure 1. Micro DTA detecting head

.1mm

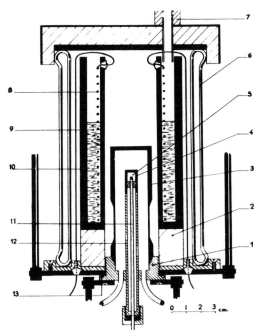

Figure 4.   Low temperature design

1, Lucite base; 2, thermal insulator; 3, 4, blackened copper sheath; 6, Dewar sleeve; 8, heating coil; 9, liquid nitrogen; 10, copper vessel; 11, six-duct sheath; 12, Lucite support; 13, O-ring

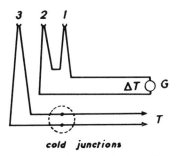

cold junctions

Figure 5.   Thermocouple assembly

temperature design of Figure 4. These figures are self-explanatory.

Thermocouple wires are BTE/CTE (Trade Mark, Aciéries d'Imphy, France, ca. 60 µv./° C.) for low temperatures, and Pallaplat (Heraeus, West Germany) or Platinel (Engelhard, USA) (both ca. 40 µv./° C.) for the higher temperatures.

DTA runs between −170° and +80° C. can be obtained with the low-temperature design, while temperatures to 1250° C. can be reached with the high-temperature apparatus.

The heating furnace 7 of Figure 2 is characterized by a fairly high power/weight ratio. This fact, coupled with the cooling brought about by the water circulating within the base, enables rates of temperature variation of 0° C./minute to ±25° C./minute to be attained. Occasionally, rates of up to 80° C./minute have been used with a slightly different arrangement (6).

In Figure 4, the heating coil 8 is used to control the rate of evaporation of the liquid nitrogen and, afterwards, the rate of warming up. In the same figure, inlet 7 is used for pressure control.

The recording set-up (Figure 5) uses a standard galvanometer coupled with a straightforward spot-follower. This set-up achieves a noise-free sensitivity of 0.4 µv./cm. (equivalent to 0.01° C./cm.) for the ΔT recording.

limited by the use of a homogenizing block as a sample holder.

EXPERIMENTAL

Apparatus and Operating Procedures. The principles involved in the present approach are embodied in the thermocouple assembly of Figure 1, showing the micro DTA detecting head used for the 1- to 200-µg.

range; Figure 2 shows the semimicro detecting head used for the 0.1- to 10-mg. range. One of the junctions holds the sample; the second, identical to the first, is the reference junction; the third, also identical to the other two, is the temperature measuring junction.

These detecting heads can be used both with the high-temperature design shown in Figure 3 or with the low-

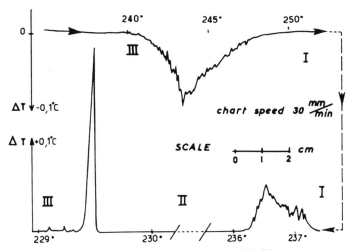

Figure 6.   µDTA of single 27-µg. fragment of Na₂SO₄ crystal

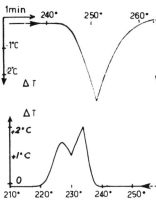

Figure 7.   DTA of 1-gram sample of Na₂SO₄ obtained with conventional apparatus

179

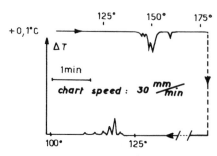

Figure 8. μDTA of 15-μg. sliver of tridymite

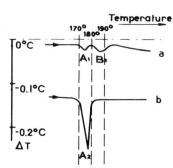

Figure 9. (a) First and (b) second heating under argon of 40-μg. sample of irradiated LiF

Two experimental difficulties arise as counterparts of the high sensitivity: "noise" in the thermocouple wires and drift of the base line. The first can be eliminated by careful welding of the thermocouples and annealing on their entire length, and by scrupulous shielding from drafts and other thermal disturbances. The second difficulty can only be eliminated by keeping the two halves of the differential thermocouple symmetrical with respect to thermal gradients.

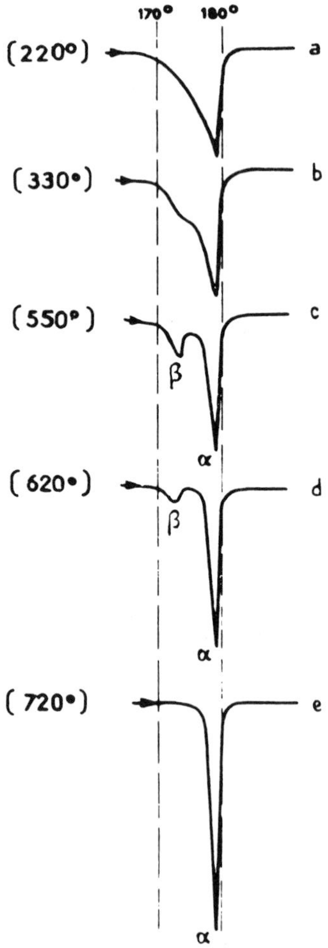

The placing of the sample in a junction crucible of Figure 1 is most easily carried out by working within the field of a binocular lens which is temporarily swung into position above the microcrucibles for this purpose. The micro sample should, preferably, be in the form of a single piece and can then be handled with short glass or piceine needles. Unfortunately, it is impractical to use the junction crucibles of Figure 1 with powdered samples because of the difficulty in packing the sample and in changing it. Moreover, certain samples badly corrode the thermocouple, especially at high temperature. The modification shown in Figure 2 meets these difficulties: packing the sample in the lining and weighing it are easier, and changing a corroded lining is no problem. Such an increase by a factor of 100 to 1000 of the mass of the sample certainly lowers the resolution. Certain existing commercial apparatus also work in this range of masses (1); however, an example given below will show the particular features of the results obtained with the present semimicro arrangement.

After the sample has been placed in the crucible, the apparatus can be thoroughly purged, if necessary, with an inert gas or evacuated.

RESULTS AND DISCUSSION

Figure 6 is a reproduction of a thermogram obtained on a single fragment of

crystalline Na₂SO₄ weighing 27 μg. Roman numerals indicate the various crystal forms. Comparison with Figure 7 showing a classical type thermogram of the same substance obtained on a ca. 1-gram mass shows the considerable increase in the power of resolution: the μDTA thermogram has a truly oscillographic character and shows the successive transformations of each crystalline domain; the irreversibility of the III→I transformation in the absence of crystalline germs can be unambiguously demonstrated. In the case of the classical thermogram of Figure 7, the overlapping of peaks makes this demonstration more obscure.

Figure 8 shows the result of μDTA performed on a thin sliver of tridymite (15 μg.) obtained from a volcanic lava. Though the thermal effect associated with the transformation is quite weak— probably less than 1 cal./gram— fragments as small as 1 μg. give peaks which are still quite visible. The successive peaks of Figure 8 correspond

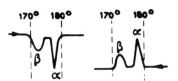

Figure 11. Same as figure 10c, 8 μg. of LiF

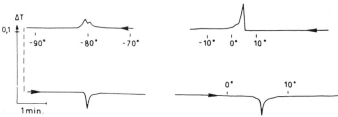

Figure 10. Successive thermograms, (under argon), of 30-μg. sample of irradiated LiF

Figures at left give highest previous temperature of treatment.

Figure 12. Low temperature μDTA of 50-μg sample of BaTiO₃

180

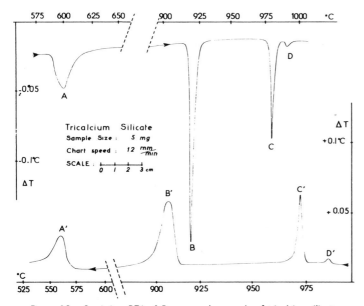

Figure 13. Semimicro DTA of 5-mg. powder sample of tricalcium silicate

shape of the thermogram is perfectly characteristic of the previous heat treatment, and it must be emphasized that only the use of a microsample makes differentiation of peaks α and β clear.

The crystallographic transformation of barium titanate shown in Figure 12 provides an example of the use of the apparatus at low temperatures.

Figure 13 shows a thermogram of a 5-mg. finely-powdered sample (particle size less than 1 micron) of tricalcium silicate, obtained with the semimicro detecting head. Determination of the temperature of transformation is still very good. Semiquantitative and relative measurement of the enthalpy changes involved in the various transitions B, C, and D can be obtained by comparison of the areas of the respective peaks. Furthermore, the area of peak B is slightly smaller than that of the peak due to the high-low transition at 574° C. of an equal mass of quartz. It should be pointed out that a very extensive and painstaking x-ray study (10) at high temperature failed to show transition B, which is the one most easily seen by DTA.

Rapid cooling—possible because of the small thermal inertia—shows the reversibility of transformation A occurring at about 600° C. on heating and a little lower on cooling; this fact, coupled with x-ray studies (10) enabled the authors to prove that A is a true polymorphic transformation and not, as was previously believed, a chemical modification.

once again to the successive $\alpha \rightleftharpoons \beta$ transformation of different domains in the microcrystal; they differ from sample to sample but are perfectly reproducible for any one sample. The hysteresis of the reverse transformation during cooling is important and is also evidenced in the figure.

An interesting and characteristic example of the use of μDTA has been the study of the small lumps of metallic Li that form within a crystal of LiF when it is subjected to thermal-neutron irradiation. X-ray studies had shown that the crystal form of the precipitated metal was not the usual one (body-centered cubic) but rather a different and abnormal one (face-centered cubic) which could be transformed into the former by heating. This result was confirmed and elaborated by μDTA of the melting and crystallization processes of the precipitated phase within the LiF matrix (3).

In Figure 9a, peak B_1, (occurring at 187° C.) shows the melting of the abnormal (face-centered cubic) and rather badly crystallized lithium, while peak A_1 is due to the melting of body-centered cubic lithium. The run shown in Figure 9b was performed on the same

sample after all the metallic lithium within it had melted and had resolidified by cooling; the curve only shows a single peak at the temperature corresponding to the melting of body-centered cubic lithium. The width of the peak is caused by the fact that the metal is in very small lumps or aggregates with a small degree of organization.

Moreover, the variation in the shape of the recorded peaks enabled the author to follow the evolution of the lumps of metallic Li during and after heat treatments at regularly increasing temperatures; in Figures 10c and d, peak α corresponds to the melting of well crystallized lumps, while β is attributed to the melting of lumps differing from the first lot in their size and degree of organization. Figures 10c and 11 show the same phenomenon, the only difference in the two analyses being in the masses used in each run: 30 μg. of irradiated LiF in the first case, and 8 μg. in the second, the quantity of metallic Li present in both cases being of the order of a few per cent. The increase in the power of resolution brought about by the use of a small sample is evident. In Figure 10, the

LITERATURE CITED

(1) Gordon, S., *J. Chem. Educ.* **40**, A87 (1963).
(2) Herold, P. G., Planje, T. J., *J. Am. Ceram. Soc.* **31**, 20 (1948).
(3) Lambert, M., Mazières, C., Guinier, A., *J. Phys. Chem. Solids* **18**, 129 (1961).
(4) Mackenzie, R. C., Mitchell, B. D., *Analyst* **87**, 420 (1962).
(5) Mazières, C., *Compt. Rend.* **248**, 2990 (1959).
(6) Mazières, C., *Ann. Chim. (Paris)* **6**, 575 (1961).
(7) Murphy, C. B., Anal. Chem. **34**, 298R (1962).
(8) Whitehead, W. L., Breger, I. A., *Science* **III**, 279 (1950).
(9) Wittels, M., *Am. Mineralogist* **36**, 615 (1951).
(10) Yannaquis, N., Regourd, M., Mazières, C., Guinier, A., *Bull. Soc. Franc. Mineral. Crist.* **85**, 271 (1962).

RECEIVED for review June 10, 1963. Accepted October 31, 1963.

24

Reprinted from *Anal. Chim. Acta.*, **52**, 397–403 (1970)

THE AUTOMATION OF THERMAL ANALYSIS INSTRUMENTATION: DIFFERENTIAL THERMAL ANALYSIS

W. W. WENDLANDT AND W. S. BRADLEY

Thermochemistry Laboratory, Department of Chemistry, University of Houston, Houston, Texas 77004 (U.S.A.)

(Received July 21st, 1970)

The modern differential thermal analysis (DTA) instrument is derived from the two thermocouple design suggested by ROBERTS-AUSTEN[1] in 1899. Many instruments have been designed and constructed since that time, each slightly different in the design of the furnace, furnace programmer, recording equipment, sample holder geometry, and so on. SMOTHERS AND CHIANG[2] in 1958 described some 225 instruments located throughout the world. This list was deleted in the second edition[3] but the latter included a bibliography of some 4248 references to DTA literature, many of them describing the instrumentation employed by the investigators. Modern DTA equipment is adequately summarized in various textbooks[3-6], while specifications on commercially available instruments are described elsewhere[7].

Present-day instruments are capable of automatic operation in that after manually inserting the sample, the temperature rise is controlled by a furnace programmer which will turn off the instrument after a preselected temperature limit is attained. After cooling the furnace back to room temperature, the pyrolyzed sample is removed from the sample holder, a new sample is introduced, and the heating cycle repeated. In this paper, an automated DTA instrument which is capable of studying eight samples in a sequential manner is described. The samples are automatically introduced into the furnace, pyrolyzed to a preselected temperature limit, and then removed. After the furnace has been cooled back to room temperature, the cycle is repeated. Operation of the sample changing mechanism, furnace temperature rise and cooling, recording, and so on, is completely automatic.

EXPERIMENTAL

General instrument features

A line drawing of the sample changing mechanism, furnace and furnace platform is shown in Fig. 1.

The powdered samples are contained in glass capillary tubes, D, of 1.6–1.8 mm i.d. (Kimax No. 34050) which are placed in the circular sample holder plate, A. The aluminum sample holder plate is of 8.0 in diameter by 1/8 in thick and has provision for retaining eight glass capillary tubes. The glass tubes are held in their respective positions by means of small spring clips. The plate is rotated by a small synchronous electric motor (Hurst Type PCSM 1/2 r.p.m.) equipped with an electromagnetic clutch. The rotation of the plate by the motor is controlled by a lamp–slit–photocell arrangement. Adjacent to each sample holder position is a 0.50 × 0.06-in slit cut in the

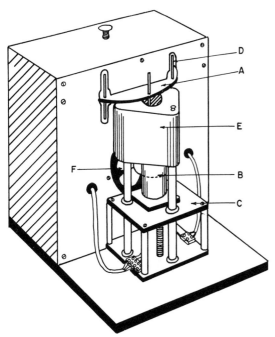

Fig. 1. General view of instrument sample changer, furnace, and furnace platform. (A) Sample holder plate; (B) furnace; (C) furnace platform assembly; (D) sample capillary tube; (E) furnace insulation; (F) cooling fan.

aluminum plate. Alignment of the plate slit between the lamp and photocell by the drive motor permits exact positioning of each capillary tube with the furnace cavity.

After the capillary tube is in position, the furnace platform, C, is raised so that the tube is positioned into the aluminum heat transfer sleeve, located on the sample thermojunction. Movement of the furnace platform is controlled by a reversible electric motor (Bodine Type KCI-S3R8, 140 r.p.m., geared down to 90 r.p.m.) con-

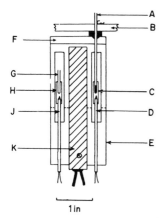

Fig. 2. Furnace and sample chamber. (A) Glass capillary tube for sample; (B) sample holder plate; (C) sample heat transfer sleeve; (D) sample thermocouple; (E) furnace block; (G) reference capillary tube; (H) reference heat transfer sleeve; (J) reference thermocouple; (K) heater cartridge.

nected to the platform by a screw-drive. Upper and lower limits of travel are controlled by two micro-switches. The furnace is insulated from the platform by a 0.25-in layer of Transite and while in the heating position, by a Marinite sleeve, E. Rotation interval for sample changing is 15 sec while it takes 50 sec to raise the furnace platform to the full upper limit.

After the sample has been heated to the upper temperature limit, the furnace is lowered, the sample holder plate rotates to a new position, and a cooling fan (Rotron muffin fan) is activated to direct air on the hot furnace. Cooling time for the furnace, from 450° to room temperature, takes about 20 min. After the furnace has been cooled to room temperature, the above cycle is repeated with a new sample.

Furnace and sample chamber

A schematic diagram of the furnace and sample chamber is shown in Fig. 2.

The cylindrical furnace, E, is of 1.5 in diameter and 3.3 in long, and is heated by a 210-W stainless-steel heater cartridge, K, (Hotwatt). The upper temperature limit of the furnace is about 500°. The sample and reference cavities are about 0.25 in diameter by 1.5 in long. Thermal contact between the sample and reference capillary tubes, A and G, is made by the aluminum heat transfer sleeves, C and H. The cylindrical sleeves are about 0.7 in long. The ends of the sleeves are drilled out so that the sample tube and the 1/16-in diameter ceramic insulator tube, D or J, fit closely within the sleeve. To minimize heat-leakage from the furnace to the sample holder plate, B, a transite cover, F, is used to enclose the top of the furnace.

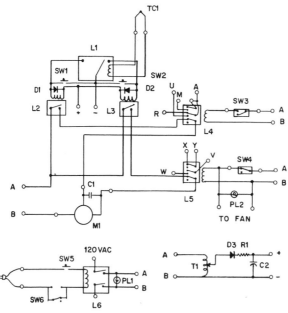

Fig. 3. Relay circuits. (L1) Meter relay, SPDT, Simpson Model 29XA, 0–50 mV; (L2 and L3) relay, SPST, 5 kΩ coil; (L4 and L5) relay, 3PDT, 120 VAC coil; (L6) latching relay, DPDT, 120 VAC coil; (C1) capacitor, 1 μfd; (C2) capacitor, 250 μfd, 25 V; (D1 and D2) diode, S93; (D3) diode, S93; (SW1, SW2, SW5) push-button switches; (SW3, SW4, SW6) micro-switches; (PL1 and PL2) pilot lamps; (M1) motor, Bodine, Type KCI-S3R8, 140 r.p.m.; (TC1) chromel–alumel thermocouple; (R1) resistor, 10 Ω, 0.5 W; (T1) variable voltage transformer, output about 13 VAC.

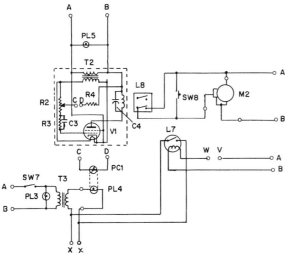

Fig. 4. Photocell and other circuits. (M2) Motor, Hurst, Type PCSM, 1/2 r.p.m.; (L7) relay, time-delay, Amperite, 115 NO5T, 5 sec delay, N.O.; (L8) relay, DPDT, 120 VAC coil; (T2 and T3) transformer, 6.3 VAC secondary; (R2) potentiometer, 500 kΩ, 3 W; (R3) resistor, 1 MΩ, 1/2 W; (R4) resistor, 100 kΩ, 1/2 W; (C3) capacitor, 500 μμF; (C4) capacitor, 8 mfd; (V1) 2D21 tube; (PC1) photocell, CL-602; (SW7) switch, SPST; (SW8) push-button switch; (PL3, PL4, PL5) pilot lamps.

Relay and other circuits

The relay and other circuits for the control of the sample holder plate and furnace platform assembly are shown in Figs. 3 and 4.

The low temperature limit contact of the meter relay, L1, activates the furnace platform motor, M1. Pushbutton switches, SW1 and SW2, permit manual control of the motor if so desired. The upper and lower limits of platform travel are controlled by the micro-switches, SW3 and SW4, which activate the two 3PDT relays, L4 and L5, respectively. The upper relay, L4, also activates the furnace programmer. The lower relay, L5, when closed, activates the time-delay relay, L8, and the photocell circuit[8,9].

When the temperature of the furnace reaches the upper preselected temperature limit of the meter relay, the furnace platform motor is activated and the platform lowered until it contacts the lower micro-switch. The latter then controls the positioning of a new sample and also turns on the furnace cooling fan. All power to the various circuits is through latching relay, L1. A micro-switch, SW6, is mounted in the sample holder plate to shut off the instrument after the eighth sample is run.

A schematic diagram of the instrument components and the furnace programmer is shown in Fig. 5.

The ΔT voltage from the differential thermocouples, TC2 and TC3, is amplified by a Leeds and Northrup microvolt d.c. amplifier, Model No. 9835-B, and is recorded on one channel of a Varian Model G-22 strip-chart potentiometric recorder. The sample temperature, as detected by thermocouple TC3 and the 0° reference junction thermocouple TC4, is recorded on the other channel. A chart-speed of 6 in h⁻¹ was employed on all of the heating runs.

The furnace temperature programmer is similar to that previously described[10].

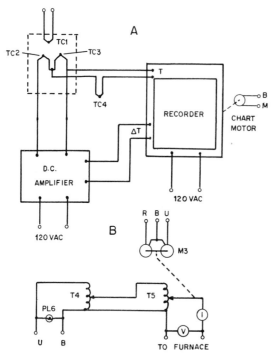

A

B

Fig. 5. (A) Schematic diagram of DTA components; (B) schematic diagram of furnace programmer. (TC2, TC3, TC4) Chromel–alumel thermocouples; (M3) motor, dual; (T4 and T5) transformer, variable voltage, Ohmite, Model No. VT3N; (PL6) pilot lamp; voltmeter, Simpson, 0–150 VAC; ammeter, Simpson, 0–5 A a.c.

Movement of the contact wiper arm of transformer T5 is controlled by a dual-speed motor unit, Bristol Type 42. The drive motor speed is 0.5 r.p.h. while the reset motor is 0.5 r.p.m. The reset limit of the wiper arm is controlled by a micro-switch connected in series with the reset motor. Variable furnace heating rates can be obtained by varying the output voltage from transformer T4. A 120 V output gives a furnace heating rate of about 7° min^{-1}; 90 V gives a heating rate of 4° min^{-1}.

Procedure

The procedure for a run consists of weighing out the samples into the eight glass capillary tubes. Sample sizes usually ranged in weight from 1 to 8 mg. The glass capillary tubes are placed in the sample holder plate, a convenient ΔT range is selected on the amplifier, and the meter relay is activated by movement of the lower temperature contact. Operation of the instrument is then completely automatic. All eight samples are heated in a sequential manner and after the eighth sample is run, the instrument shuts off all power to the various components via the latching relay.

RESULTS AND DISCUSSION

Two sequential runs of copper(II) sulfate pentahydrate samples are shown by the DTA curves in Fig. 6. The DTA curve and the temperature of the sample are shown

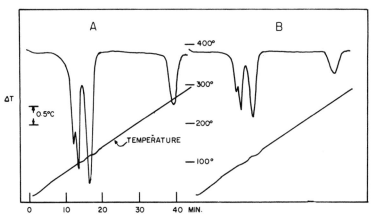

Fig. 6. DTA curves of two samples of $CuSO_4 \cdot 5H_2O$. Sample A, 5.72 mg; Sample B, 3.32 mg.

by the two curves for samples A and B. The only difference in the samples is their weight. The curves show the dehydration reactions very clearly. The first shoulder peak indicates the evolution of liquid water by the reaction,

$$CuSO_4 \cdot 5H_2O \text{ (s)} \rightarrow CuSO_4 \cdot 3H_2O \text{ (s)} + 2H_2O \text{ (l)} \tag{1}$$

while the second endothermic peak is due to the vaporization of the liquid water[11]. The formation of the monohydrate is indicated by the third endothermic peak, as shown by the reaction,

$$CuSO_4 \cdot 3H_2O \text{ (s)} \rightarrow CuSO_4 \cdot H_2O \text{ (s)} + 2H_2O \text{ (g)} \tag{2}$$

The fourth endothermic peak is caused by the dehydration of the monohydrate, according to,

$$CuSO_4 \cdot H_2O \text{ (s)} \rightarrow CuSO_4 \text{ (s)} + H_2O \text{ (g)} \tag{3}$$

A large number of inorganic salt hydrates have been studied by the automated instrument. The convenient size of the recorded curve permits the chart paper to be cut to fit a E-Z sort punched card and hence filed for easy reference at some future date. The instrument should find a wide use for routine DTA examination of a large number of samples, both inorganic and organic. The automated features of the instrument should permit convenient computer interfacing so that reaction temperatures, peak areas, and so on, can be easily calculated or tabulated.

The financial support of this work by the U.S. Air Force, Office of Scientific Research, through Grant No. 69–1620 is gratefully acknowledged. Many of the mechanical aspects of the instrument were designed and built by Mr. RALPH MARTIN. The Sun Oil Co. is acknowledged by W. S. B. for a scholarship.

SUMMARY

An automated DTA instrument is described which is capable of studying eight individual samples, each contained in a glass capillary tube, in a sequential manner. The samples are automatically introduced into the furnace, pyrolyzed to a preselected temperature limit, and then removed. After the furnace has been cooled back to room temperature, the cycle is repeated. Operation of the sample changing mechanism, furnace temperature programming, recording, etc. is completely automatic.

RÉSUMÉ

On décrit un appareil automatique pour l'analyse thermique différentielle, capable d'examiner huit échantillons à la suite. Les substances à analyser sont introduites automatiquement dans le four, pyrolysées à une température préalablement fixée et évacuées. Après refroidissement du four, le cycle se répète. Les manipulations de l'échantillon, la programmation de la température du four, l'enregistrement, etc. sont entièrement automatiques.

ZUSAMMENFASSUNG

Es wird ein automatisiertes DTA-Gerät beschrieben, mit dem aufeinanderfolgend acht einzelne, jede für sich in einem Glaskapillarrohr befindliche Proben untersucht werden können. Die Proben werden automatisch in den Ofen eingeführt, bis zu einer vorgewählten Temperaturgrenze pyrolysiert und dann entfernt. Nachdem der Ofen auf Raumtemperatur abgekühlt ist, wird der Cyclus wiederholt. Probenwechsel, Ofentemperatursteuerung, Aufzeichnung, etc. erfolgen völlig automatisch.

REFERENCES

1   W. C. ROBERTS-AUSTEN, *Metallographist*, 2 (1899) 186.
2   W. J. SMOTHERS AND Y. CHIANG, *Differential Thermal Analysis: Theory and Practice*, Chemical Publishing Co., New York, 1958, pp. 294–399.
3   W. J. SMOTHERS AND Y. CHIANG, *Handbook of Differential Thermal Analysis*, Chemical Publishing Co., New York, 1966.
4   W. W. WENDLANDT, *Thermal Methods of Analysis*, Interscience–J. Wiley, New York, 1964, Chapter 6.
5   P. D. GARN, *Thermoanalytical Methods of Investigation*, Academic Press, New York, 1965, Chapter 4.
6   E. M. BARRALL AND J. F. JOHNSON, in P. E. SLADE AND L. T. JENKINS, *Techniques and Methods of Polymer Evaluation*, M. Dekker, New York, 1966, Chapter 1.
7   Anon., *Industrial Research*, Nov. 1969, p. 25.
8   C. GROOT AND V. H. TROUTNER, *U.S. At. Energy Comm.*, HW-41007, Jan. 20, 1956.
9   W. W. WENDLANDT, *Anal. Chem.*, 30 (1958) 56.
10 W. W. WENDLANDT, *J. Chem. Educ.*, 38 (1961) 571.
11 W. W. WENDLANDT, *Thermochim. Acta*, 1 (1970) 11.

25

Reprinted from *Anal. Chem., 30,* 867–872 (Apr. 1958)

Differential Thermal Analysis

C. B. MURPHY

General Engineering Laboratory, General Electric Co., Schenectady, N. Y.

A N EFFORT has been made in this review to cover the significant developments in differential thermal analysis, from its advent to about October 1, 1957. This survey does not include an exhaustive coverage of the literature with respect to materials. A critical study of the method has been made by Norton (*83*). Two bibliographies have also appeared on this subject. An excellent review was written by Grim (*38*). More recently, an extensive review has been given by Lehmann, Das, and Paetsch (*64*). Because of the increased activity of chemists in this technical area, it was considered appropriate to review some of the problems involved in the application of this technique, to indicate recent attempts to improve and standardize it, and to illustrate the widespread application of this procedure.

EARLY HISTORICAL DEVELOPMENT

Differential thermal analysis is a technique developed by ceramists and mineralogists for studying the phenomena occurring when materials are heated. Le Chatelier (*23, 24*) was the first to use the thermal transformations occurring in matter as an analytical procedure. Other investigators (*5, 70, 99, 125, 135*) took up this technique.

The usual practice involved charging a material into a platinum crucible and following the response of a single, centrally embedded thermocouple on heating. This technique advanced to where a single recording gave results from two galvanometers, one for the material and the other for a reference substance (*99*).

Austen (*7*) devised the differential thermocouple circuit from which the process derives its name. In this case, the differential thermocouple was used to measure the difference in temperature between the sample and ambient furnace conditions. Burgess (*20*) elaborated on the principles of this circuit and suggested practical working thermocouple systems. Fenner (*30*) used the differential thermocouple to study silicate minerals, the first application outside metallurgy. Differential thermal analysis, as we know it today, was first applied by Houldsworth and Cobb (*49*). In their study of clays, the temperature of a reference material, rather than the furnace, was used for comparison.

THERMOCOUPLES

Many varieties of thermocouples have been used in differential thermal analyses. In general, they fall into two classifications, base metal and noble metal. Copper-constantan, (*51, 89*), Chromel-Alumel (*11*), and nickel–nickel chromium (*48*) are some of the base metal types that have been used The platinum–platinum rhodium system is a very commonly used noble metal thermoelement. Gold palladium–platinum rhodium has been used (*61*).

Iron-constantan (*122*) and Chromel-Alumel (*32*) thermopiles offer greatest advantage where the thermal effects must be amplified.

The choice of thermocouples for differential thermal analysis apparatus is governed by the temperature limitations to be imposed, the thermal response desired from the thermocouple, and the chemical reactivity of the materials to be investigated. In general, the noble metal type are operable at higher temperatures and are less sensitive to chemical attack. A tungsten - molybdenum thermocouple has been reported (*19*) for use up to 2200° C. Thermocouples for high temperature application have recently been reviewed by Wisely (*133*). The base metal thermocouples have to be applied with more caution because of their reactivity. However, they will give a much greater differential electromotive force response to thermal excitation than the noble metal type, and are less expensive. This latter permits more frequent replacement, and reduces the possibility of spurious effects produced by contaminated thermocouples. Base metal thermocouples also have lower thermal working limits.

Opposing physical and electrical factors are encountered in thermocouple circuits. A large thermocouple bead requires more heat to raise the temperature of its mass. However, a large bead is much more efficient in maintaining the electromotive force to give a significant differential thermal analysis peak. Moreover, these two factors make it essential that both beads in the differential thermocouple circuit be identical in size, to reduce the tendency for base line drift.

Wire size must also be considered. Large wires have high thermal conductivity and low electrical resistance. It is obvious that there must be a compromise in these two properties. Boersma has calculated (14) that the heat leakage through thermocouple wires can reduce a peak area to less than 50% of its theoretical value. This would indicate the desirability of having fine thermocouple wires. These would offer high electrical resistance to the small current generated in the thermocouple circuit and would require frequent replacement, because they could not withstand many exposures to high temperatures.

The platinum–platinum–10% rhodium thermocouple is the most widely used noble metal thermocouple and is so well characterized that it is used to define the International temperature scale from 630° to 1063° C. The platinum–6% rhodium–platinum–30% rhodium thermocouple, introduced recently in the field of high temperature measurements, appears to have a number of advantages over the platinum–platinum–10% rhodium type. The alloy has greater mechanical strength than the pure metal, and can be used to higher temperatures. The thermal response, somewhat lower at low temperatures, is equal to that of the platinum–platinum–10% rhodium thermocouple at high temperatures.

A study has been made of the effects of wire and bead size in the platinum–platinum–10% rhodium system (45). With illite under identical conditions, a bead 1.43 mm. in diameter gave the more significant thermograms, compared with beads 0.80 and 0.90 mm. in diameter. A comparison of wire size effects with beads 1.43 mm. in diameter showed that 28-mm. wire gave more significant thermographic effects than 22-mm. wire. Similar experiments were conducted with kaolin.

The thermocouple circuits have been employed (Figure 1). In *A* temperature measurement is controlled by a separate thermocouple. In *B*, the tempera-

ture measurement is made with a part of the differential circuit. Both methods give similar results. However, from a maintenance standpoint, the *A* circuit is more easily repaired than *B*. By using a platinum foil sample holder, it has been demonstrated by West (128) that the temperature-measuring couple can be completely encapsulated by the foil and used continually. This same technique, applied to the reference material, introduces a degree of constancy in all thermograms. In this way, calcination of the reference material is repeated with every run.

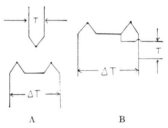

Figure 1.   Thermocouple circuits

The centering of thermocouples is also important. A thermocouple is subjected to excessive thermal transmission from the sampler holder if not centered. This results in a gradually increasing thermal difference between the two thermocouples, and a base line drift in the thermogram. Smyth (111) has discussed the problem of temperature distribution during an endothermic reaction, and has shown that a marked variation in peak shape will result from noncentered thermocouples. His calculations show that if the differential temperature is plotted against the temperature of the center of the reference specimen, neither the point of departure from the base line nor the peak temperature corresponds to the inversion temperature. Smyth's work indicates that if the differential temperature is recorded as a function of the surface temperature of the sample or the temperature of the center of the sample, the point of departure in the former or the peak temperature in the latter will give the temperature of the inversion.

SAMPLE HOLDERS

Sample holders may be classified in several ways. The simplest is based on thermocouple insertion. Side-inserted thermocouples have been employed by Pask and Warner (86), McConnell and Earley (67), and Berkelhamer (13). Systems with top-mounted thermocouples have been used by West (128), Vold (122), and Gordon

and Campbell (35). Thermocouples inserted through the bottom of the block have been used by Stone (112), Parodi (85), and Carthew and Cole (22). Top - mounted thermocouple systems have found special application in the study of waxes, greases, and liquid-forming systems. The thermal behavior of substances into the liquid state can be followed in other types of systems by using an inert material as an adsorbent (78). This prevents the material from running out of the sample holder.

Multiple sample holders have been used by Kulp and Kerr (63), McConnell and Earley (67), and Kauffman and Dilling (58).

The variation in sample holders has ranged from inexpensive and simple to complex and costly. Insley and Ewell (50) used a divided platinum cylinder. A simple refractory block was used by Houldsworth and Cobb (49). Refractory blocks with replaceable refractory cups (40) and platinum cylinders (46) have also been employed. Nickel blocks have been used by Grim and Rowland (39), Norton (83), and many others. The most simple holder developed consists of holes drilled into sections of insulating brick (81). Platinum foil has been wrapped about thermocouple beading to form the sample holder by West and Sutton (129) and Kissinger (59). Pask and Warner (86) have used blocks made of nickel, graphite, and platinum. Platinum microcrucibles (42) in some cases have been incorporated as part of the thermocouple system (110). An aluminum sample holder has been developed for moderately high temperature application in the field of soaps and greases (27).

Normal procedure in differential thermal analysis has been to arrange the thermocouple circuit so that junctions are embedded in the sample and reference material. In a new type of sample holder developed by Boersma (14) the thermocouple junctions are inserted in insulated nickel block sections without physical contact between the thermocouple junction and the materials. This system is claimed to alleviate the dependency of the peaks on sample packing and thermal conductivity of the sample. Another obvious advantage is that thermocouples do not have to be replaced because of contamination by reactants.

Most equipment in use today has metal blocks of nickel, Inconel, or stainless steel. A ceramic block, because of its low thermal conductivity, would give more intense thermographic peaks. A smaller temperature differential would exist between the programing thermocouple and the furnace in a metal holder because of the high thermal conductivity of the metal block. This would result in better control of the furnace with a metal block.

The comparative performance of nickel and porous alumina sample holders of identical dimensions was investigated by Webb (*127*), studying the thermal decomposition of calcium hydroxide and calcium carbonate. Lower decomposition temperatures were observed with the ceramic holder. When vitreous silica inserts were used with the porous black, this effect disappeared. It was reasoned that permitting the escape of gaseous decomposition products led to a more rapid completion of reaction. The nickel block also gave a smaller peak area, showing less sensitivity than the porous alumina. The temperature difference between the furnace atmosphere and the center of the inert reference material was smaller in the nickel holder than the porous material. Similar experiments conducted by Mackenzie (*68*) on the exothermic peak for kaolinite confirm these results. Mackenzie has also shown that the nickel sample holder, without a cover, approaches the same maximum temperature value obtained with the porous holder. But, again, the peak area was smaller than that obtained with the latter material. It was also shown that low porosity alumina compares favorably with the nickel as a sample block material.

Using the same mass of kaolin, the nickel sample block and the platinum foil type holder have been compared by Patterson (*87*). The larger peaks with the platinum foil sample holder are attributed to the heat capacity differences in the two sample holders used. The platinum foil type holder has a tendency to show marked drift in the base line, because the specific heats of the sample and reference material become significant. More attention to this variation might develop a comparative method for the determination of specific heats. However, using this type of holder, and replacing the platinum foil after each run, have the distinct advantage of preventing contamination in subsequent experiments. Scrap platinum can be salvaged to reduce operational costs.

Until recently, sample holders have been constructed to conform with the dimensions of available furnaces. Geometric design was applied to the bulk material that could be accommodated. The design of the sample holder involves several factors. Most important is that the material of construction should not react with the substances investigated or the furnace atmosphere to be used. The sample cell should be small to minimize the thermal gradient through the material, yet sufficiently large to give maximum differential thermal effects. These and other factors have been considered at some length.

One of the first investigators to determine the relationship of the physical dimensions of the sample holder and the resulting thermograms was Hauser (*45*). Using platinum foil sample holders, $1/8$, $3/16$, and $1/4$ inch in diameter and $3/4$ inch long, he obtained best results with the $3/16$-inch holder. It gave peaks more intense than the smaller unit, and more distinct than the larger cell.

A more complete investigation was published (*86*) on the construction of nickel blocks, and confirmed with platinum and graphite sample holders. Using cell diameters of $1/4$, $3/8$, and $1/2$ inch, much larger peaks were obtained with the larger cells, but much of the detail was lost. Excessively large peaks completely obliterated small thermal reactions. The maximum number of reactions was shown by the smallest holder. Pask and Warner (*86*) also found that the variation of the ratio of the bulk volume of the sample holder to the volume of the sample holes, for values of 2.7, 5.8, and 10.6 produced only minor effects in the thermograms of ball clay. These numbers apply with a small total mass of sample block.

Mackenzie and Farquharson (*69*), reporting on the results obtained with standard clay samples from 32 laboratories, have concluded that ceramic sample holders give higher peak temperatures than metal types up to 650° C. Comparable results are obtained to 750° C. Above this last temperature reproducibility appears to be better with the metal sample holder.

A recent report by Walton (*126*) illustrates the use of pressed pellets of clay into which the thermocouples are inserted. No sample holder is used. Morita (*71*, *72*) has used 150-mg. samples between 200–mg. portions of calcined alumina compressed at 200 pounds per square inch to ensure a compacted mass for examination.

RATE OF HEATING

Rates of heating have been employed from 0.5° (*27*) to 100° C. per minute (*9*). Because a rate of heating is employed, the temperatures recorded for various phenomena are not those of thermodynamic conditions. A time lag occurs. The slower the rate of heating, the smaller the variation. However, with a slow rate of heating, the temperature differential between the reference material and the substance examined will become less. This will result in rounded peaks that tend to be more insignificant as the heating rate is lowered. With a rapid rate of heating, peaks become intense and considerable detail is lost. Larger deviations from thermodynamic temperatures will also result (*79*).

Most published differential thermal analysis results have been made at heating rates of from 8° to 15° C. per minute. Recommendations made at the International Geological Congress in London, in 1948, called for a heating rate for the sample of 10° C. per minute, with a maximum variation of 1° C. per minute (*69*).

The development of modern apparatus has tended toward controls permitting variability in the heating rate (*89*, *119*, *132*). Such apparatus has been applied to the determination of reaction kinetics (*59*).

EFFECT OF PARTICLE SIZE

The effect of particle size on the nature of the thermogram has been considered by several investigators. Norton (*83*) has pointed out that finer particles give up their heat more rapidly and have a distinct effect on the thermogram. It has been said (*41*) that the temperature at which reactions begin and end thermographically will be lowered with increasing fineness of the material studied. Recent results published by Berg (*10*) show increased peak area as well as peak intensity with increased fineness of the clay particles examined. The areas of 0.8-gram samples of various particle size fractions of kaolinite were measured by Carthew (*21*); variation of area with particle size was found.

It is unfortunate that the bulk of this work has been conducted with clays. These are such complex chemical structures that there is no assurance that various fractions have the same composition. Further efforts in this direction should be conducted on materials of unquestioned structure and purity.

In a simple cubic crystal, the unit cell rupture involves the breaking of three bonds per atom. In gross crystals, six bonds are ruptured per atom. From a theoretical point of view, it would appear that differences in energy would result with increasing fineness. However, particles sizes employed to date have by no means approached this limit.

Also, with fine samples, the packing of the sample should be greater. This would lead to an increased thermal conductivity in the denser material. A quicker thermocouple response would result, lowering the temperature at which reaction occurs, as reported by Grimshaw, Heaton, and Roberts (*41*).

ATMOSPHERE CONTROL

The original differential thermal analysis systems provided for the programed heating of samples in the air present in the furnace atmosphere. This technique resulted in thermograms that were difficult to interpret because thermal effects were not always separable from oxidative effects. The desire to separate these two effects first led to the development of equipment operable under vacuum. Such apparatus has been described by Whitehead and Breger (131), West (128), Stone (112), Haul and Heystek (42), and others.

In vacuum techniques, endothermic reactions resulting from phase transitions would show little or no variation because of the lack of pressure dependency of such reactions. However, endothermic reactions originating from the volatilization of decomposition products—i.e., release of gases such as carbon dioxide, water, sulfur dioxide, etc.—would show peaks at lower temperatures under vacuum. This would result from the lowered partial pressure of these materials in vacuum systems. Oxidation reactions would not be likely to occur, and peaks for such phenomena would disappear from thermograms obtained using vacuum.

A greater understanding of the thermal properties of materials could be obtained by studying the effects of various atmospheres on resulting thermograms. Equipment to serve such a purpose was developed by Rowland and Lewis (100). Greater significance could be attached to results obtained over a wide variation of pressures. Stone (112) has developed equipment capable of operation from vacuum up to 6-atm. pressure, and uses either a single gas or a mixture of two gases under dynamic conditions. This shows promise as being the most versatile type of equipment devised to date.

FURNACES

For the most part, furnace construction has consisted of a ceramic tube wound with a resistance wire element. The tube has been mounted vertically and horizontally; no particular advantage has been found in either method. Nichrome, Kanthal, and platinum wire have been used most frequently as resistance heating elements. Globars have also been used. More recent furnace construction has incorporated design tending to reduce induced e.m.f. in the thermocouples. Stone's (112) equipment has six symmetrically located, electrically balanced coils as the heating element; this eliminates induced effects in thermocouples.

A sufficiently large uniform zone of heating should exist, so that the sample block and thermocouple system can be accommodated at the same temperature. This ensures that the differential temperatures observed are a function of materials examined rather than furnace construction.

Brewer and Zavitsanos (18) developed an inductive heated differential ther-

mal analysis apparatus and applied it to a study of the germanium-germanium dioxide system. The sample was held in silica sample holders inserted in a molybdenum block. Temperatures were measured with platinum-platinum rhodium thermocouples. The results agreed satisfactorily with those from conventional differential thermal analysis equipment. This same equipment was applied to the silicon-silica system (17). It was suggested that this technique, using pyrometric temperature determinations, should permit differential thermal analysis to temperatures of 3000° C.

MULTIPURPOSE EQUIPMENT

Many investigators have used other techniques to obtain information to complement differential thermal analysis results. Typical examples are resistance analysis used by Reisman, Triebwasser and Holtzberg (96), x-diffraction applied by Stone and Weiss (115), and thermogravimetric analysis utilized by Johnson (52). The desire to obtain supplementary information led Powell (94) to the development of equipment giving simultaneous thermogravimetric and differential thermal analysis curves from one sample.

APPLICATIONS

Phase Equilibria. Differential thermal analysis registers the thermal transformations occurring in matter, and, therefore, can be applied to the detection of phases formed when materials are heated. The hydrates of many compounds, including those of nickel nitrate (51), chromium orthophosphate (116), cupric chloride (56), magnesium sulfate (55), and chromic chloride (84), have been studied. A study of the system of magnesium oxide-magnesium chloride-water (26, 29) has shown four hydrated oxychlorides. Katz and Kedesdy (57) have used the method to show the existence of a new hydrate of aluminum arsenate. Phase equilibria in greases has been studied by a number of investigators (27, 122–124). Perloff (90) has used the method to distinguish between two forms of anhydrous gallium orthophosphate.

Heats of Reaction. The relationship of the peak area in a thermogram to the heat of reaction has been discussed by Borchardt and Daniels (16), Ramachandran and Bhattacharyya (95), and many others. Borchardt and Daniels have determined the heat of decomposition of benzenediazonium chloride to be −37.0 kcal. per mole and the heat of reaction of ethyliodide with N,N-dimethylaniline to be −20.4 kcal. per mole. The application of differential thermal analysis as a microcalorimeter has been

demonstrated by Wittels (134), Allison (4), Sabatier (101), and Talibudeen (118). Sabatier (101) has given heats of dissociation in calories per gram: gypsum 153, zinc carbonate 114, magnesium carbonate 302, and calcium carbonate 404.

Stone (113), using the dynamic gas method of differential thermal analysis (112), determined the heat of dissociation of magnesite to be 10.1 kcal. per mole. The thermograms showed increasing decomposition temperature with increasing carbon dioxide pressure. When log p is plotted against $1/T$, the Clausius-Clapeyron equation, the slope of the plot, $-\Delta H/R$, can be used to determine the heat of reaction. Borchardt (15) has recently considered the kinetic effects in this method of determining heats of reaction, and has concluded that it is probably a valid procedure.

A practical application of this aspect of differential thermal analysis has been made with coal. Payne (88) has determined the ΔH values of semibituminous coals by differential thermal analysis, and has found that they compare favorably with values obtained by peroxide bomb technique. Glass (33) has been able to correlate the differential thermal analysis curves of coking coals with standard rank classification. Clegg (25) has presented an extensive report on experimental factors that modify thermograms of bituminous coal.

QUANTITATIVE ANALYSIS

Quantitative differential thermal analysis has been reported by a number of investigators. The relationship of the peak area to the heat of reaction has been shown by Kracek and associates (34, 62), Alexander, Hendricks, and Nelson (2), Schafer and Russell (103), Berg (8), and many others. Recently, de Jong (53) published a detailed verification of the usage of peak area for quantitative measurements from an instrumental aspect. Apart from this, the thermograms of a material are often affected by the impurities present. Variations in dolonite thermograms have been determined as a function of contaminants (36, 105). Impurities have marked effects on the thermograms of kaolin (75), calcium carbonate (117), ammonium nitrate (1), and mullite (78). Lehmann and Fischer (65) have pointed out that the determination of quartz in clays may have interference from dehydration unless the sample is rapidly heated.

Berkelhamer (12) has determined quartz in calcium and Schedling (104) reported that quartz can be determined in dusts at concentrations of approximately 1%. Carthew (21) recently studied the 600° endothermic peak of kaolinite. He found that an empirical relationship between the slope ratio of the peak and the ratio of area to width of the peak at half amplitude gives a quantitative measurement of kaolinite, independent of particle size distribution and degree of crystallinity. Heystek (47) has applied differential thermal analysis to the qualitative determination of gangue minerals in chrome ores. Under favorable conditions, the thermograms can lead to quantitative data.

Recent trends in clay analysis involve complex formation with organic molecules. Sand and Bates (102) have used ethylene glycol to form a complex with endellite in the presence of halloysite and kaolinite. The endothermic peak of the endellite complex allowed quantitative determination of this component to within 3% of the weighed amount. Konta (60) has used this technique for montmorillonite identification. Allaway (3) has investigated the differential thermal analysis of piperidine-clay complexes.

SOLID STATE REACTIONS

The number of solid state reactions studied by this technique has been very extensive. Haul and Schumann (43) investigated the reaction between manganese dioxide and iron pyrites. Although manganous sulfate was formed, incompleteness of reaction eliminated consideration as a commercial process. Differential thermal analysis has also been used to study the formation of zinc-chromium spinel (80). Greenberg (37), Kalousek, Logiudice, and Dodson (54), and Newman (82) have applied the method to studies of reactions in the calcium hydroxide–silica-water system. The reaction of iron, alumina, and sodium hydroxide has been investigated by Smothers and Chiang (108). The reaction of alkali monohydrogen sulfate with monohydrogen phosphate was shown (6) to produce sodium sulfate and dihydrogen phosphate. Differential thermal analysis has been applied to solid state reactions occurring in phosphor formation (81, 98). Crandall and West (28) have used differential thermal analysis to study the oxidation of cobalt by alumina. The oxidation of uranium dioxide by air has been shown by Johnson (52). Differential thermal analysis has also been applied to the decomposition of heteropoly acids of molybdenum and tungsten (130).

ORGANIC APPLICATIONS

In general, application of differential thermal analysis to organic materials has not progressed very rapidly. The study of greases by Vold and associates

(122-124) was one of the first in this field. Pirisi and Mattu (93) have applied this technique to a number of organic compounds, including benzoic acid and sodium benzoate, o-, m-, and p-aminobenzoic acids (92) and succinic acid, ammonium succinate, succinamide, and succinimide (91). Morita and Rice (74) first applied this method to polymeric materials, investigating the thermal phenomena associated with poly(vinyl chloride) and vinyl chloride–vinyl acetate copolymers. Subsequent investigations by Morita have demonstrated the applicability of this technique to the study of polyglucosans (73) and starch and related polysaccharides (72).

Murphy and coworkers (77) have recently applied differential thermal analysis to the study of cure in a series of Vibrin 135 resins prepared with different heat treatments. This technique has differentiated between two fundamentally different curing exotherms. Different polymerization catalysts produced markedly different effects in the final polymer.

CATALYST STUDIES

Stone and Rase (114) have shown that differential thermal analysis can be used to measure the activity of silica-alumina cracking catalysts. Fischer catalysts have been studied by Trambouze (120). In this work, it has been shown by differential thermal analysis patterns of variously prepared nickel precipitates that a close relationship exists between the conditions of precipitation and catalytic activity. Hauser and le Beau (44) have used differential thermal analysis to assist in studies of reactivity of alumina and silica gels. Murphy (76) has used a differential thermal system to study the recombination coefficient of oxygen atoms on various surfaces. The system used incorporated differential thermocouples in the apparatus developed by Linnett and Marsden (66).

KINETICS

To extend the usefulness of differential thermal analysis, many investigators (79, 106, 107) have attempted to obtain kinetic data. Borchardt and Daniels (16) have applied the method and obtained kinetic data for the decomposition of benzenediazonium chloride and the reaction of ethyl iodide with N,N-dimethylaniline. Data obtained were in good agreement with those reported using other techniques. Kissinger (59) recently has derived an expression relating peak temperature at different heating rates to kinetic quantities. Borchardt (15) has applied kinetics to magnesium carbonate decomposition.

LITERATURE CITED

(1) Alekseenko, L. A., Boldyrev, V. V., *Zhur. Priklad. Khim.* **29**, 529 (1956).
(2) Alexander, L. T., Hendricks, S. B., Nelson, R. A., *Soil Sci.* **48**, 273 (1939).
(3) Allaway, W. H., *Soil Sci. Soc. Amer., Proc.* **13**, 183 (1948).
(4) Allison, E. B., *Silicates Ind.* **19**, 363 (1954).
(5) Ashley, H. E., *J. Ind. Eng. Chem.* **3**, 91 (1911).
(6) Audrieth, L. F., Mills, J. R., Netherton, L. E., *J. Phys. Chem.* 482 (1954).
(7) Austen, R., Fifth Rept. Alloys Research Comm., *Proc. Inst. Mech. Eng.* **1899**, 35.
(8) Berg, L. G., *Compt. rend. acad. sci. U.R.S.S.* **49**, 648 (1945).
(9) Berg, L. G., Rassonskays, I. S., *Doklady Acad. Nauk S.S.S.R.* **73**, 113 (1950).
(10) Berg, P. W., *Ber. deut. Keram. Ges.* **30**, 231 (1953).
(11) Berkelhamer, L. H., U. S. Bur. Mines, Rept. Invest. **3762** (1944).
(12) *Ibid.*, **3763**.
(13) Berkelhamer, L. H., U. S. Bur. Mines, Tech. Paper **664** (1954).
(14) Boersma, S. L., *J. Am. Ceram. Soc.* **38**, 281 (1955).
(15) Borchardt, H. J., *J. Phys. Chem.* **61**, 827 (1957).
(16) Borchardt, H. J., Daniels, F., *J. Am. Chem. Soc.* **79**, 41 (1957).
(17) Brewer, L., Greene, F. T., *J. Phys. Chem. Solids* **2**, 286 (1957).
(18) Brewer, L., Zavitsanos, P., *Ibid.*, **2**, 284 (1957).
(19) Budnikov, P. P., Tresvyatskiĭ, S. G., *Ogneupory* **20**, 166 (1955).
(20) Burgess, G. K., U. S. Bur. Standards, Bull. **5**, 199 (1908–9).
(21) Carthew, A. R., *Am. Mineralogist* **40**, 107 (1955).
(22) Carthew, A. R., Cole, W. F., *Australian J. Instr. Technol.* **9**, 23 (1953).
(23) Chatelier, H. le, *Bull. soc. franç. mineral.* **10**, 204 (1887).
(24) Chatelier, H. le, *Z. physik. Chem.* **1**, 396 (1887).
(25) Clegg, K. E., Ill. State Geol. Survey, Rept. Invest. **190** (1955).
(26) Cole, W. F., Demediuk, T., *Australian J. Chem.* **8**, 234 (1955).
(27) Cox, D. B., McGlynn, J. F., ANAL. CHEM. **29**, 960 (1957).
(28) Crandall, W. B., West R. R., *J. Am. Ceram. Soc.* **35**, 66 (1956).
(29) Demediuk, T., Cole, W. F., Hueber, H. V., *Australian J. Chem.* **8**, 215 (1955).
(30) Fenner, C. N., *Am. J. Sci.* **36**, 331 (1913).
(31) Gale, R., Rabitin, J., unpublished results.
(32) Gamel, C. M., Smothers, W. J., *Anal. Chim. Acta* **6**, 442 (1952).
(33) Glass, H. D., *Fuel* **34**, 253 (1955).
(34) Goranson, R. W., Kracek, F. C., *J. Phys. Chem.* **36**, 913 (1932).
(35) Gordon, S., Campbell, C., ANAL. CHEM. **27**, 1102 (1955).
(36) Graf, D. L., Ill. State Geol. Survey, Rept. Invest. **161** (1952).
(37) Greenberg, S. A., *J. Phys. Chem.* **61**, 373 (1957).
(38) Grim, R. E., *Ann. N. Y. Acad. Sci.* **53**, 1031 (1951).
(39) Grim, R. E., Rowland, R. A., *J. Am. Ceram. Soc.* **27**, 65 (1944).
(40) Grimshaw, R. W., *Trans. Brit. Ceram. Soc.* **44**, 72 (1945).
(41) Grimshaw, R. W., Heaton, E.,

Roberts, A. L., *Ibid.*, **44**, 87 (1945).

(42) Haul, R. A. W., Heystek, H., *Am. Mineralogist* **37**, 166 (1952).

(43) Haul, R. A. W., Schumann, H. J., *J. S. African Chem. Inst.* **8**, 80 (1955).

(44) Hauser, E. A., le Beau, D. S., *J. Phys. Chem.* **56**, 136 (1952).

(45) Hauser, R. E., thesis, N. Y. State College of Ceramics, Alfred University, Alfred, N. Y., 1953.

(46) Herold, P. G., Plange, T. J., *J. Am. Ceram. Soc.* **31**, 20 (1948).

(47) Heystek, H., *Bull. Am. Ceram. Soc.* **35**, 133 (1952).

(48) Hiller, J. E., Probsthain, K., *Erzmetall* **7**, 257 (1955).

(49) Houldsworth, H. S., Cobb, J. W., *Trans. Brit. Ceram. Soc.* **22**, 111 (1922-23).

(50) Insley, H., Ewell, R. H., *J. Research Natl. Bur. Standards* **14**, 615 (1935).

(51) Jaffray, J., Rodier, N., *J. recherches centre natl. recherche sci. Lab. Bellevue (Paris)* No. 31, 252 (1955).

(52) Johnson, J. R., *J. Metals* **8**, 660 (1956).

(53) Jong, G. de, *J. Am. Ceram. Soc.* **40**, 42 (1957).

(54) Kalousek, G. L., Logiudice, J. S., Dodson, V. H., *Ibid.*, **37**, 7 (1954).

(55) Kamecki, J., Palej, S., *Rocznik Chem.* **29**, 691 (1955).

(56) Kamecki, J., Trau, J., *Bull. acad. polon. sci., Classe III* **3**, 111 (1955).

(57) Katz, G., Kedesdy, H., *Am. Mineralogist* **39**, 1005 (1954).

(58) Kauffman, A. J., Jr., Dilling, E. D., *Econ. Geol.* **45**, 222 (1950).

(59) Kissinger, H. E., *J. Research Natl. Bur. Standards* **57**, 217 (1956).

(60) Konta, J., *Universitas Carolina, Geologica* **1**, 29 (1955).

(61) Kracek, F. C., *J. Phys. Chem.* **33**, 1281 (1929).

(62) *Ibid.*, **34**, 225 (1930).

(63) Kulp, J. L., Kerr, P. F., *Science* **105**, 413 (1947); *Am. Mineralogist* **33**, 387 (1948).

(64) Lehmann, H., Das, S. S., Paetsch, H. H., *Tonind.-Ztg. u. Keram. Rundschau*, Beiheft 1 (1954).

(65) Lehmann, H., Fischer, P., *Ibid.*, **78**, 309 (1954).

(66) Linnett, J. W., Marsden, D. G. H., *Proc. Roy. Soc. (London)* **A234**, 489 (1956).

(67) McConnell, D., Earley, J. W., *J. Am. Ceram. Soc.* **34**, 183 (1951).

(68) Mackenzie, R. C., *Nature* **174**, 688 (1954).

(69) Mackenzie, R. C., Farquharson, K. R., *Compt. rend. XIX Session Congr. Geol. Inst. (Algeria)* **18**, 183 (1952).

(70) Mellor, J. W., Holdcroft, A. D., *Trans. Brit. Ceram. Soc.* **10**, 94 (1911).

(71) Morita, H., ANAL. CHEM. **27**, 336 (1955).

(72) *Ibid.*, **28**, 64 (1956).

(73) *Ibid.*, **29**, 1095 (1957).

(74) Morita, H., Rice, H. M., *Ibid.*, **27**, 336 (1955).

(75) Mukherjee, A. K., *Science and Culture (India)* **21**, 36 (1955).

(76) Murphy, C. B., unpublished results.

(77) Murphy, C. B., Palm, J. A., Doyle, C. D., Curtis, E. M., *J. Polymer Sci.*, in press.

(78) Murphy, C. B., West, R. R., unpublished results.

(79) Murray, P., White, J., *Trans. Brit. Ceram. Soc.* **54**, 204 (1955).

(80) Nagura, S., Kato, E., Niboshi, I., *Nippon Kagaku Zasshi* **76**, 1492 (1956).

(81) Nagy, R., Lui, C. K., *J. Opt. Soc. Amer.* **37**, 37 (1947).

(82) Newman, E. S., *J. Research Natl. Bur. Standards* **59**, 187 (1957).

(83) Norton, F. H., *J. Am. Ceram. Soc.* **22**, 54 (1939).

(84) Pamfilov, A. V., Gumenyuk, N. N., *Zhur. Obshchei Khim.* **23**, 1065 (1953).

(85) Parodi, J. E., unpublished results.

(86) Pask, J. A., Warner, M. F., *Bull. Am. Ceram. Soc.* **33**, 168 (1954).

(87) Patterson, R. C., *Ibid.*, **32**, 117 (1953).

(88) Payne, N., *Econ. Geol.* **46**, 846 (1951).

(89) Penther, C. J., Abrams, S. T., Stross, F. H., ANAL. CHEM. **23**, 1459 (1951).

(90) Perloff, A., *J. Am. Ceram. Soc.* **39**, 83 (1956).

(91) Pirisi, R., Mattu, F., *Ann. chim. (Rome)* **43**, 574 (1953).

(92) Pirisi, R., Mattu, F., *Chimica (Milan)* **9**, 10 (1954).

(93) Pirisi, R., Mattu, F., *Rend. seminar. fac. sci. univ. Cagliari* **22**, 81 (1952).

(94) Powell, D. A., *J. Sci. Instr.* **34**, 225 (1957).

(95) Ramachandran, V. S., Bhattacharyya, S. K., *J. Sci. Ind. Research (India)* **13A**, 365 (1954).

(96) Reisman, A., Triebwasser, S., Holtzberg, F., *J. Am. Chem. Soc.* **77**, 4228 (1955).

(97) Rengade, E., "Analyse Thermique et Metallographique Microscopique," Vol. 1, Hachette, Paris, 1909.

(98) Rice, A. P., *J. Electrochem. Soc.* **96**, 114 (1949).

(99) Rieke, R., *Sprechsaal* **44**, 637 (1911).

(100) Rowland, R. A., Lewis, D. R., *Am. Mineralogist* **36**, 80 (1951).

(101) Sabatier, G., *Bull. soc. franç. mineral.* **77**, 953 (1954).

(102) Sand, L. B., Bates, T. F., *Am. Mineralogist* **38**, 271 (1953).

(103) Schafer, G. M., Russell M. B., *Soil Sci.* **53**, 353 (1942).

(104) Schedling, J. A., *Chim & ind. (Paris)* **69**, 1066 (1953).

(105) Schwob, Y., *Compt. rend.* **224**, 47 (1947).

(106) Segawa, K., *J. Japan. Ceram. Assoc.* **56**, 7 (1948).

(107) Sewell, E. C., *Clay Minerals Bull.* **2**, 233 (1955).

(108) Smothers, W. J., Chiang, Y., *Proc. Intern. Symposium Reactivity of Solids, Gothenburg, 1952*, 501.

(109) Smothers, W. J., Chiang, Y., Wilson, A., "Bibliography of Differential Thermal Analysis," University of Arkansas, Fayetteville, Ark., 1951.

(110) Smyth, F. H., Adams, L. H., *J. Am. Ceram. Soc.* **45**, 1167 (1923).

(111) Smyth, H. T., *Ibid.*, **34**, 221 (1951).

(112) Stone, R. L., *Ibid.*, **35**, 76 (1952).

(113) *Ibid.*, **37**, 46 (1954).

(114) Stone, R. L., Rase, H. F., ANAL. CHEM. **29**, 1273 (1957).

(115) Stone, R. L., Weiss, E. J., *Clay Minerals Bull.* **2**, 214 (1955).

(116) Sullivan, B. M., McMurdie, H. F., *J. Research Natl. Bur. Standards* **48**, 159 (1952).

(117) Takanaka, J., Yajima, S., *J. Sci. Hiroshima Univ., Ser. A*, **17**, 257 (1953).

(118) Talibudeen, O., *J. Soil Sci.* **3**, 251 (1952).

(119) Theron, J. J., *Brit. J. Appl. Phys.* **3**, 216 (1952).

(120) Trambouze, Y., *Compt. rend.* **230**, 1169 (1950).

(121) Vigneron, G., "Analyse Thermique pendant ces dix dernières années 1939-1948," F. Jacobs, Brussels, 1949.

(122) Vold, M. J., ANAL. CHEM. **21**, 683 (1949).

(123) Vold, M. J., Hattiangdi, G. S., Vold, R. D., *Ind. Eng. Chem.* **41**, 2320 (1949).

(124) *Ibid.*, p. 2539.

(125) Wallach, H., *Compt. rend.* **157**, 48 (1913).

(126) Walton, J. D., Jr., *J. Am. Ceram. Soc.* **38**, 438 (1955).

(127) Webb, T. L., *Nature* **174**, 686 (1954).

(128) West, R. R., personal communication.

(129) West, R. R., Sutton, W. J., *J. Am. Ceram. Soc.* **37**, 221 (1954).

(130) West, S. F., Audrieth, L. F., *J. Phys. Chem.* **59**, 1069 (1955).

(131) Whitehead, W. L., Breger, I. A., *Science* **111**, 279 (1950).

(132) Wilburn, F. W., *J. Soc. Glass Technol.* **38**, 371 (1954).

(133) Wisely, H. R., *Ceram. Age* **66**, 15 (1955).

(134) Wittels, M., *Am. Mineralogist* **36**, 615, 760 (1951).

(135) Wohlin, R., *Sprechsaal* **46**, 719, 733, 749, 767, 781 (1913).

Part II

COMMENTARY ON
THERMOGRAVIMETRY

Editor's Comments
on Papers 26 Through 36

The art and science of weighing was known in Egypt as early as 3000 B.C. However, it was not until the Middle Ages that the balance was used to follow the course of a chemical reaction (2). The cupellation procedure was already in general use in the fourteenth century and many records indicate that it was in statutory use for the examination of gold (1). Philip VI of France described the analysis procedure, which is perhaps the earliest example of a standard method of analysis, and stated that "the balance used for the test should be of good construction, precise, and should not pull to either side. The test should be carried out in a place where there is neither wind nor cold, and whoever carries out the test must take care not to burden the balance by breathing upon it."

Berzelius used the analytical balance in gravimetric analysis and was quite concerned with its construction and the weights that were employed. The smallest weight then in use was a 5-mg rider, but some balances were sensitive to 1 mg of a 10-gram load. It should be noted that Berzelius was the first chemist to use the metric system (1) in weighing. As a result of the work of Berzelius and others, gravimetric analysis was very well developed by 1847; the methods and techniques used were almost the same as those used today.

The effect of heat on analytical precipitates was initially studied to improve the accuracy of gravimetric procedures. The determination of drying-temperature ranges and the composition of the precipitates were of great interest. To do this, a balance had to be constructed which permitted the heating of the precipitate simultaneously with weighing. According to Szabadvary (1), the first such instrument of this type was constructed and used by Nernst and Riesenfeld (3) in 1903. They used a Nernst quartz torsion microbalance equipped with an electric furnace to study the mass losses on heating of Iceland spar, opal, and zirconia. Two years later, in 1905, Brill (4) recorded what was probably the first TG curve by the continuous weighing of calcium carbonate up to 1200°C. Also, in 1907, Truchot (5) heated pyrite and pyrrhotite using a method resembling present-day thermogravimetry. In 1912, Urbain and Boulanger (6) constructed an electromagnetic compensation balance which was used to study the dehydration of metal salt hydrates. In 1922, Urbain remarked to Duval, then a young researcher in his laboratory, "the results depend too much on the heating conditions"(7).

The starting point of our papers on thermogravimetry is the paper by Honda (Paper 26), which describes the development of the first instrument to be called a *thermobalance*. The balance was a rather crude affair, but it enabled him to determine the TG curves of $MnSO_4 \cdot 4H_2O$, $CaSO_4 \cdot 2H_2O$, $CaCO_3$, and CrO_3. He employed the quasistatic heating technique, in which the furnace was maintained isothermal during the region of mass change. This greatly increased the amount of time needed to complete a TG curve; some runs required from 10 to 14 hours to reach a maximum temperature of 1000°C. Honda was an exceedingly modest man and he concluded his paper by stating: "All of the results given are not altogether original; the present investigation with the thermobalance has however revealed the exact positions of the change of structure and also the velocity of the change in respective temperatures. The investigation also shows the great convenience of using such a balance in similar investigations in chemistry."

Honda's thermobalance was taken over by his student, Saito, who during the 1920s made numerous improvements as well as elucidating the effect of experimental variables on the resulting curves. Saito's work, as well as that of other Japanese workers, has been summarized (*8,9*).

The French school of workers was started by Guichard, who in 1923 began a series of studies "directed principally to the realization of a linear elevation of temperature with respect to time" (*10*). His pioneering efforts in this field are summarized by the statement: "The method of studying chemical systems through their changes in weight when subjected to temperatures varied in a regular fashion may be widely employed, provided certain precautions are taken. It will be usefully applied to certain operations of quantitative analysis" (*11,12*).

Although a number of thermobalances were described following the pioneering efforts of Guichard, the thermobalance developed by Chevenard and his co-workers in 1936 had the greatest influence on the French school and others. Chevenard was a metallurgist who was interested in the oxidation behavior of stainless steels, which had just been developed in the factory where he worked. In order to withstand the constant vibrations common in steel mills, he adopted a bifunicular suspension for the balance beam which was similar to that found in the Kelvin electrodynamometer. Also, to avoid air convection currents, he used an inverted furnace geometry. Because of World War II, publication of his paper was delayed for six years, finally appearing in print in 1941 (Paper 27). The photographically recording model of the thermobalance become commercially available in France in 1945.

The man who had the greatest impact on modern analytical thermogravimetry was Duval, who with the recently developed Chevenard thermobalance studied the thermal properties of over 1000 analytical precipitates as well as the development of an automatic method of analysis based on this technique (Paper 28). He recalls that on a visit to Ludwigshafen on the Rhine in 1946, he was impressed by the automatic execution of organic combustion analyses based on the method developed by Zimmermann. From that time on, he was never free of the idea of developing rapid, automatic inorganic analyses which would require no preliminary separation and whose results would be given in graphical form, as is done with an infrared spectrum. He was also concerned with writers who were so meticulous in specifying precipitation conditions and stated little if nothing about the subsequent treatment of the precipitate beyond the simple direction of "heat and bring to constant weight." Other loose phrases, such as "heat without exceeding dull red"; "ignite in such a fashion that the bottom of the crucible becomes red"; and so on, were entirely inexcusable.

The rapid development of thermogravimetry in the late 1950s and early 1960s was due in large part to the availability of good-quality thermobalances. This is similar to the development of differential thermal analysis, discussed in Part I. Commercial instrumentation permitted the thermal analyst to concentrate on the applications of the technique rather than building thermobalances, a task that he was not especially qualified to do. In 1960, Gordon and Campbell (Paper 31) reviewed the state of the art for automatic and recording balances and noted the increased activity of instrument manufacturers in this field. At that time there were 18 commercially available thermobalances—six were manufactured in the United States, one each in England and Switzerland, three in France, and seven in Western Germany.

Many of the commercial thermobalances were built around a readily available microbalance such as the Cahn Electrobalance. This automatic recording vacuum ultramicrobalance, based on an elastic ribbon–suspension torque motor, was first described by Cahn and Schultz in 1962 (Paper 33). It had a sensitivity of 0.1 μg and permitted a precision of 1 part in 10,000 of the mass change. Since it could be enclosed in a glass system, TG curves could be obtained easily in controlled atmospheres or at low pressures.

Peters and Wiedemann in 1959 (Paper 30) described a high-quality thermobalance from which the sophisticated Mettler Thermoanalyzer I was to develop (Wiedemann, 1964). Next to the Derivatograph developed by Paulik et al. (*13*), this is perhaps the most widely used thermobalance in the world. An early model of the thermobalance is illustrated in Figure 1.

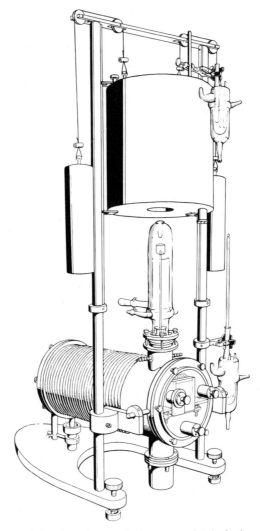

Figure 1 Early model of a thermobalance, which led to the development of the Mettler Thermobalance. (*Courtesy of H. G. Wiedemann.*)

A look into the future is provided by the automation of the TG technique by Bradley and Wendlandt in 1971 (Paper 36). The instrument not only permitted automatic control of the heating rate and furnace atmosphere but sample changing also. It provided operator-free operation after the eight samples were loaded into the sample changer.

Perhaps the most comprehensive and complete guide to the interpretation of TG curves were the papers by Newkirk in 1960 (Paper 32) and Simons and Newkirk in 1964 (Paper 34). Newkirk pointed out the errors that were frequently overlooked in thermogravimetry, such as changing air buoyancy and convection, temperature measurement, and the effects of furnace atmosphere, heating rate, and other parameters. He stated that these limitations seem to indicate that the principal value of TG is in the determination of approximate quantitative behavior of a system and not, by itself, for the determination of precise constants. In a later paper, they pointed out that the behavior of $CaC_2O_4 \cdot H_2O$ under controlled conditions in a thermobalance could provide an unusually versatile reference substance to guide the interpretation of TG measurements.

Although nonisothermal kinetics methods were known since the 1920s, attention was refocused on this branch of kinetics by Freeman and Carroll in 1958 (Paper 29). This was one year later than the application of DTA to kinetics by Borchardt and Daniels (see Paper 13) who had described a nonisothermal method for kinetics determinations in solution. Freeman and Carroll's method consisted of a simple graphical evaluation of the activation energy and order of reaction from a single TG curve.

REFERENCES

1. F. Szabadvary, *History of Analytical Chemistry*, Pergamon Press, Oxford, 1966, p. 16.
2. R. Vieweg, in *Progress in Vacuum Microbalance Techniques*, Vol. 1, Th. Gast and E. Robens, eds., Heyden & Son Ltd., London, 1972, p. 1.
3. W. Nernst and E. H. Riesenfeld, *Ber.*, **36,** 2086 (1903).
4. O. Brill, *Z. Anorg. Chem.*, **45,** 275 (1905).
5. P. Truchot, *Rev. Chim. Pure Appl.*, **10,** 2 (1907).
6. G. Urbain and C. Boulanger, *Compt. Rend.*, **154,** 347 (1912).
7. C. Duval, *Inorganic Thermogravimetric Analysis*, Elsevier, Amsterdam, 1963, p. 4.
8. H. Saito, in *Thermal Analysis*, R. F. Schwenker and P. D. Garn, eds., Academic, New York, 1969, p. 11.
9. H. Saito, *Thermobalance Analysis*, Technical Books Pub. Co., Tokyo, 1962.
10. M. Guichard, *Bull. Soc. Chim. Fr.*, **33,** 258 (1923).
11. M. Guichard, *Bull. Soc. Chim. Fr.*, **2,** 539 (1935).
12. Reference 7, p. 7.
13. F. Paulik, J. Paulik, and L. Erdey, *Z. Anal. Chem.*, **160,** 241 (1958).

26

Reprinted from *Sci. Rep. Tohoku Imp. Univ.*, Ser. 1, **4**, 97–103 (1915)

On a Thermobalance.

BY

KÔTARÔ HONDA.

With two plates.

---

1. The ordinary method of following a chemical change taking place in a compound at high temperatures is to heat the compound to various temperatures and every time, it is cooled to the room temperature, to measure its weight by means of an ordinary balance. As the process is very troublesome, it is highly desirable to measure, if possible, the change of weight at high temperatures without cooling to the room temperature. For this purpose, the author has constructed a thermobalance, which admits us to follow continuously the change in its weight at gradually varying temperatures. The following is a brief account of the instrument:—

Fig. 1. Elevation.

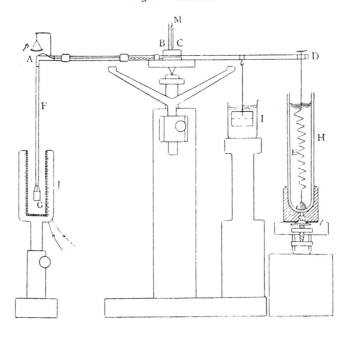

Fig. 2. Plan.

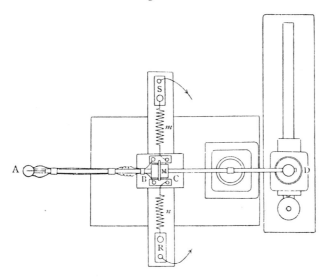

AB, CD, Figs. 1, 2, are the two arms of the balance made of silica glass, whose expansion-coefficient is negligibly small. They are fixed to a wooden block BC, which rests on an agate plane with an agate knife-edge. This block also holds a vertical mirror M to reflect the image of a vertical scale into an observing telescope. At one end A of the arm, a thin porcelain tube is vertically fixed. A short cylindrical vessel made of platinum or magnesia hangs from the lower end of this tube by means of three platinum wires; the specimen to be tested is placed in it. From the other end D of the arm CD, a very weak steel spiral E is stretched vertically into the bottom of a Dewar vessel H filled with oil. This arrangement protects the rapid change of temperature in the spiral, and also prevents the oscillations of the beam during the heating. The Dewar vessel can slowly be raised or lowered by means of a lever system (Fig. 3), the amount of which can be read on the screw head l, as in the case of a spherometer. Since the damping of the spiral alone is not sufficient, another damper I is suspended from the arm CD; it has a cylindrical form with a partition in its middle part.

To measure the temperature of the specimen, one junction of a platinum platinum-rhodium element is suspended into the cylindrical vessel

G through the porcelain tube. The element extends from G to B, there to form the cold junction with the interposition of lead wires of copper. These wires are stretched from BC to the screws R and S in the form of very weak spirals *m* and *n*; the screws are then electrically connected with a galvanometer of Siemens & Halske for the measurement of the thermocurrent. The temperature of the cold junction is read by a mercury thermometer, and the corresponding correction always applied. The calibration of the thermoelement is effected in the usual way by the melting points of pure metals.

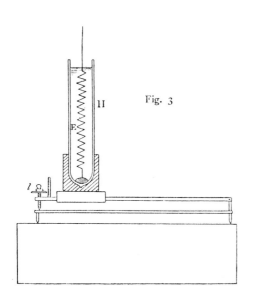

Fig. 3

The heating is made by an electric current. On a thick porcelain tube 2.5 cm. in internal diameter and 15 cm. long, a platinum wire of 0.4 mm. in diameter is wound noninductively at a rate of four turns per centimeter; this coil is then thickly covered with caolin and then with asbestus paper. To improve the uniformity of temperature along the heating coil, a thick tube of nickel 12 cm. long, which fits inside the coil, is placed in it. The heating coil J can be vertically raised or lowered by means of a screw motion.

2. The measurement was carried on in the following way:—The specimen weighing about 0.6 grams was put in the vessel G; the beam was then adjusted to the horizontal position by raising or lowering the Dewar vessel, and the zero position of the beam was read by means of the scale and telescope. The sensibility of the instrument was then determined, by placing a weight of 1 centigram in a small pan at one end of the arm AB and observing the deflection of the image of the scale in the telescope. With a scale-distance of about 1 meter, the deflection amounted to

about 17 mm, so that a deflection of 1.7 mm. corresponds to a change of weight of 1 mg. The accuracy of the instrument was therefore nearly the same as that of an ordinary balance; but the sensibility can further be increased, if desired.   A gradually increasing current was then passed through the heating coil, and the change of deflection of the scale was constantly observed.   If the deflection increased more than about 20 mm, a suitable weight was placed on the small pan *p* at the end A of the arm AB to bring back the image of the scale nearly to its original position, or the Dewar vessel was raised or lowered by a suitable amount. The further change of deflection was then observed, and so on.   This plan also permitted us to calculate the sensibility at different stages of heating; the sensibility was found to be nearly constant.   Since the velocity of a chemical change is usually very small, the heating must be done very slowly.   When a change began to take place, the temperature was kept quite constant, till the weight of the substance ceased to change.   This required usually from 1 to 4 hours.   The whole time of heating up to 1000°C lasted from 10 to 14 hours.   As most of the chemical change was irreversible in a condition as investigated by myself, the change of the weight during the cooling was not usually observable, and therefore the cooling to the room temperature was effected in one or two hours.

Notwithstanding the damping arrangement above described, the small oscillation of the beam due to the convection current in the heating coil was not entirely absent at high temperatures above 300°C.   But as the heating was very slow, the small oscillation of the image of the scale did not cause any sensible error in determining its mean position.   The effect of the convection current is equivalent to a small diminution of weight and almost proportional to the temperature.   This effect can easily be seen from the inclination of the deflection-temperature curve to the axis of temperature, when there is no chemical change.

If a greater accuracy is required, a null method can be used.   The deflection of the scale is always compensated for by raising or lowering h: Dewar vessel by means of the screw motion, the amount being read on the screw-head.   The calibration of the reading can easily be done, by placing a weight of 1 or 2 cgr. in the pan *p* and by turning the screw head to bring back the image of the scale to its zero position.

3. In what follows, some of the results of my experiments are given to show the working of the instrument. In all figures, the increase of the ordinates means a decrease in the weight of the substance, and the origin of the ordinates is quite arbitrary.

(a) Manganous sulphate $Mn\,SO_4.\,4H_2O$: Pl. I, Fig. 1.

Fig. 1 shows that the water of crystallisation is given off at two different ranges of temperature, that is, three molecules in 70°–110°C and the remaining one molecule in 230°–260°C. The mean velocities of separation are respectively 0.347 gr. and 0.0762 gr. per hour, the total initial weight being 0.629 gr. From 280°C upward, the sulphate exists quite free from the water of crystallisation, till at about 820°C, it begins to decompose and change into Mn_3O_4. This rate of decomposition always increases with temperature, and terminates at about 950°C. The mean velocity of the change between 900°–950° is 0.328 gr. per hour. The following table contains the calculated and observed results accompanying these changes:

	Separation of 3 H_2O	Separation of H_2O	Decomposition of $Mn\,SO_4$ into Mn_3O_4
Cal. weight	0.154 gr.	0.050 gr.	0.210 gr.
Obs. weight	0.156	0.051	0.200

The coincidence between the calculated and observed weights is thus very satisfactory. During the cooling, no change of weight was observed.

The slight inclination of the curve, which is observable in the figure, when there is no chemical change, shows the effect of convection-current during the heating or cooling.

(b) Calcium sulphate $CaSO_4.\,2H_2O$: Pl. I, Fig. 2.

The water of crystallisation begins to escape at about 130°C and the whole is given off at about 260°C. The mean velocity of separation is about 0.0992 gr. per hour. The theoretical and observed change of weight are respectively 0.112 gr. and 0.116 gr., which coincide well with each other. The initial weight was 0.535 gr.

(c) Calcium carbonate $CaCO_3$: Pl. II, Fig, 3.

By heating, the compounds begins gradually to decompose and change into CaO at about 500°C. The rate of decomposition always increases with the rise of temperature, and at 820°C, it attains a maximum. The decomposition terminates at about 880°C. The calculated and observed change of weight coincide fairly well with each other, being respectively 0.201 gr. and 0.196 gr. for the intial weight of 0.455 gr.

(d) Chromic anhydride CrO_3: Pl. II, Fig. 4.

The change of structure of chromic acid at high temperatures was formerly investigated by Mr. T. Soné and the present author.[1] By measuring the magnetic susceptibility of chromic anhydride at different high temperatures, the following changes were found :—

$$Cr\,O_3 \xrightarrow{280°} Cr_6O_{15} (\rightarrow Cr_5O_9) \xrightarrow{420°} Cr_2O_3 .$$

Between 300° and 400°, the strong paramagnetic compound Cr_6O_{15} and the ferromagnetic one Cr_5O_9 exist in mixture. Above 500°C, all the substance is transformed into chromic oxide Cr_2O_3.

The present investigation with the thermobalance also confirmed the above results. Fig. 4 shows two distinct diminutions in weight at 300° and 435°C, and the two others in 80°–100°C and 330°–420°C, which are rather gradual. The first change of weight in 80°–100°C is due to moisture and the second to the decomposition $CrO_3 \rightarrow Cr_6O_{15}$; the third gradual change is probably the effect of two partial decompositions $Cr_6O_{15} \rightarrow Cr_5O_9 \rightarrow Cr_2O_3$, and the fourth is due to the decomposition $Cr_5O_9 \rightarrow Cr_2O_3$. The total initial weight was 0.520 gr. The following table contains the observed result and that calculated according to the above view :—

	Separation of moisture	$CrO_3 \rightarrow Cr_6O_{15}$	$Cr_5O_9 \rightarrow Cr_2O_3$
Cal. weight	—	0.040	0.080
Obs. weight	0.018	0.038 (*ab*)	0.091 (*cd*)

(1) Sci. Rep. **3**, p. 223, 1914.

The quantitative coincidence between the calculated and observed results is not so good as in the other cases, probably because of the double decompositions. The further heating of the substance up to 1000°C did not cause any further change in the structure of the chromic oxide.

All the results above given are not altogether original; the present investigation with the thermobalance has however revealed the exact positions of the change of structure and also the velocity of the change in respective temperatures. The investigation also shows the great convenience of using such a balance in similar investigations in chemistry.

In conclusion, I wish to express my hearty thanks to Messrs. T. Matsushita, T. Takayanagi, M. Yamada, S. Konno and T. Hongô, who were so kind as to take observations with the balance.

Pl. I.

Fig. 1. $Mn SO_4. 4H_2 O.$

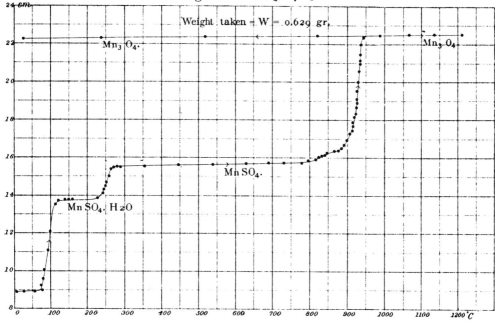

Fig. 2. $Ca SO_4. 2H_2 O.$

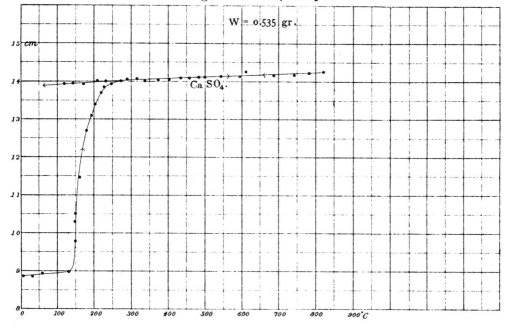

Pl. II.

Fig. 3. Ca CO₃.

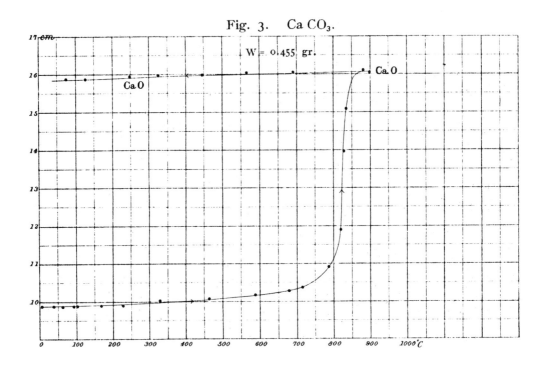

Fig. 4. Cr O₃.

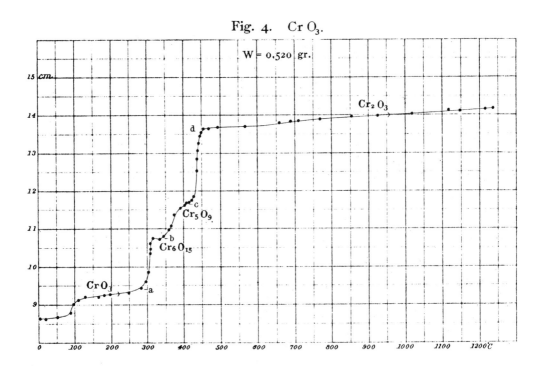

27

Reprinted from *Bull. Soc. Chim. Fr.*, **11**, 41–47 (1944)

ÉTUDE DE LA CORROSION SÈCHE DES MÉTAUX AU MOYEN D'UNE THERMOBALANCE

P. Chévenard, X. Waché, et R. de la Tullaye

Pour étudier la corrosion sèche des métaux et des alliages, les auteurs ont réalisé une *thermobalance* qui enregistre sur papier sensible, en fonction du temps, le gain de poids d'un échantillon maintenu à température stationnaire dans l'air, dans un gaz ou dans une vapeur corrosive.

En prévision d'une longue série d'expériences, la construction a été étudiée de manière à obtenir un appareil sensible, fidèle, robuste et d'emploi commode. Le couteau est remplacé par une suspension funiculaire; des amortisseurs à huile rendent l'appareil insensible aux trépidations, etc... Les courbes enregistrées permettent des pointés à 0,2 mg près. Elles sont assez nettes pour supporter le tracé des *dérivées*, élaboration qui fournit la vitesse d'oxydation pour toute valeur du temps et accroît sensiblement le pouvoir de résolution des diagrammes.

Utilisée pour étudier l'oxydation du nickel, l'appareil a permis de retrouver en gros les résultats classiques : en particulier la loi en $e^{-\frac{A}{RT}}$ a été vérifiée avec une excellente précision. Mais l'élaboration des graphiques a montré que les courbes « oxydation-temps » du nickel ne sont pas exactement des paraboles.

Les résultats sont autrement complexes dans le cas des alliages inoxydables au nickel-chrome : l'allure des courbes dépend au plus haut point de l'état de surface de l'échantillon et de l'allure du chauffage. Les auteurs espèrent élucider l'influence de ces facteurs à l'aide de la thermobalance.

La corrosion sèche des métaux et des alliages, c'est-à-dire l'attaque par les gaz et par les vapeurs à différentes températures, offre un double aspect aux yeux du métallurgiste.

Le forgeage et le laminage à chaud des lingots et des billettes, le recuit des fils ou des feuilles aigris par un travail à froid, la chauffe des aciers pour la trempe ou le revenu entraînent des *pertes au feu*. Il y a un intérêt évident à les réduire, d'autant plus qu'elles ne se bornent pas au métal transformé en oxyde. Il s'y ajoute le métal altéré, c'est-à-dire décarburé en surface, contaminé par l'oxygène ou par le soufre et qu'il faudra écrouter à la meule ou à l'outil.

Un gain notable peut être obtenu, dans quelques cas, en agissant sur l'atmosphère du four. Par exemple, le recuit « blanc » des fils et des feuillards, entre les passages à la filière ou au laminoir à froid, supprime presque le décapage chimique : il en résulte une économie d'acides, on évite de jeter à la rivière quelque 5 0/0 du métal élaboré et on s'épargne bien des contestations avec le voisinage.

Un deuxième problème est de réaliser des alliages dits *réfractaires* ou *inoxydables à chaud*, c'est-à-dire susceptibles de se conserver longtemps quand ils sont maintenus à haute température dans l'air ou dans les gaz corrosifs. Les applications en sont innombrables : fils et rubans pour résistances électriques de chauffage, supports pour tubes de chaudières, soles et chaînes convoyeuses pour fours de recuit, récipients de synthèse sous pression, tubes de récupérateurs pour fours à chaleur régénérée, etc. Tantôt ces pièces sont homogènes, l'alliage étant anobli dans sa masse par des additions convenables : nickel, chrome, aluminium, silicium, zirconium, etc.; tantôt leur résistance chimique n'est développée qu'à la surface par un placage, une cémentation métallique, etc.

Comme on le sait depuis longtemps, et notamment depuis les belles mises au point de la question faites par M. Portevin, par M. Chaudron et ses élèves, etc., les alliages inoxydables doivent leur résistance, non pas à une indifférence chimique, privilège des métaux nobles, mais à la protection d'une *pellicule d'oxyde* continue, adhérente, et quasi imperméable. Il est donc important d'étudier la formation de cette pellicule, et d'en caractériser les qualités protectrices. La thermobalance qui va être décrite a été créée dans ce but : trois exemplaires sont en service dans notre laboratoire de recherches installé à Imphy.

I. — *Méthodes d'étude.*

La première idée qui vient à l'esprit est de suivre, en fonction du temps, la variation de masse d'un échantillon porté à température stationnaire dans un réactif gazeux, dont la pression et la composition demeurent constantes. De fait, cette méthode très simple a été la plus usitée. Les observateurs ont mesuré, tantôt le gain de poids de l'échantillon : Utida et Saito, Pilling et Bedworth, Dunn, Krupkowski, Valensi, etc., tantôt la perte après enlèvement mécanique ou chimique de la pellicule formée : von Schwartz, Ziegler, Dickenson, Jominy et Murphy, etc.

Une autre méthode, plus sensible mais plus délicate, consiste à mesurer le volume de gaz combiné. La pression dans le tube-laboratoire est maintenue constante et on évalue la contraction : par exemple, MM. Portevin, Prétet et Jolivet (1) manœuvrent un réservoir à mercure dont le déplacement est proportionnel à la quantité de gaz absorbé ; il serait facile de réaliser une manœuvre automatique et de rendre l'appareil enregistreur. On peut aussi, comme l'a fait un de nos anciens collaborateurs M. Keitzer, maintenir la pression constante en faisant croître la température du gaz contenu dans un récipient auxiliaire calorifugé : cette élévation de température est enregistrée au moyen d'un pyromètre du type classique.

Bien entendu, on associe à ces techniques d'autres méthodes d'investigation. La transformation progressive d'un fil métallique en oxyde peut être caractérisée par les changements de la résistance électrique, ou par les variations de température, si le fil est chauffé électriquement sous tension constante. La diffusion mutuelle du métal et de l'oxyde est étudiée par micrographie et par macrographie. La composition moyenne de la couche oxydée est du ressort de l'analyse chimique, tandis que la spectrographie par rayons X a permis à M. Chaudron et à ses élèves de caractériser les variations de cette composition suivant l'épaisseur de la pellicule. La méthode par diffraction d'électrons convient à l'examen de fines pellicules d'oxyde, trop minces pour donner des couleurs d'interférence.

Pour nos recherches et nos travaux de contrôle industriel, nous avons adopté la première méthode dont la mise en œuvre est très simple, et dont les résultats sont relativement faciles à interpréter.

II. — *Thermobalance.*

Avant d'établir le projet d'un appareil, il faut discuter les données du problème. Dans le cas des alliages dits « inoxydables », c'est-à-dire qui se recouvrent d'une pellicule oxydée protectrice, les gains de poids tendent plus ou moins rapidement vers une limite relativement petite : aussi l'appareil doit-il être *sensible.*

Les facteurs du phénomène sont multiples : composition chimique du métal ; composition, température et pression de l'atmosphère corrosive ; états de surface et forme des échantillons, etc. Il faut donc prévoir de nombreux essais, et les résultats récoltés pendant plusieurs années doivent former un ensemble cohérent : la *fidélité* sera une qualité indispensable à la thermobalance.

Pour être significatifs, pour traduire en particulier les modifications plus ou moins intenses et plus ou moins tardives de la pellicule oxydée, les observations doivent être sans lacune, et les essais de longue durée : 3 à 7 jours. D'où nécessité d'un appareil *automatique.* D'ailleurs, l'enregistrement du gain de poids offre un autre avantage que l'économie de temps et de main-d'œuvre. Si les courbes sont assez fines et assez vigoureuses pour être susceptibles d'une élaboration graphique précise, par la méthode des dérivés par exemple, les moindres singularités dans la marche de l'oxydation seront mises en évidence : la méthode aura un haut *pouvoir de résolution.*

Enfin, la thermobalance va fonctionner dans un laboratoire industriel. Elle subira des trépidations et des chocs ; elle sera exposée à des champs magnétiques parasites et à des variations parfois importantes de la température extérieure ; elle sera confiée à des manipulateurs d'une formation scientifique sommaire, etc. Il faut donc construire un appareil *robuste.*

Sans prétendre à l'originalité, chose difficile pour un appareil aussi simple qu'une thermobalance, l'appareil qui va être décrit nous paraît satisfaire aux conditions énoncées. La fidélité est assurée par la suppression de tout frottement solide, par l'utilisation de matériaux amagnétiques et peu dilatables, par l'emploi d'un régulateur thermostatique précis, par une construction soignée ; la sensibilité est obtenue par l'enregistrement photographique de la masse de gaz fixée ; le pouvoir de résolution des diagrammes résulte d'un bon réglage de l'amortissement et de la finesse des courbes enregistrées.

Description. — L'éprouvette est soit une mince plaquette de 60×17 mm de côté, soit une série d'épingles à cheveux obtenues en partageant un fil de 120 cm de longueur et de 0,5 mm de diamètre ; la surface est égale à 20,4 cm² dans le premier cas, et à 18,9 cm² dans le second. L'emploi d'un fil complique les calculs quand on veut tenir compte de la réduction progressive de la surface « métal-oxyde » (2). Mais l'élaboration quantitative des diagrammes enregistrés est le plus souvent inutile dans l'étude des alliages « inoxydables » industriels, dont l'oxydation obéit à des lois compliquées. Un fil est non seulement plus facile à préparer qu'une plaquette, mais il offre une surface plus grande pour un moindre poids. En outre, le départ des gaz au cours de la chauffe déforme les épingles à cheveux, et on recueille ainsi une indication complémentaire intéressante. Pour ces raisons, l'éprouvette filiforme est la plus usitée.

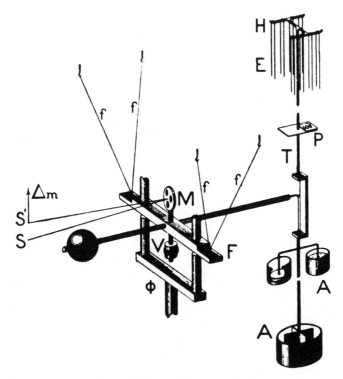

Fig. 1. — Schéma de la thermobalance.

L'éprouvette E (fig. 1), plaquette ou fil, est suspendue à une herse H en silice fondue, portée par une hampe de silice T convenablement lestée. La herse est engagée dans une cloche ouverte vers le bas, contenue dans un four à régulateur. Un tube accédant au sommet de la cloche y conduit le gaz corrosif quand l'essai n'est pas effectué dans l'air. Pour le moment, on opère à la pression atmosphérique et, si on veut réduire la pression partielle du gaz actif, on le dilue par de l'argon. Une modification est à l'étude pour travailler à des pressions différentes de l'ambiante.

Dans notre appareil (fig. 2 et 3), le four est placé au-dessus de la balance. Cette disposition complique un peu la construction, mais elle épargne au fléau l'effet perturbateur des courants d'air ascendants. Ce fléau F, en duralumin et en silice fondue, est léger et non magnétique; la longueur du bras est insensible aux changements de la température. Il est porté par deux paires ff, ff de minces rubans

[*Editors' Note:* Figures 2 and 3 have been omitted because the original halftones were unavailable.]

d'invar. Cette suspension funiculaire est préférable au classique couteau d'agate : les frottements solides sont exclus, de même que les déplacements sous l'effet de trépidation ; la construction est peu coûteuse, le réglage aisé et la fidélité parfaite.

C'est également un ruban flexible qui suspend l'équipage à l'extrémité du fléau. Des amortisseurs A à huile fluide s'opposent à toute oscillation parasite. Le gaz fixé par l'échantillon force le fléau à s'incliner d'un angle proportionnel au gain de poids Δm. Cette déviation amplifiée par le miroir M, est enregistrée sur un papier sensible porté par un tambour de chronographe. Une vis-contrepoids V permet de régler la sensibilité, qu'on vérifie de temps à autre au moyen d'une masse marquée placée sur le plateau P : 100 mg correspondent à 50 mm environ.

Le four électrique, muni d'un régulateur à fil dilatable, assure à 1° *près* la constance et l'uniformité de la température, jusqu'à 1100°. L'hélice de chauffe est enroulée en bifilaire pour éliminer le champ magnétique du courant et les pièces métalliques sont reliées à la terre pour annuler les attractions électrostatiques. La netteté et la régularité des courbes enregistrées (fig. 4, 7, 8 et 9) prouvent l'efficacité de ces précautions.

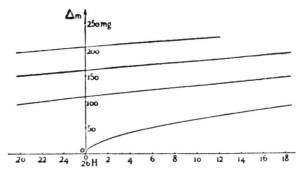

Fig. 4. — Courbe « oxydation-temps » d'une feuille de nickel chauffée à 950° dans l'air.

Applications de la thermobalance.

1° *Etude de l'oxydation du nickel.* — L'oxydation d'une feuille de nickel obéit à des lois simples et connues : l'étude qui va être résumée a été entreprise pour vérifier le bon fonctionnement de notre appareil.

Chauffé au contact de l'air, le nickel donne un oxyde moins dense que lui et se recouvre par conséquent d'une pellicule continue. Pour se combiner, l'oxygène et le métal doivent diffuser à l'intérieur de cette pellicule, et la vitesse de cette diffusion conditionne la vitesse de la réaction. L'hypothèse la plus simple, formulée

en 1923 par Pilling et Bedworth, est de supposer cette vitesse à l'instant $t : \dfrac{d\frac{m}{s}}{dt}$

réciproque à l'épaisseur, c'est-à-dire à la masse Δm de l'oxygène déjà combiné :

$$\frac{d\frac{m}{S}}{dt} = \frac{K_1}{2 \gimel m}$$

S étant la surface de contact métal-oxyde, constante dans le cas d'une plaquette.

D'où :

$$\left(\frac{\Delta m}{S}\right)^2 = Kt$$

D'après Dunn (1926), la réaction, considérée du point de vue de l'activation, s'accélère avec la température selon la loi :

$$\frac{d \, L K}{d \, T} = \frac{A}{RT^2}$$

D'où :

$$L K = L C - \frac{A}{R T} \; ; \; K = C e^{-\frac{A}{RT}} \; et \; \frac{\Delta m}{S} = \sqrt{Ct}.e^{-\frac{A}{2 \, RT}}$$

R est la constante des gaz parfaits. La chaleur d'activation A est exprimée en cal/mol et la constante $\sqrt{C}$ en g. cm^{-2} min$^{-\frac{1}{2}}$.

D'après cette théorie, les courbes (Δm, t) seraient des paraboles : le moyen le plus simple de le vérifier est de transformer les courbes enregistrées par *anamorphose logarithmique*. La figure 5 montre le résultat de cette élaboration pour cinq courbes obtenues avec des plaquettes de nickel commercial, étudiées à la thermobalance entre 850° et 1050°. On obtient bien des droites, mais un peu. plus inclinées que l'oblique OA dont le coefficient angulaire est 1/2 : les pentes de ces cinq droites s'échelonnent entre 0,514 et 0,535.

Un semblable résultat a été retrouvé avec des feuilles et des fils de nickel de provenances variées : on a obtenu des paraboles transcendantes à exposant un peu supérieur à $\frac{1}{2}$. Peut-être l'allure de la réaction, légèrement plus rapide que ne le voudrait la loi de Pilling et Bedworth, a-t-elle pour cause la superposition de deux phénomènes : à l'oxydation superficielle par tranches plates, prévue par la théorie, s'ajoute en effet l'oxydation en profondeur, pénétrant par les joints de grains.

Quant à la loi en $e^{-\frac{A}{RT}}$, elle se trouve vérifiée de manière excellente comme le prouve le diagramme (Log K, $\frac{1}{T}$) de la figure 6.

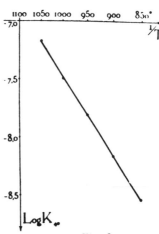

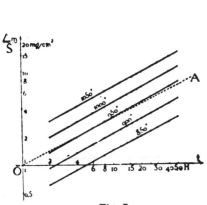

Fig. 5.

Fig. 6.

FIG. 5. — Anamorphose logarithmique des courbes « oxydation-temps » du nickel, enregistrées à différentes températures.
Si c des courbes étaient exactement des paraboles, les droites résultant de leur anamorphose seraient parallèles à OA, droite de coefficient angulaire 1/2.

FIG. 6. — Variation linéaire du logarithme $L K_{40}$ relatif au nickel en fonction de l'inverse 1/T de la température absolue.
K_{40} est la constante de l'oxydation mesurée au bout de 40 heures.

Le tableau suivant rapproche les valeurs de A et de $\sqrt{C}$ données par Krupkowski (3) de celles qui résultent de nos propres expériences.

Nickel étudié	A cal/mol	$\sqrt{C}$ g.cm⁻².mm⁻¹
Nickel commercial étudié à Imphy à l'état de feuilles C = 0,10; Si = 0,08; Mn = 0,19; Fe = 0,20; Cu = 0,14............................	47.220	2,27
Nickel Wiggin à 99,5 Ni étudié à Imphy à l'état de fil ...	45.500	1,35
	44.150	1,27
Nickel Mond refondu, tréfilé et étudié à Imphy à l'état de fil C = 0,01; Si = traces; Mn = 0,16; Fe = 0,5	39.500	0,96
Nickel étudié par Krupkowski	43.420	0,95
Nickel étudié par Pilling et Bedworth, résultats calculés par Krupkowski	45.700	2,04

D'un échantillon de nickel à l'autre, les variations de A et de $\sqrt{C}$ dépendent de la pureté du métal. D'ailleurs, nous avons cru constater une corrélation entre ces grandeurs et le coefficient de thermorésistivité, paramètre rapidement affecté par les éléments incorporés; nous comptons revenir sur la possibilité de contrôler la pureté du nickel d'après la marche de l'oxydation. Pour le moment, nous nous bornerons à souligner le bon fonctionnement de la thermobalance, prouvé par la netteté parfaite des courbes enregistrées, par la reproductibilité des résultats, par la cohérence des données recueillies à des températures différentes.

2° *Oxydation des alliages fer-nickel-chrome.* — Les choses se compliquent dès que l'alliage renferme plusieurs éléments dont les oxydes concourent à former la couche protectrice. La proportion de ces oxydes varie suivant l'épaisseur de la pellicule et cette répartition dépend de toutes les conditions de l'expérience. Un fil d'alliage nickel-chrome pour résistances chauffantes, recuit sous le vide, se colore en vert par formation d'oxyde de chrome aux dépens du chrome et de l'oxygène du métal; le résultat est le même si le recuit a lieu dans une atmosphère d'argon impur. Ainsi, sous faible pression partielle d'oxygène, l'oxyde qui tend à dominer est celui du métal le plus oxydable, en l'espèce le chrome.

Dès la température ordinaire, les ferronickels chromés sont recouverts d'une pellicule protectrice, variable selon la manière dont on a usiné et nettoyé l'échantillon. Cette première pellicule oppose un obstacle à l'action chimique de l'air chaud quand l'échantillon est introduit dans le four. La courbe $(\Delta m, t)$ s'élève d'abord lentement, puis il se produit une accélération au delà de laquelle on observe parfois une allure parabolique.

Tel est le cas, par exemple, pour un ferronickel chromé relativement riche en nickel : 35 0/0, et pauvre en chrome : 10 0/0 (fig. 7). Les échantillons identiques

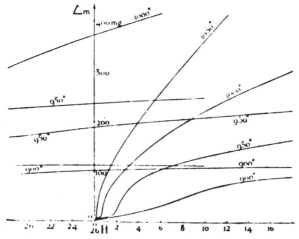

Fɪɢ. 7. — Courbes « oxydation-temps » d'un ferronickel chromé à 35 0/0 Ni et 10 0/0 Cr, enregistrées à différentes températures.

provenaient d'une feuille de 0,5 mm d'épaisseur, laminée à froid, nettoyée au papier émeri fin, et dégraissée à l'éther. On voit la double inflexion rétrograder vers l'origine au fur et à mesure que la température s'élève. La pellicule finale, relativement riche en nickel et pauvre en chrome, demeure assez perméable à l'oxygène pour permettre la diffusion : c'est pourquoi les courbes affectent une allure parabolique une fois franchie la singularité initiale.

Si l'alliage est riche en chrome et pauvre en nickel, la couche d'oxyde devient rapidement quasi imperméable, et la vitesse d'oxydation diminue bien plus vite que ne le voudrait la loi parabolique. Ainsi, après 20 heures de chauffe à 1000° dans l'air, la courbe d'un alliage inoxydable du type 18/8 (fig. 8) se confond presque avec une horizontale. La transformation de la pellicule initiale ne paraît pas s'effectuer en tous les points à la fois.

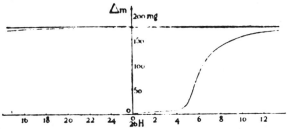

Fig. 8. — Courbe « oxydation-temps » d'un alliage inoxydable 18/8 enregistrée à 1000°

Si tel est le rôle de la première pellicule, on peut prévoir que la *préparation de la surface et l'allure de la chauffe* jouent des rôles essentiels dans la marche de l'oxydation sèche. Les trois courbes de la figure 9 le montrent clairement. Elles ont été obtenues avec des fils d'un alliage à 20 0/0 Ni et 20 0/0 Cr étudiés à 1050° à l'air.

La courbe 1 concerne des fils nettoyés au papier émeri fin, dégraissés au trichloroéthylène, et introduits dans le four chaud. Les fils auxquels correspond la courbe 3 ont subi la même préparation, mais ils ont été introduits dans le four froid et chauffés progressivement. Enfin, les fils qui ont donné la courbe 2 ont été décapés dans l'acide chlorhydrique pur à 60°, soigneusement rincés, séchés et introduits dans le four chaud.

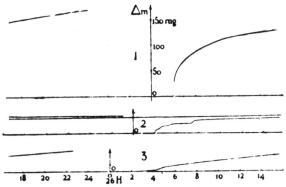

Fig. 9. — Courbe « oxydation-temps » enregistrée à 1050° avec des fils d'alliages à 20 0/0 N et 20 0/0 Cr, ayant subi des préparations différentes ou chauffés à des allures différentes!

L'extrême discordance de ces trois courbes souligne la complexité des phénomènes : l'oxydation par tranches plates s'ajoute en effet à la pénétration d'oxygène par les joints des grains, et de toute évidence cette dernière action dépend beaucoup du mode de décapage.

De longues recherches sont encore nécessaires pour éclairer le difficile et important problème de la formation d'une pellicule protectrice à la surface des alliages inoxydables au nickel et au chrome : le programme des essais est tracé, et nous espérons que la thermobalance, qui s'est révélée un appareil sensible, fidèle et pratique, nous permettra de le mener à bien.

(1) PORTEVIN, PRETET et JOLIVET, *Revue de métallurgie*, 1934, p. 101. « Méthodes d'études de la corrosion des métaux et alliages par les gaz à température élevée et leurs applications ». On trouvera à la fin de ce mémoire (p. 235) une bibliographie très complète de la corrosion sèche.
(2) G. VALENSI, « Cinétique de l'oxydation des fils métalliques ». *C. R.*, 1935, **201**, 602.
(3) A. KRUPKOWSKI et J. JASZCZUROWSKI, « Vitesse d'oxydation des métaux à température élevée ». Congrès international des Mines, de la Métallurgie et de la Géologie appliquée Paris, octobre 1935.

28

Reprinted from *Anal. Chem., 23*(9), 1271–1286 (1951)

Continuous Weighing in Analytical Chemistry

CLÉMENT DUVAL

The Sorbonne, Paris, France

With the collaboration of Pr. Jean Besson, Pierre Champ, Monique De Clercq, Thérèse Dupuis, Raymonde Duval, Pierre Fauconnier, Yvette Marin, Josette Morandat, Pr. André Morette, Simonne Panchout, Simonne Peltier, Janine Stachtchenko, Suzanne Tribalat, and Nguyen Dat Xuong

Translated by RALPH E. OESPER, *University of Cincinnati, Cincinnati, Ohio*

By registering, photographically or by pen, the curve representing the loss or gain in weight undergone by a heated substance, important facts are brought to light, which are not revealed by the classic method of discontinuous weighing. In this way it is possible to determine gravimetrically many materials, either alone or in mixture, by an automatic process which involves no personal error in the weighing. The new method lends itself to numerous determinations of humidity and ash (grains, plaster, paper pulp, etc.). By this means it is possible to set up a hierarchy among the methods in use; this order of excellence can usually be established in a short time, and the results will be recorded in an indisputable permanent form. The greatest advance realized through the thermobalance is the fixing of the exact limits of temperature between which a given precipitate acquires a constant weight, a condition indispensable to the making of a sure weighing.

NOTABLE advances in analytical chemistry, especially in gravimetry, can be realised with the aid of the Chevenard thermobalance. This instrument automatically traces on photographic or ordinary paper a curve showing as a function of the temperature or time the gain or loss in weight of a material while it is being heated.

The pyrolysis curves of almost a thousand precipitates (described in the bibliography) have been discussed and interpreted. On these curves, the horizontals indicating constant weight and marked out by precise temperatures, and registered by the apparatus itself, show what temperature domain the chemist should finally employ to be sure of obtaining a weight that, if not constant, is nevertheless correct.

New compounds have been discovered by this method, and it has also yielded about 80 new methods of determination. It has drawn attention to numerous errors of interpretation, aided in the selection of methods, furnished sublimation curves and diagrams of isotherms, disclosed new methods of separation, made possible a critical study of all the inorganic microgravimetric procedures and a comparison of the precipitants that yield colloidal hydroxides, followed the progress of the washing of precipitates, and estimated the quantity of adsorbed materials and the temperature at which they are desorbed. Finally, the methods of automatic determination have now been selected for all the ions, after a patient and fastidious study of the choice of filtering crucibles. It is possible to make some determinations on mixtures without previous separation. In any case, the analyst, who does not run the risk of making a false weighing, has a permanent record on which the measurement of the weight is reduced to a length, a measurement which, if need be, may be made by the operator long after the record is actually made.

A glance at the current reviews of analytical chemistry, or simply at the semimonthly "Analytical Chemistry" section of *Chemical Abstracts*, will show that gravimetry, at least in inorganic chemistry, is being superseded by methods which seemingly are more rapid, such as titrimetry, colorimetry, polarography, and spectrography. These branches of analytical chemistry have made wonderful progress in the past 20 years and it may well be asked whether the funnel, the paper filter, the crucible, and the balance with its knife-edges and discontinuous weighings, heritages of a long tradition, are not destined to disappear soon, and give way to other more modern devices capable of equal precision.

While visiting Ludwigshafen on the Rhine in 1946, the author was impressed by the automatic execution of organic combustion analyses by the methods that had recently been published by W.

Zimmermann. From that time, he was never free of the idea of developing rapid, automatic gravimetric mineral analyses, which would require no preliminary separations, and whose results would be given by graphical inscription, as is done for infrared absorption spectra. Fortunately, the Chevenard thermobalance had progressed beyond the trial stage, but nonetheless a tremendous amount of work was required before the new methodology was adapted to routine analyses that could be made by a technician. Three years of hard work went into these preparatory labors. This paper gives an account of certain facts, diverse in nature, which were encountered in the course of the daily experiences with our thermobalances, and which should convince the reader of the incontestable advantages afforded by continuous weighing.

APPARATUS

The thermoponderal method is so simple that many investigators have thought it could be carried out with a more or less worn analytical balance of the usual precision, by modifying the instrument somewhat—e.g., by drilling a hole in the base of the balance; one of the pans is replaced by a rod extending into a heating device placed below, etc. These easy solutions had numerous defects, the following being especially bad. The rod, which extends into the furnace, rubs if the orifice is too narrow, or it is subject to convection currents if the opening is too large; the effect becomes evident, particularly above 600° C. Moreover, the traditional knife-edge and agate plate of a balance can no longer be recommended if there is continuous contact between them, instead of the intermittent contact employed in the ordinary weighing procedure. The vibrations—so often an unavoidable problem in industrial laboratories—dull the knife-edges. The bifunicular suspension, which involves no friction between solids, is far superior to a knife-edge in providing an invariant axis of orientation. The bell-shaped furnace, closed at the top and just covering the substance, which is placed above the balance, makes the convection currents inoffensive. The bifilar winding, in the furnace, also contributes to removing any action of the field on heated magnetic materials.

Figure 1 is a diagram of the thermobalance, as developed by Chevenard, Waché, and De la Tullaye (6), and used in the author's studies. He has dealt particularly with improvements and details that are of special interest to the problem of using this instrument in chemical analysis.

The nonmagnetic beam, *f*, 23 cm. long and constructed of Duralumin, is supported by two sets of tungsten wire (0.02 mm.

in diameter). As may be seen in Figure 3, one end of the beam carries a perfectly vertical silica rod, F, which is well ballasted. The material being studied or determined is contained in a suitable crucible, which is placed at top of the rod. The other end of the beam is fitted with a counterpoise, Cp, and carries a concave mirror, M, which catches the well diaphragmed incident ray sent out from a small automobile light bulb. The reflected ray strikes a photographic paper (24 × 30 cm.) carefully rolled on a cylinder which is revolved at a uniform rate about a perfectly vertical axis by means of a clockwork or synchronous motor.

The temperatures are measured to within 1° (within 0.5° below 250°) by means of a platinum-rhodium plated platinum couple, whose hot junction, housed in a silica sheath, reaches to the level of the material being heated.

The calibration in weights is made by placing an overload of 50 mg., consisting of an aluminum wire analogous to a microbalance weight, on a small platform fastened to the silica rod. In all recordings, a gain or loss of 50 mg. in ordinates corresponds to a distance of 25 mm. on the paper after drying.

Figure 2 is a photograph of the balance, properly speaking, surmounted by the furnace capable of reaching about 1100° C., which is generally sufficient for any analytical problem. (Only once was a temperature above 1035° required—namely, to obtain zinc oxide free of carbonate.) The furnace is provided, at its base, with a recent device, which enables the furnace to function in a desired cycle, to keep it at a constant temperature, etc. The balance is connected with the case carrying the photographic film holder containing the drum.

The auxiliary equipment is not shown: a 110/8-volt transformer, a 0–7-ampere ammeter, a 0–110-volt voltmeter, the relays permitting the conversion of the furnace into a thermostat, the automatic rheostat, and the temperature-recording device. In the latest model of the thermobalance, the photographic housing is removable and the record is made on ordinary paper with a pen and by usual technique.

Figure 3 is a photographic reproduction of the base of the thermobalance, with its protective parts removed for the sake of clarity. Sometimes it may be of interest to heat a precipitate in some gas other than air (operations in a Rose crucible) or to replace the oxygen by nitrogen in order to gain an insight into the singularities of a curve.

Figure 4 is a diagram of the silica sheath which fits the center of the furnace. The hydrogen, for example, is introduced from above at a rate that is strictly regulated; a silica baffle above the

substance being heated effectively prevents the production of any parasitic convection currents.

METHOD OF OPERATION

These thermobalances have made it possible to record the thermolysis curves from 20° to 1000° (provided the material did not explode previously) of 933 precipitates which had been proposed, up to July 14, 1950, for use in inorganic gravimetry. These precipitates have been prepared in strict accord with the directions given by their respective sponsors. In every case, the material was slightly impregnated with the wash liquid before the heating was begun. On the other hand, for reasons which will appear later, the author has avoided keeping them in filter paper or in contact with paper pulp or asbestos.

Figure 2. Photograph of Balance and Casing

Actual length 1.32 meters
Bo. Cover of noncorrosive alloy carrying sheath of hot junction
C₁. Leveling screws
Bu. Shoulder check for furnace
Ce. Flange protecting beam
D₁. Metal cover protecting oil of damping device against dust
Ce₁, Ce₂. Vertical metallic guides for furnace
L. Metallic tongue closing slot of *D*
Ma. Metallic sleeve permitting adjustment of balance and casing
P. Silica rod supporting crucible (another silica rod and porcelain crucible are on table)
R. Metallic platform supporting photographic chassis
So. Cold source
Tu₁, Tu₂. Metal sheaths protecting damping device, extremity of beam, and silica rod

The detailed results derived from all these curves have been or probably will be published in *Analytica Chimica Acta*.

At first, no electrolytic deposits were included, but in response to several requests the study was extended, to thallium, lead, nickel, copper, cobalt, cadmium, zinc, gold, silver, and mercury.

The ordinates of the curves represent the variations in weight, and the abscissas either the time, or more frequently, the temperatures. If, then, a part of the curve runs parallel to the lower

Figure 1. Diagram of Thermobalance

b. Balance
Ca. Photographic housing
Ch. Chassis containing rotating cylinder
Ce. Columns supporting furnace
f. Beam
h. Furnace, movable vertically, range 20° to 1100° C.
t. Device for regulating rate of heating furnace

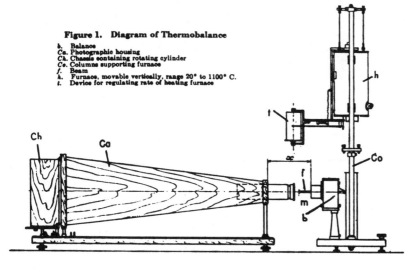

edge of the sheet, the occurrence of this horizontal may be taken as evidence of constant weight. The author accepts this conclusion without question, but not all chemists share his opinion.

When the extremities of the horizontal are known within 1°, it is apparent that the literature of analytical chemistry is being enriched with many data. Admittedly, the results are reproducible only when the material is not explosive; in this case the observed explosion temperature falls when the rate of heating is increased.

From the beginning, it has been possible to separate the methods worthy of consideration from those of little or no value; to quote an apt expression by F. Feigl: "I have written—it seems—the chapter on the pathological anatomy of analytical chemistry." The author has been especially successful in correcting certain formulas and drying temperatures that have been suggested in the textbooks—for example, magnesium pyrophosphate reaches a constant weight from 477° on; therefore it is useless to maintain it for 3 hours at the full temperature of the Meker burner or the blast lamp.

The author has always wondered why writers are so meticulous in specifying precipitation conditions (volume of the liquid to be determined, maximum concentration, temperature, volume and titer of the reagent, pH of the milieu, length of time the precipitate should age, grade of the filtering crucible or ashless paper, preparation and volume of the wash liquid), while frequently even the larger classic texts state nothing about the subsequent treatment of the precipitate, beyond a simple direction "heat and bring to constant weight," or such loose phrases as "heat without exceeding dull red," "ignite in such fashion that the bottom of the crucible becomes red," or "heat with the full flame of a Meker burner."

What was excusable in the time of Berzelius or Rose should no longer be tolerated after Le Chatelier invented the thermoelectric couple some 70 years ago. This lack of progress is extremely regrettable, since in this respect, analytical chemistry, which has made such remarkable advances in the domain of aqueous systems, voluntarily preserves its empirical character for which it has been so much ridiculed by the "pure chemists."

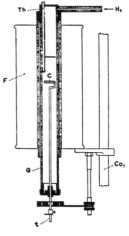

Figure 4. Diagram of Silica Sheath Installed in Center of Furnace

F. Electric furnace
Th. Cavity for housing hot junction of thermocouple
t. Silica rod ending in ring *C* (crucible carrier)
Co₁. Column supporting furnace (see Figures 2 and 3)

PRECISION

Precision is a function almost entirely of the thickness of the line that constitutes the curve. Two kinds of paper have been used, both with mat surface. One paper, very fast, especially designed for recording, registers the slightest details of the losses or gains in weight, particularly at the instant of sudden decompositions—e.g., nitro iderivatives, oximes, cupferronates, picrates; the other paper s much slower and serves for automatic determinations when the exposure time is much longer. The inscribed record is barely visible but very fine; the mean thickness of the line is 0.1 mm., which corresponds to about 0.2 mg. As the paper is 240 mm. high, which is equivalent to a maximum gain or loss of 480 mg., it is apparent that the relative error in a "weighing" is 4×10^{-4} at the minimum. The author's experience has been that this error is between 3×10^{-3} and 4×10^{-4}. It is recommended that the photographic paper be allowed to dry spontaneously, and that the recording be started in such manner that a given line extends toward the two edges, right and left, of the sheet. The other precautions regarding the use of the thermobalance and its installation are described in the pamphlets published by the manufacturer (Société de Commentry, Fourchambault et Decazeville, 84 Rue de Lille, Paris).

STUDY OF PRECIPITATES TO FIX HEATING LIMITS

Example I. The author has tried unsuccessfully to use the precipitate of the triple periodate, which supposedly has the formula $KLiFeIO_6$, for the determination of lithium (44), inasmuch as the factor for this ion is extremely favorable and the precipitation occurs in the presence of other alkali metal ions, if they are not present in too great excess. Actually, the precipitate [which is used in titrimetry (97) and in spot test analysis (51)] is not suitable for gravimetric purposes because it does not have a constant composition, and its pyrolysis curve (Figure 5) shows a continuous descent. The excess of potassium which it would need to contain in order to exist gives a variable weight in the residue above 947° C.

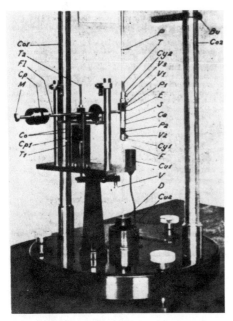

Figure 3. Base of Thermobalance

Actual diameter, 30 cm.
Bu. Support for furnace
Ca. Framework supporting rod and supported by wire *S*
Co₁, Co₂. Columns supporting furnace
Cu¹, Cu². Cups containing two dampers and some oil
D. Excentric permitting vertical adjustment of silica rod
Cy¹, Cy². Adjusting screw of tungsten wire, 0.02 mm. diameter permitting articulation between beam and rod
V. Set screw of balance
V₁, V₂. Heads of *Cy₁, Cy₂*
P. Rod supporting upper damper
S. Tungsten wire
P₁, P₂. Clamps holding wire *S*
E. End of beam
V. Screw holding collar of silica rod
T. Brass sleeve of silica rod
P. Platform carrying weights for calibration
Fl. Beam
M. Mirror
Cp. Counterpoise
T₁, T₂. Adjusting rods
Co. Column supporting balance
Cm. Movable member serving for transporting balance. Two tungsten wires supporting beam are not visible

Example II. Wenger, Cimerman, and Corbas (*109*) proposed the determination of cobalt with the aid of anthranilic acid. They recommend that the precipitate be dried at 120° to 130°C. The curve obtained with this anthranilate (Figure 6) indicates the loss of moisture up to 108°C; then a horizontal begins and extends to 290°C. (The loss on 201.5 mg. between the extremities of this horizontal is not even 0.1 mg.) Accordingly, the temperatures given by these authors are correct. The destruction of the organic matter is indicated by the change in direction at 403°C. Above 609°C, we have the equally good horizontal which pertains to the oxide, Co_3O_4 (*57*). A very simple calculation, which starts with weights of this latter oxide, shows that the weights of the anthranilate (measured on the paper in millimeters) agree within at least 0.1%. We are dealing here with one of the better gravimetric methods.

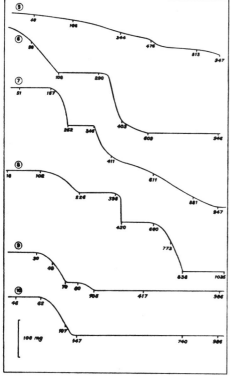

Figures 5 to 10. Thermolysis Curves

　　　5.　Iron potassium lithium periodate
　　　6.　Cobalt anthranilate
　　　7.　Uranium oxinate
　　　8.　Calcium oxalate
　　　9.　Artificial calcium sulfate
　　10.　Gypsum

Example III. Another type of curve is given by the most precise method (in the author's opinion) for determining hexavalent uranium. The technique adopted was that of Hecht and Donau (*65*) or Claassen and Visser (*7*). While it is well known that these workers produced a precipitate carrying one molecule of oxine of crystallization, the thermogravimetric curve (Figure 7) of this precipitate revealed a new fact. The usual complex involved in this analysis, $UO_2(C_9H_6ON)_2.C_9H_7ON$, is stable up to 157°; it is wetted so little by water that it dries almost instantly, especially if the final washing is with alcohol and ether. Hence the temperatures between 110° and 140°, suggested hitherto for drying this material, are correct. If now, the heating is carried beyond 157°, the extra molecule of oxine sublimes, and a new horizontal begins at 252° and extends to 346°, where a sharp decomposition occurs. The compound, corresponding to this horizontal, conforms strictly to the formula $UO_2(C_9H_6ON)_2$.

Complete carbonization is rather difficult; it is not accomplished at 1000° and there should be no thought of weighing the oxide at this temperature because decomposition will have set in. Moreover, the advantage of the favorable analytical factor will be lost. Among all the oxinates studied (28 in all), that of uranium is the only one that exhibits this behavior, and the presence of this horizontal between 252° and 346° should permit the determination of the uranium in a mixture containing other metal oxinates (*59*).

Example IV. Figure 8 presents (*87*) the more complicated conditions attending the heating of hydrated calcium oxalate, $CaC_2O_4.H_2O$, which has been precipitated and washed in accord with accepted analytical procedures. This curve is very instructive (*87*). Because of its four parallel horizontals, this material is used to adjust the thermobalances after they have been set up, cleaned, or repaired.

From room temperature to 100° we have the domain of the monohydrate, and the curve tends slightly downward, the actual slope depending on how much "moisture" is held by the precipitate. If one begins with a sample that has been allowed to stand in the open air for several days, this branch of the curve is perfectly horizontal. The precipitate loses its bound water between 100° and 226°. Accordingly, in the microdetermination of calcium it is not advisable to dry the hydrated precipitate at 105° because there is danger of low results. A second perfectly horizontal section is observed from 226° to 398°; it corresponds to the anhydrous calcium oxalate. This provides a second method for determining calcium, a method which is too seldom used. At 420° the anhydrous oxalate suddenly loses carbon monoxide in accord with the equation: $CaC_2O_4 \rightarrow CO + CaCO_3$, which on the photographic paper is easily verified with a precision of 0.1%.

A third horizontal, absolutely parallel to the preceding, extends from 420° to 660°. It was with this section that the writer successfully accomplished the first automatic determination (discussed later) and also the laboratory exercise for determining the atomic weight of carbon. We have here a third form of the gravimetric determination of calcium. The familiar dissociation of calcium carbonate begins at this point: $CaCO_3 \rightarrow CaO + CO_2$. It ends, under the conditions of our method of heating (open crucible, free escape of the carbonic gas), around 840°. Finally, starting at 840°, we have the horizontal corresponding to quicklime. The heating was stopped when the temperature reached 1025°.

Among all the metal oxalates, the only ones capable of yielding the corresponding oxide via a stable carbonate are those of calcium, strontium, and barium. In other words, they lose their carbon monoxide and dioxide in separate stages.

STUDIES IN PURE INORGANIC CHEMISTRY

In the course of this study, which was intended to be essentially analytical, some purely inorganic facts were noted, which were simply touched on in earlier publications, as the author hoped to study them in detail when time permitted.

Gypsum and Calcium Sulfate. When precipitated calcium sulfate, $CaSO_4.2H_2O$, is heated, 1.5 moles of water come off up to 70°, while from 70° to 80° there is a short horizontal corresponding rigorously to the appearance of the hemihydrate. The last half-molecule of water then disappears between 82° and 105° (Figure 9). The presence of this short horizontal was somewhat surprising (*87*) in view of the previous studies of this matter, and that is the reason for also recording the curve for gypsum (Figure 10). The specimen used was of high purity; like the precipitate of Figure 9, it was ground to pass a 100-mesh sieve, and the materials gave identical Debye and Scherrer spectra. Figure 10 therefore shows no discontinuity in the course of the dehydration, which agrees with a well-known study by Jolibois and Lefèvre (*69*). The loss of 1.5 H_2O is indicated on this curve at exactly 107°. Furthermore, this is the temperature previously reported by van't Hoff, whereas the dilatometric method showed that this transformation occurs at 107.2°.

Rammelsberg Reaction. This name is given to the passage of an alkali earth iodate into the corresponding paraperiodate (*94*)—for example, $5Ca(IO_3)_2 \rightarrow Ca_5(IO_6)_2 + 4I_2 + 9O_2$. For the first time, the temperatures at which this reaction occurs have been determined precisely (*89*). The course of the pyrolysis of calcium iodate is given in Figure 11:

Ca salt. **Between 550° and 887°**; very rapid between 680°
and 750°
Sr salt. **Between 600° and 748°**; very rapid between 650°
and 700°
Ba salt. **Between 476° and 720°**; very rapid at 610°

The other iodates which were heated do not seem to yield the paraperiodate.

Oxidation of Platinum Sponge. Ammonium chloroplatinate, $(NH_4)_2PtCl_6$, begins to decompose at 181° (*54*). The reaction, which gives rise to isomeric platinodichlorodiamines, proceeds slowly up to 276°, and then accelerates until it becomes almost explosive. The decomposition is complete at 407°. A constant weight of platinum is obtained up to 538° (Figure 12); then the metal takes up oxygen, doubtless coating itself with PtO; the maximum weight is attained at 607°. Thereafter, the oxide decomposes and the initial weight of the metal is regained at 811°. These variations in weight (which may amount to 5 mg. in 123 mg.) have probably not always been taken into account when the molecular weights of amines were being determined by the classical chloroplatinate method. The phenomenon evidently does not occur in hydrogen. It is necessary to take due note of these changes when the reaction is being followed in a Gooch-Neubauer crucible.

Uranium "Peroxide." When uranyl nitrate is treated with 12-volume hydrogen peroxide, the resulting precipitate is usually given the formula $UO_4.2H_2O$, which is correct from the stand-

point of percentage composition, but incorrect when written in this form (*40*). Reference to Figure 13 shows that up to 90°, the water of imbibition is given off readily, and only at this temperature are we dealing with a compound of the formula just given. In the interval 90° to 180°, it loses precisely one molecule of hydrogen peroxide. Only then does the residue have the formula $UO_3.H_2O$ (uranic acid). This slowly gives up a molecule of water, to produce between 560° and 672° the horizontal pertaining to UO_3. In its turn, this oxide is stable between 800 and 946°, which is the useful interval with respect to the analyst.

These findings therefore indicate that the formula should be written $UO_3.H_2O_2.H_2O$. This point of view obviously does away with the notion of the octavalence of uranium (which is no longer accepted). It is rather surprising to discover that a compound containing a molecule of hydrogen peroxide is relatively stable in the vicinity of 100°. This finding is also justified physically. The measurement by Lecomte and Freymann, and those which the writer has repeated with Lecomte, show, without doubt, that the "peroxide," as a powder, gives an infrared absorption spectrum between 6 and 15 microns, which greatly resembles that of dry or moist UO_3, and exhibits, in particular, a very strong infrared absorption band at 920 cm.$^{-1}$, which is characteristic of the UO_2 group and all the so-called uranyl salts which contain this radical. The compounds of the type XO_4 have an entirely different spectrum.

Silver Chromate. As shown by Figure 14, this salt is stable and dry from 92° to 812° (*22*). It gives a perfect horizontal between these temperatures, and consequently there is no need for drying it at precisely 135° as is directed by some writers. From 812° to 945°, each molecule of silver chromate loses exactly one molecule of oxygen, leaving a mixture of silver and its chromite, a reaction which has not been reported before. It can be written: $2Ag_2CrO_4 \rightarrow 2O_2 + 2Ag + Ag_2Cr_2O_4$. The silver is easily dissolved out by nitric acid, leaving the green chromite, which is not affected by the acid.

Chromium Hydroxide and Chromous Acid. When ammonium hydroxide is added, with no special care, to the solution of a chromic salt, no definite product results. When a very slow current of air, charged with ammonia gas, is passed into a water solution of chrome alum that has not previously been treated with ammonium salts, a crystalline precipitate of chromic hydroxide starts to appear at pH 5.5. The passage of the gas is stopped when the pH reaches 11, and the precipitate is filtered off at once and washed. The curve (Figure 15) obtained from this product is interesting on account of the inflection to the horizontal tangent between 440° and 475°. This inflection corresponds exactly to the formula $Cr(OH)_3$. To the authors' knowledge, this hydroxide has never been so well characterized with respect to its purity. This part of the curve is rather too long to permit the recommendation that chromium be weighed in this form.

If, however, a chromic salt is treated with potassium cyanate or if a chromate is allowed to react with thiosemicarbazide (*22*), a granular product results, which is readily filtered. This produces the curve (Figure 16) on which there is, between 320° and 370°, a horizontal conforming to the composition $Cr_2O_3.H_2O$ or $H_2Cr_2O_4$, which doubtless is chromous acid—i.e., the acid from which chromites are derived. (It gives the same infrared absorption spectrum as oxalic acid, $H_2C_2O_4$.)

The Question of Arsenic Anhydride. There is some lack of agreement as to the dehydration temperature of orthoarsenic acid. However, this oxide might serve as the basis of a possible method of determining arsenic, when the element or one of its derivatives is treated with fuming nitric acid—in other words, if arsenic acid is heated. Although Bäckström gives 435° to 450° for the limits of existence of the anhydride As_2O_5, and Victor Auger gives 400° as the upper limit, the curve shown in Figure 17 shows clearly that for a compound of the initial approximate composition $As_2O_5.3/2H_2O$, the anhydride is obtained at 193° and begins to break down into arsenic trioxide and oxygen at 246° (*30*).

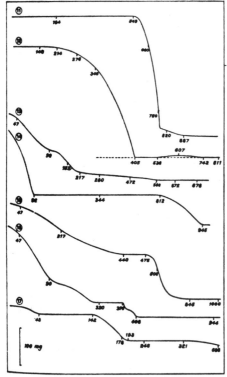

Figures 11 to 17. Thermolysis Curves
11. Calcium iodate
12. Platinum sponge
13. Uranium "peroxide"
14. Silver chromate
15. Chromium hydroxide by ammonia
16. Chromium hydroxide by thiosemicarbazide
17. Arsenic acid

Suboxide of Bismuth. Inorganic chemists have often argued the question whether bismuth exhibits a valence below 3. By following the directions of Mawrow and Muthmann (*76*) the author obtained, from bismuth oxychloride and 50% hypophosphorous acid (specific gravity 1.27), a black precipitate, which filtered rather poorly and showed a marked tendency to assume the colloidal condition. In that case it is necessary to heat the filtrate to reprecipitate the product. The curve (Figure 18) has a horizontal from 92° to 191°, which indicates the existence of a material with the empirical formula Bi_2O. However, as the Debye and Scherrer spectrum demonstrated, we were not dealing with a definite compound, but with metallic bismuth, which carried adsorbed oxygen, and in fortuitous amounts. Above 230°, the mixture oxidized rapidly and yielded Bi_2O_3 quantitatively beyond 840°. Obviously, the determination of bismuth by this procedure is not recommended.

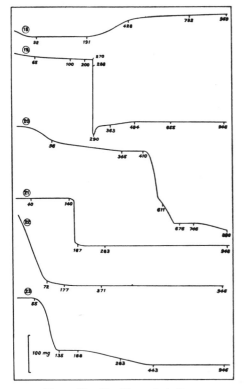

Figures 18 to 23.　Thermolysis Curves

　　18.　Colloidal bismuth
　　19.　Copper oxalate
　　20.　Cerium iodate
　　21.　"Reduced silver"
　　22.　Cobaltous acid
　　23.　Boric acid

Pyrolysis of Copper Oxalate. When a boiling, neutral solution of a cupric salt is treated with an oxalate, the precipitation is incomplete and the product is hard to filter. Its behavior toward heat is rather singular (*74, 75*). Figure 19 shows that, after dehydration, the anhydrous copper oxalate is stable from 100° to 200°. It decomposes abruptly at 288° but, contrary to all expectations, the pyrolysis does not proceed: $CuC_2O_4 \rightarrow CO + CO_2 + CuO$. The final product is cuprous oxide, doubtless be-

cause of the reducing action of the carbon monoxide. Consequently, the following equation is more in line with the facts: $2CuC_2O_4 \rightarrow CO + 3CO_2 + Cu_2O$. However, the cuprous oxide is not within its zone of stability and so it reoxidizes very rapidly. From 494° on, we have the black oxide, CuO (which, in turn, regenerates the red oxide above 1030°).

DISCOVERY OF NEW COMPOUNDS

Frequently a thermolysis curve includes a level portion which corresponds to a compound whose existence has not been recorded in the literature. Its molecular weight can be obtained by a simple calculation from the formula of the initial compound and that of the residue. Then, as the extreme temperatures of the horizontal are known, it is permissible to heat another sample to a temperature in the center, for instance, of the horizontal in question. The following illustrative examples have been selected from among the eighty cases studied.

Cerium Iodate. Chernikhov and Uspenkaya (*5*) proposed that cerium be precipitated as $2Ce(IO_3)_4.KIO_3.8H_2O$, which could be weighed after drying with ether. Actually, the pyrolysis curve (Figure 20) shows a steady fall, and then a sudden drop between 410° and 650°, which corresponds to a loss of oxygen and iodine (analogous to the case of the alkali earth iodates). However, this is probably not a Rammelsberg reaction. A mixture of potassium iodide and cerium metaperiodate, $Ce(IO_4)_4$, is present between 650° and 746°. The residue at 880° contains cerium peroxide. Though this precipitate cannot be used in gravimetry, its pyrolysis curve gives promise of initiating new studies in pure inorganic chemistry (*56*).

Reduced Silver. A more or less oxidizable deposit of metallic silver is obtained from a silver nitrate solution by electrolysis, or by the reducing action of ammoniacal cuprous chloride, hypophosphorous acid, cadmium, aluminum, hydroxylamine, vitamin C, etc. However, if we select the classic instance of the silvering of glass—i.e., the reduction by formaldehyde and 15% ammonia—no metallic silver is deposited in the cold. The black precipitate gives a curve (Figure 21) which, after a very short horizontal, has a slow fall from 40° to 160°, where sudden decomposition occurs. The constant weight for the silver is obtained from 500° on, and drying above this temperature is advisable when this method is used for quantitative purposes.

The original precipitate has attracted attention. It is known that the NH groups are specific in the detection of silver. Under the present experimental conditions, one molecule of ammonia combines with two molecules of formaldehyde, producing dihydroxymethylamine, $HOCH_2NHCH_2OH$. Replacement of the central hydrogen atom by silver leads to a precipitate, $HOCH_2NAgCH_2OH$, the so-called reduced silver (molecular weight 184), which has been decomposed by heat (*75*).

Cobaltous Acid. The thermolysis curve obtained from the product of the action of potassium persulfate or sodium hypochlorite on a cobaltous salt, gives no indication of a hydrated cobaltic oxide. The break in this curve (Figure 22) at 72° corresponds to $Co_2O_3.H_2O$ or $H_2Co_2O_4$, the parent acid of the cobaltites. This is analogous to chromous acid and is the only compound of which there is positive evidence on the graph. It subsequently loses water and changes into Co_2O_4 at 371°.

Metaboric Acid. When orthoboric acid is heated progressively, it retains the composition H_3BO_3 up to 55°. No part of the dehydration curve (Figure 23) indicates the existence of pyroboric acid, $H_2B_4O_7$. On the contrary, the horizontal between 135.5° and 168° represents the existence region of metaboric acid, HBO_2. This finding provided a simple method of isolating the latter in a pure state. The molecular weight indicated by the graph is 43.8 (calculated 43.828). Above 168°, the material loses water steadily, and the perfectly level portion of the curve corresponding to boric anhydride, B_2O_3, begins at 443° (*41*).

Metavanadic Acid. If slightly moist ammonium metavanadate is heated (*59*), a horizontal is obtained between 45° and 134° (Figure 24); this agrees closely to the composition NH_4VO_3. From 134° to 198°, the loss of weight corresponds to the elimination of all the ammonia, and the residue consists of metavanadic acid, HVO_3. This finding suggests a method for the preparation of this acid, but the latter is not stable. It begins to lose water at 206°, and the anhydride, V_2O_5, is obtained quantitatively at 448°. This oxide is very stable, up to 1000°, at least.

NEW FORMS OF GRAVIMETRIC DETERMINATION

The author has studied the pyrolysis curves of almost one thousand precipitates of interest to chemical analysis. Although

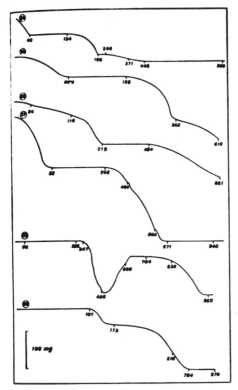

Figures 24 to 29. Thermolysis Curves

24. Ammonium metavanadate
25. Benzidinium sulfate
26. Cesium cobaltinitrite
27. Mercurous chromate
28. Cuprous thiocyanate
29. Iron neocupferronate

after this critical study he suggests that only about two hundred of these methods should be retained, he has also recommended the revival of certain procedures that have fallen into disuse, usually because their inventors did not indicate proper temperatures or because they destroyed the precipitate without ascertaining that it could be weighed to advantage. Of course, a systematic choice can be made only on the basis of continuous weighing. Up to the present, he has suggested 89 new methods.

It should likewise be noted, for the future, that if an investigator discovers a new gravimetric method, a simple experiment requiring 3 hours—and furnishing a supporting document—may save weeks of work. Two things must be determined. (1) If the curve given by the precipitate has no horizontal, the method is of no value; it is useless to continue with it, even though the precipitation is total and the precipitate does not run through the filter. Sometimes, it is possible to dissolve the precipitate in a suitable solvent and then finish the determination by some other gravimetric procedure or by a titrimetric or colorimetric method. (2) If there is at least one horizontal, all desiccations of the precipitate should be carried out at some stated temperature between the extremities of this horizontal. In general, curves have proved that in the ordinary laboratory practice the analysts are entirely too prone to overheat precipitates and to prolong the heating period unduly. Several new techniques are discussed here.

Benzidinium Sulfate. Sulfate is often determined by the procedure devised by Müller (80), in which the precipitated $C_{12}H_{12}N_2.H_2SO_4$ is finally titrated with standard base. The author wished to ascertain whether this precipitate was suitable for gravimetric purposes (32). The curve (Figure 25) has a good horizontal extending from 72° to 130°, which satisfactorily corresponds to the formula just given. From then on, the compound breaks down completely into water, sulfur trioxide, carbon dioxide, nitrogen, and carbon. Only the portion of the curve up to 610° is shown here.

Cesium Cobaltinitrite. In contrast to the analogous products given by potassium, rubidium, and thallium, the result of the action of sodium cobaltinitrite on cesium chloride contains no sodium; the precipitate conforms to the formula $Cs_2[Co(NO_2)_6].H_2O$. The water of crystallization is lost at 111° with no observable change in the slope of the curve (Figure 26). In other words, the loss of the molecule of water and of three NO_2 groups is not reflected by any change in the curve. The temperature 110° recommended for drying is accordingly correct, but the domain of existence of the anhydrous salt is extremely limited. For example, for an initial weight of precipitate of 451.89 mg., the loss at 110° is 10.80 (theoretical); at 127°, it amounts to 13.00 mg. From 219° to 494°, the curve has a good horizontal, demonstrating the existence of a compound or a mixture of definite compounds. Qualitative analysis of the residue that remains beyond this horizontal reveals the presence of CoO and cesium nitrate, with not even a trace of nitrite. The explanation of this new reaction follows clearly from the fact that the cobaltous oxide freely fixes oxygen of the air, either in the atomic form or in the form Co_3O_4. At the temperatures under consideration, the oxygen which is made available in this fashion would be capable of quantitatively converting the cesium nitrite into nitrate. For the analyst, the practical result is a new method of determining cesium. Destruction of the cobaltinitrite between 219° and 494° produces a mixture $CoO + 3CsNO_3$ with the "molecular weight" 659.7 and containing 60.44% of cesium.

If the contents of the crucible are treated with hot water after the operation and centrifuged, the resulting solution yields cesium nitrate in a high state of purity (55).

Mercurous Chromate. Various writers, including Treadwell advocate that the chromium content of solutions, which must contain only chromates and nitrates, be determined by precipitating the former as mercurous chromate, Hg_2CrO_4, which is then decomposed "under a well-drawing hood" and the chromium trioxide weighed. The author believes that it is useless to employ this method if no advantage is taken of the high atomic weight of mercury. In fact, the curve of mercurous chromate (Figure 27) shows that, after drying, the compound remains perfectly stable from 52° to 256°. The chromium trioxide, in its turn, appears from 671° on. Therefore, he proposes that the anhydrous mercurous chromate be used as the weighing form. This is produced between 52° and 256°: weight of chromium equals weight of precipitate × 0.100; hence the interest of the procedure for microgravimetry. The use of an efficient hood is thus rendered unnecessary (32).

Cuprous Thiocyanate. Cuprous thiocyanate, CuCNS, was precipitated in the cold, according to the directions of Kolthoff and van der Meene (71). In general, most authors do not advise direct weighing of the thiocyanate; they prefer to convert it into cupric oxide or cupric sulfate. However, the thermolysis curve (Figure 28) indicates that cuprous thiocyanate is stable up to 300°. The portion of the curve corresponding to this compound is admittedly not entirely horizontal, but rises slightly. The gain of oxygen is less than $1/_{100}$ in weight. Therefore, it is feasible to weigh the copper in this form of combination.

Above 300°, there is decomposition with release of sulfur and cyanogen, and a minimum, which satisfies the formula Cu_2S, is found at 440°. Going over to the curve given later (Figure 51) which refers to thiosulfate—i.e., to the oxidation into a mixture of CuO + CuSO₄—finally, around 950°, nothing but the black cupric oxide, CuO, is present (75).

Cupferronates and Neocupferronates of Iron and Copper. Since the classic studies by Baudisch, it has been known that cupferron (and neocupferron) precipitate a great many metals, that the resulting complexes are usually unstable, and when heated suitably they break down into the corresponding metal oxides.

The author was greatly surprised to find that the iron and copper cupferronates and neocupferronates are the most stable of all, and that they resist decomposition up to temperatures around 100°. This form obviously provides a more favorable gravimetric factor than the oxide for the determination of these two metals. Figure 29 refers to iron neocupferronate, which is stable (horizontal section) up to 101°. The organic matter decomposes and disappears from 101° to 754°; beyond 754° we have the horizontal corresponding to the sesquioxide (52). An easy separation of iron and titanium has been developed from these facts.

Determination of Germanium. When an excess of 5,7-dibromo-oxine hydrochloride is added to a hydrochloric acid solution of ammonium germanomolybdate on the boiling water bath, the germanium is quantitatively precipitated. The product has the composition $H_4[Ge(Mo_{12}O_{40})].(C_9H_4Br_2ON)_4.H_2O$ with a molecular weight of 3097, which corresponds to a gravimetric factor of 0.0339 for germanium (14).

The curve shown in Figure 30 traces the fate of this precipitate under the action of heat. The horizontal, observed up to 200°, agrees with the composition as just given. Next, the organic matter is given off up to 410°, and a second horizontal extends from 410° to 880°. This corresponds closely to a mixture GeO₂ + 12MoO₃. At higher temperatures molybdic anhydride is lost by sublimation.

Gallium Camphorate. The curve (Figure 31), traced with gallium camphorate, has made it possible to improve the final step of the ordinary determination (20). When the precipitate, $Ga_2[C_8H_{14}(CO_2)_2]_3$, is washed with ethanol, acetone, and ether, it dries rapidly and retains this composition up to 125°. It is sufficient to keep it between 94° and 110° for 30 minutes. The gravimetric factor is 0.189. This is the most precise method that the author has found for gallium; however, it cannot be used if iron or indium is present.

The decomposition of the organic matter begins somewhat above 125° and is complete at 478°, where the horizontal corresponding to Ga₂O₃ starts.

Determination of Gold in Thiophenolate. The forty gravimetric methods that have been proposed for the determination of gold can be divided as follows (4):

Citarin, for instance, precipitates metallic gold directly; the weight of the metal does not change, its thermal record is a straight line.

Pyrogallol, ferrous sulfate, sulfur dioxide, pyrocatechol, hydroxyhydroquinone, etc., precipitate very finely divided gold which, from room temperature up to 950° to 980°, takes up oxygen and releases it on cooling.

Certain reagents, such as resorcinol, hydrogen sulfide, etc., precipitate a mixture of gold and a more or less broken down compound; around 500° there remains only metallic gold, which does not oxidize.

Thiophenol, which is a class by itself, acts in a very special fashion (11). It precipitates gold quantitatively, as a single definite compound, which has been recognized as aurous thiophenolate, C₆H₅SAu. This white material, which turns yellow after prolonged exposure to light, dries at once. When heated, its weight remains constant up to 157° (Figure 32) and agrees within ¹/₃₀₀th with the composition just given. The residual gold appears above 187°, but it is better not to decompose the salt, because the unpleasant odor of burning rubber is avoided and advantage is taken of the more favorable gravimetric factor. Obviously, this finding should lead to a detailed study of the precipitation of gold by homologous thiophenols.

Critical Study of Some Precipitants of a Hydroxide. Metal hydroxides often come down in the colloidal condition, and they

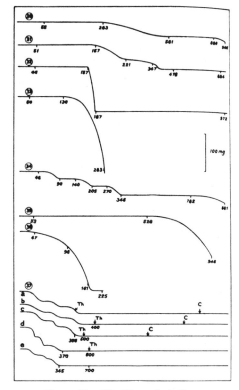

Figures 30 to 37. Thermolysis Curves

30.	Germanomolybdate of dibromo-oxine
31.	Gallium camphorate
32.	Aurous thiophenolate
33.	Mercurous chloride
34.	Ammonium molybdate
35.	Lead chloride
36.	Gallium oxinate
37.	Isotherms of ammonium molybdate
Th.	Placed in thermostat
C.	Obtaining of constant weight

must be aged, sometimes over a long period, before their particles become large enough to be retained by the filter. Continuous recording provides a means for judging the quantity of impurities adsorbed by the precipitate, and also reveals the minimum temperature at which the oxide reaches a constant weight. Table I gives a comparison of the efficacy of various precipitants for alumina (19).

Among all these methods, the author's preference goes to the use of ammonia gas and potassium cyanate.

Table I. Efficacy of Alumina Precipitants

Precipitation by	Minimum Temp., °C.	Precipitation by	Minimum Temp., °C.
Ammonia	1031	Ammonium sulfide	414
Ammonia gas	475	Ammonium carbonate	409
Urea	672	Ammonium bicarbonate	514
Urea succinate	611	Hydrazine carbonate	524
Mercuric amido chloride	676	Potassium cyanate	510
Urotropine	473	Ammonium nitrite	480
Pyridine	473	Sodium thiosulfate	675
Ammonium acetate	475	Potassium iodide-iodate	580
Ammonium formate	539	Sodium bisulfite	413
Ammonium succinate	509	Carbon dioxide	945
Ammonium benzoate	607	Bromine	380
Sodium salicylate	680	Tannin	808

He has made analogous critical studies of the precipitants for iron, copper, zirconium, and beryllium.

The following noteworthy fact applies to the precipitation of beryllium hydroxide, Be(OH)$_2$ (14). If the precipitate is brought down by means of a slow stream of air charged with ammonia vapors, the filtration may be begun soon. The precipitate does not run through the filter, and only 418° is required to obtain the beryllium oxide at constant weight. However, a study of the Debye and Scherrer spectra of the hydroxide precipitated in this manner reveals that it is not crystalline. Contrary to the accepted notion, excellent filterability of a precipitate does not depend on the perfection of its crystalline state but rather on the shape of its molecules.

SUBLIMATIONS

The author has come to the belief that vapor tension measurements tend to lose precision as the temperature rises, because he has found very great divergences in the temperatures at which sublimation begins, particularly of chlorides. The precision with which the thermobalance operates has made it possible to correct the figures reported in many cases.

Mercurous Chloride. The weight of this precipitated salt stays constant up to 130° (Figure 33). By progressive heating, the author arrived at the following losses of weight (in an open crucible and under atmospheric pressure), relative to an initial weight of 417.6 mg. of dry calomel (31).

t, °C.	130	167	283	410
Mg.	0	5.8	184.0	366.6

Molybdic Anhydride. This compound serves as the final product in numerous gravimetric determinations. Most authors agree that the precipitates which contain this oxide must be heated with caution, but their figures vary from 425° to 650° with respect to the maximum permissible ignition temperature. Recently, Pavelka and Zucchelli (85) claimed that molybdic anhydride begins to lose weight at 300°.

The anhydride derived from dissociation of ammonium molybdate shows a horizontal between 343° and 782°. In addition to the information furnished by Figure 34, the following losses suffered by 239.8 mg. of anhydride can be cited:

t, °C.	782	827	840	852	866	881
Mg.	0	0.98	2.94	4.90	8.82	17.64

Dupuis (13) has made a systematic study of the temperature at which this anhydride begins to volatilize for all possible cases of gravimetric determinations encountered in ordinary practice. The materials studied were either alone or mixed with such anhydrides as silica, phosphorus pentoxide, germanium dioxide, etc. In practice, it is necessary to count on a minimum temperature of 800°.

Lead Chloride. Figure 35 shows that this salt begins to lose weight toward 528° (42). Up to this temperature, there is a slight gain in weight (probable superficial formation of oxychloride). The losses, as measured on the record, sustained by 328 mg. of lead chloride are:

t, °C.	528	675	788	826	859	915	928	946
Mg.	0	4	30	51	80	180	220	292

The residue at 946° consists of chloride contaminated with a slight quantity of lead oxide, apparently due to incipient decomposition.

Gallium Oxinate. Several authors have proposed that gallium oxinate be weighed after drying between 110° and 150°. An examination of the curve (Figure 36) shows clearly that this complex compound loses weight from room temperature upward. No horizontal appears below 180°, and at this temperature the crucible is already full of carbon. The compound is manifestly sublimable and the region 110° to 150° corresponds precisely to the highest rate of decomposition. Therefore, the author is dubious about the feasibility of determining gallium in the form of its oxinate (20).

CONSTRUCTION OF A SYSTEM OF ISOTHERMS

As was pointed out earlier, it is possible at any time to use the furnace as a thermostat. This is why the thermobalance rendered such excellent service in reviewing the studies by Pavelka and Zucchelli (85)—i.e., in tracing the isotherms relating to 300°, 400°, 500°, 600°, and 700° with respect to the molybdic anhydride (obtained from ammonium molybdate). Each trial lasted at least 2 hours. In the loss of weight–time diagram it is easy to see, for all these temperatures (Figure 37, a, b, c, d, and e) with the ordinary precision of analysis, some straight lines parallel to the abscissas. The error calculated from the thickness of the lines traced on the photographic paper employed justifies the conclusion that a relative loss of weight of ^{1}/$_{500}$th would be easily observable. The ordinary weighing errors do not enter in here because the crucible is never taken out of the furnace. No variation in weight was observed even for the isotherm at 600°, where Pavelka and Zucchelli reported a maximum loss of 0.435% at the end of 2 hours (corresponding to a deflection of 1 mm. on the photographic paper).

If the isotherms of molybdic anhydride shown in a, b, c, d, and e are carefully examined, it is very important to note, with respect to chemical analysis, that their parallelism to the time axis is attained more quickly the higher the temperature (points C). This is apparent in the following tabulation:

Temp., ° C.	300	400	500	600
Time, minutes	90	60	45	0

For instance, a sample weighing 371 mg., which seemed to have reached constant weight from 360° on, would have to lose 0.6 mg. more to come to its theoretical weight. These losses in weight observed by these workers may have been due to this phenomenon which certainly is not appreciable on an ordinary pan balance (13).

SHOULD A PRECIPITATE BE DRIED OR IGNITED?

This is a question which has already consumed much paper and ink, because many writers have wished, according to their temperaments, to give a categorical answer. Some, such as Spacu and his associates, propose that the precipitate be washed with ether and then dried in a desiccator under reduced pressure. Others, like Winkler, prefer to use the temperature provided by their ovens (in the present case 132°) on all occasions, while others—e.g., Carnot, Treadwell, etc.—apparently are unable to finish a determination to their satisfaction without applying the heat of a Meker burner or blast lamp.

Continuous recording of the weight makes it possible for everybody to reach agreement and avoid ambiguity. For each precipitate there exists a drying or desiccation zone, whose location is furnished by the authors' curves and which must be respected if correct results are desired. The following six examples support this viewpoint.

Example I. According to Spacu and Dima (101), arsenates may be determined by precipitating TlAg$_2$AsO$_4$, which is weighed after drying in vacuo at room temperature. The curve, or rather the straight line, which constitutes Figure 38 shows a constant weight from 20° to 846°. In one run, the author (30) found initial and final weights of 360.73 and 360.05 mg., respectively. Hence, any temperature between 20° and 846° may be selected.

Example II. Figure 39 presents the decomposition of calcium picrolonate, Ca(C$_{10}$H$_7$O$_5$N$_4$)$_2$.7H$_2$O (and not 8 H$_2$O). This salt is ordinarily dried by means of a dust-free air current. Sometimes the method cannot be used in the tropics; in fact, the curve reveals that the decomposition of calcium picrolonate begins slightly above 30° (87). Thorium picrolonate is incomparably more stable.

Example III. Spacu and Spacu (103) recommend that mercurous iodate be weighed after drying in vacuo at room temperature, while Gentry and Sherrington (*63*) advocate heating at 140° for an hour. The curve (Figure 40) shows that this salt is stable up to 175°, as it gives a perfect horizontal up to that point. The subsequent decomposition occurs in four stages up to 642°, where the crucible is completely empty. The portion of the curve between 230° and 449° corresponds to mercuric iodide, which sublimes and dissociates (*53*).

Example IV. When a zinc salt is treated with sodium carbonate in contact with the air, a more or less carbonated hydroxide results. In the course of the dehydration, there is a sudden change in curvature around 100° (Figure 41); then the loss of water becomes slower. At 200° we are dealing with a relatively stable basic carbonate, which does not release the whole of its carbon dioxide below 1000°. Consequently, the crucible may be returned to the furnace as often as one wishes, at 950°, for instance. Though the weight will always be constant, the result will be false (*8*).

Example V. The author (*79*) has shown that, because of the complications which it introduces into the determination of antimony, the familiar Treadwell hot-air bath and its modifications, which are commonly used to bring antimony sulfide into a suitable weighing form, can no longer be justified. In fact, after the loss of water and sulfur, the curve (Figure 42) has a horizontal between 176° and 275°, which agrees rigorously with the composition Sb_2S_3 and the existence of a black, homogeneous product. The author advises drying in an electric furnace for 10 minutes at 176°, and especially urges the use of ammonium thiocyanate rather than hydrogen sulfide for the precipitation.

Example VI. Cadmium anthranilate produces a curve (Figure 43) which rises slowly until the temperature reaches 222°, where decomposition sets in (*43*). The gain in weight is slight, as can be seen from the following figures, which refer to 118.25 mg. of dry anthranilate:

Temp., ° C.	51	87	150	222
Loss, mg.	0.6	0.9	1.8	2.0

However, in view of the ease with which the leaflets of cadmium anthranilate dry, particularly after washing with ethanol, it seems preferable to keep the compound in a desiccator below 40° before weighing.

The question which was asked in the heading of this section can therefore not be answered categorically. Whenever a chemist attempts by trials alone to find the optimum drying temperature for a particular precipitate, he is in much the same situation as a blind man who has lost his cane.

A CURIOUS OXIDATION PHENOMENON

In several instances a rise in the curve has been observed on progressive heating of certain oxidants (nitrates, chlorates, bromates, iodates, chromates, stannates, etc.), which are expected to give off oxygen at a proper temperature. The initial gain in weight is due to the oxygen of the air. In some instances the phenomenon is reversible; in others it is irreversible. In the latter case, it is thought that the weighing made in the cold, after cooling without having dissociated the heated compound, gives false results. In the former case, there is a chance that this weighing may be good, but the automatic determination has become impossible. Therefore, when a compound is to give up oxygen, it begins by taking a small quantity of this element from the air to form an unstable peroxide or persalt, which doubtless is capable of starting a chain decomposition.

Anhydrous lead chromate is stable from 91° to 904°. From this temperature on, it loses oxygen, but previously it gains slightly in weight (1 mg. per 400 mg.) notably from 673° on (Figure 44). Is this oxygen carried on the chromium, on the lead, or is it absorbed by the chromate? The thermobalance cannot answer this question, but merely confirms the fact (*22*).

When nitron nitrate is heated up to 256° (Figure 45), a temperature close to its decomposition point, and then allowed to cool at the same rate at which it was heated, the loss of oxygen during the cooling is equal to the gain observed during the rise in temperature. Accordingly, why heat this compound before weighing it?

It is logical to discuss now the different uranium oxides which the writer has been able to present on the same graph (Figure 46),

despite the slight variation in the oxygen content which accompanies the passage from one oxide to the other (*40*).

This curve represents the pyrolysis of the pale green uranium oxalate, $U(C_2O_4)_2.6H_2O$, which, first of all, loses 4 molecules of water between 50° and 78°. The dihydrate is shown by the short horizontal extending from 78° to 93°. Then the anhydrous oxalate is present from 126° to 158°, after which it begins to decompose, ordinarily losing carbon monoxide and carbon dioxide, notably at 250°, while the residue takes up some oxygen. Actually, it is very surprising to note that the residual oxide does not have the expected composition UO_3; it corresponds instead to U_3O_8. The latter oxide is not within its zone of stability. it combines with the quantity of oxygen necessary to produce UO_3 quantitatively at 542°. Finally, the latter oxide regenerates the U_3O_8 above 700° and at 942° the oxide UO_2 begins to appear. These successive oxidations and reductions may be verified quantitatively on the photographic paper with about 100 mg. of oxide.

This gives a good idea of the precision that may be attained by means of the Chevenard thermobalance.

THE PROGRESS OF WASHING A PRECIPITATE

The following example illustrates a problem that otherwise can be solved only by cut and try methods (*8*).

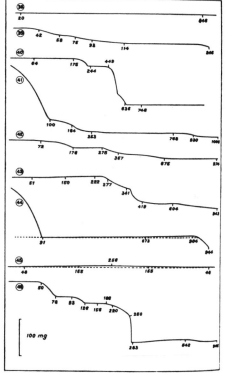

Figures 38 to 46.   Thermolysis Curves

38. Silver thallium arsenate
39. Calcium picrolonate
40. Mercurous iodate
41. Carbonated zinc hydroxide
42. Antimony sulfide
43. Cadmium anthranilate
44. Lead chromate
45. Nitron nitrate
46. Uranyl oxalate

Merritt and Walker (77) recommended that zinc be precipitated with 8-hydroxyquinaldine and weighed after drying at 130° to 140°. The curve produced by this precipitate (Figure 47) shows a good horizontal extending from 110° to 216°, but it corresponds to a zinc complex, contaminated with reagent, which is removed with difficulty even by hot water. When the temperature is continuously raised, the theoretical weight is not attained below 355°, but the quinaldinate decomposes at once. However, if the precipitate is washed 1, 2, 3, 4, and 5 times with hot water and alcohol, and then kept in a constant temperature oven (not above 220°) the theoretical weight of quinaldinate given by Merritt and Walker is gradually reached.

In this carefully controlled instance, the use of the thermobalance allows the operator to evaluate the extent of the adsorption occasioned by a precipitate and to judge the efficacy of successive washings.

THE THERMOBALANCE IN GASOMETRY

When the gas being determined is the sole product evolved on heating a solid, the thermobalance can obviously record the loss with no need for using a gas collecting tube, whose manipulation at temperatures around 1000° is not always a simple matter.

A sample of gypsum gave the data:

	%		%
Water (moisture)	0.3	SO_4	59.9
Water (combined)	20.2	CO_2	1.5
Calcium	23.4	$Al_2O_3 + Fe_2O_3$	Trace
Silica	0.1		

The pyrolysis curve (Figure 48) of this material is compared with that of Figure 10, which pertains to a specimen of gypsum that contains no calcite. The dissociation of the calcium carbonate becomes evident around 820°; the loss of 4 mg. of carbon dioxide measured on the graph corresponds to 1.69% in terms of CO_2. (The figure 1.5% had been found by the classical volumetric method.)

THERMOBALANCE FOR DISCOVERING METHOD OF SEPARATION

The separation of gallium and iron is known to be one of the most difficult problems of analytical chemistry. Iron is always found in commercial gallium when the samples are examined spectrographically or tested with α,α'-bipyridine. The thermobalance likewise reveals this contamination, as the numerous curves produced with gallium hydroxide show a rise (Figure 49) at high temperatures, notably from 800° on, which signals the gain of oxygen capable of converting the iron into the Fe_2O_3 state. This finding has made it possible to compare the efficacies of the various methods of separation that have been proposed. All of them are unsatisfactory (20). On the other hand, treatment with sodium sulfite (or bisulfite) is the only known procedure that yields a perfectly horizontal section in the gallium hydroxide curve (Figure 50). This proves that this reagent leaves the iron in solution (provided there is not more than 1% present initially). When a trace of iron is added to the gallium oxide remaining after this procedure, a curve analogous to that of Figure 49 is obtained.

The sulfite method leaves less than 10^{-4} iron in the gallium which originally contained 1%, and furthermore this separation is achieved in a single operation and by means of a reagent which is readily available.

EXPLANATION AND CORRECTION OF ERRORS IN ANALYTICAL CHEMISTRY

Hardly 200 of the 933 known gravimetric procedures are worthy of inclusion in future texts on analytical chemistry; this declaration suggests another quip by F. Feigl—namely, "the thermobalance is capable of distinguishing between the good and the poor chemists." The writer has no desire to start a debate and merely points out some ten cases, to which their authors have doubtless given all necessary care.

Bell-Shaped Curve of Cuprous Sulfide. When a cupric solution is treated above 80° with an excess of sodium thiosul-

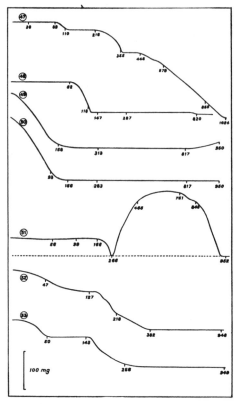

Figures 47 to 53.   Thermolysis Curves

47.   Zinc 8-hydroxyquinaldinate
48.   Carbonated gypsum
49.   Gallium hydroxide by aniline
50.   Gallium hydroxide by sodium sulfite
51.   Cuprous sulfide
52.   Luteocobaltivanadate in acid medium
53.   Luteocobaltivanadate in neutral or alkaline medium

fate, cuprous sulfide and sulfur are formed. Figure 51 shows that this sulfur is removed by heating, and that $Cu_2S = 159$ (sometimes already oxidized) is obtained at 226°. If the heating is continued, deep-seated oxidation occurs and the color goes from black to white:

$$Cu_2S + 2^1/_2O_2 = CuO + CuSO_4$$

Then the cupric sulfate decomposes

$$CuSO_4 \longrightarrow SO_3 + CuO$$

so that at 952° the residue consists of $CuO + CuO = 159$, which has the same molecular weight as the initial cuprous sulfide, since the atomic weight of sulfur is double that of oxygen. This fact has led to the assertion that the method is very exact; from case to case the percentage of copper is the same. However, what is weighed is not cuprous sulfide; the residue contains no sulfur. The ignition ought to be made at 950° and the calculations should be based on the black oxide, CuO (74).

In passing, attention is called to the irregularity between 751° and 845°. This corresponds to a golden yellow basic sulfate, $2CuO.SO_3$, which the author has isolated.

Determination of Vanadium with Luteocobaltichloride. Parks and Prebluda (85) treated a vanadate solution with luteocobaltichloride, and claimed to have obtained different precipitates according to the nature of the reaction theater: acid, neutral, basic. The author (59) has proved that the same product is obtained in neutral or alkaline surroundings—namely, [Co(NH₃)₆](VO₃)₃, which can be used as the weighing form after drying between 58° and 143° (Figure 53). On the other hand, the formula [Co(NH₃)₆](V₆O₁₇)₃ proposed for the product obtained in an acid medium, is off by almost 1.7%; the curve (Figure 52) has no horizontal in the region of 100°. More exact figures are obtained if the product is dried at 127°, but decomposition sets in at once.

Determination of Tungsten by Oxine. Jilek and Rysanek (68) assign the incorrect formula, WO₂(C₉H₆ON)₂, to the precipitate, which is finally ignited to the oxide as weighing form. The precipitate gives a good horizontal up to 218°, but obviously it does not conform to the formula just given. The composition varies from trial to trial, and the precipitate always contains less than one molecule of oxine for one tungstic anhydride. In Figure 54, the apparent molecular weight according to the horizontal is 436 instead of 503.92 required by the formula. The oxide WO₃ appears from 674°.

Determination of Copper by Thiocyanate of Copper-Benzidine Complex. Spacu and Macarovici (102) thought that they had prepared a complex, [Cu(Bzd)](SCN)₂, by treating a copper salt with thiocyanate and then with benzidine (or toluidine, or o-dianisidine). The product, for which they reported no analysis, did not seem to agree with this composition when the author submitted it to thermolysis (74). In fact, its qualitative analysis and the calculation from the curve (Figure 55) exclude the presence of organic matter. Furthermore, the three curves are superposable no matter whether benzidine, toluidine, or o-dianisidine, which are homologs of each other, was used. The author has thus proved that these curves pertain to one and the same compound—cupric thiocyanate, Cu(CNS)₂.¹/₂H₂O. This material, which admittedly is not well known among inorganic chemists, is stable up to 168°. (None of the usual spot tests gave a positive reaction for benzidine with this material.) Hence it may be concluded that the organic material serves merely to facilitate the preparation of the cupric thiocyanate; beyond this, it has no function.

When pyridine is used, the organic matter is retained. According to Spacu and Dick (100), the complex [Cu(C₅H₅N)₂]-(SCN)₂ is readily formed and can be used in the analysis. The same is true of the complex prepared with isoquinoline by Spakowski and Freiser (104)—namely, [Cu(C₇H₇N)₂](SCN)₂—which can be weighed after drying at 100°. (This latter complex is highly thixotropic.)

The precipitate obtained by Vejdilek and Vorisek (107) does not contain nitrobenzimidazole. The curve (not reproduced here) is entirely identical with that given by ammonia.

In the author's opinion (82), the precipitate discovered by Solodovnikov (98) and formulated as BI₃[(CH₃)₆N₄I] includes not a trace of hexamethylenetetramine; it is nothing but bismuth iodide.

Purpureo-Cobaltic Molybdate. After a critical study, Congden and Chen (10) stated that Carnot's method (5) is the poorest way to determine cobalt gravimetrically. The formula of the initial precipitate is not known, but the residue at 110° has the composition Co₂O₃.10NH₃.6MoO₃ according to Carnot, and 2CoO.7MoO₃ according to Congden and Chen.

The authors' results may be summed up as follows: The conversion of bivalent cobalt into purpureocobaltic chloride is never quantitative; consequently, the determination cannot be exact, no matter what formula is accepted.

The authors then formed the precipitate by starting with ammonium molybdate and pure purpureocobaltic chloride. The pink precipitate has a composition which conforms as well as possible to [CoCl(NH₃)₅]₂Mo₇O₁₆. It loses weight (loss of water and chlorine) up to 167° (Figure 56). From 167° to 206° there is

a horizontal, which, in agreement with Carnot, conforms to the composition 3MoO₃.¹/₂Co₂O₃.5NH₃ = 600, with the gravimetric factor 0.098 for the cobalt. Then, the ammonia is released, and a second horizontal extends from 338° to 836°. This is more certain than the preceding one, and represents a mixture, 3MoO₃.¹/₂Co₃O₄ = 512.12, with the factor 0.115 for the cobalt. Molybdic anhydride is lost rapidly above this temperature.

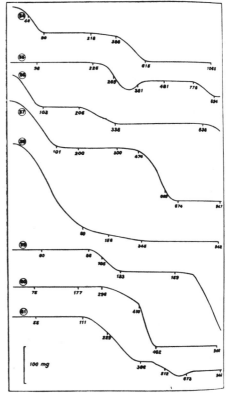

Figures 54 to 61. Thermolysis Curves

54. Oxine tungstate
55. Cupric thiocyanate
56. Purpureocobaltimolybdate
57. Thorium iodate
58. Titanium phosphate
59. Magnesium oxinate
60. Cuprous iodide
61. Antimony oxinate

It may therefore be concluded that the formula given by Carnot for the material weighed is good, whereas that of Congden and Chen is false. The selected temperature, 110°, is too low. As the method is not quantitative, it should be abandoned as a means of determining bivalent cobalt (58).

Determination of Thorium as Iodate. Chernikhov and Uspenkaya (5) suggested the complicated formula, 4Th(IO₃)₄.KIO₃.18H₂O, for this precipitate. However, the author's analysis agrees better with the formula given by Moeller and Fritz (78). The compound in question is the normal iodate, Th(IO₃)₄, which is unstable and readily hydrolyzes during washing. The curve (Figure 57) shows a rapid loss of water up to about 100°. The anhydrous iodate is present, with almost constant weight between 200° and 300°. Beyond this range, iodine and oxygen are evolved. The residue, produced from 674° on, consists of thorium oxide and is free of both iodine and potassium (*24*).

Determination of Titanium as Phosphate. Jamieson and Wrenshall (*67*) advise that titanium phosphate be brought to "the highest temperature possible." Figure 58 shows that this phosphate is obtained pure from 400° on. The formula $Ti_2P_2O_9$ is more suitable with respect to valence than $TiPO_4$, but its gravimetric factor is less correct.

Determination of Magnesium by Oxine. Even though it does not give satisfactory results, this method is in wide use, especially in biological chemistry. In practice two methods of precipitation are employed: The solution being analyzed is treated successively with ammonia and oxine; these same reagents are added in the reverse order. Berg, Hahn, and Wieweg state that the air-dried precipitate has the composition $Mg(C_9H_6ON)_2.4H_2O$; this compound is supposed to give the dihydrate on drying at 100° to 105°, and the anhydrous salt when it is kept at 130° or 160°.

The curves obtained with the thermobalance show that the precipitate dried in the air never has the composition of a tetrahydrate. Figure 59 shows a rapid fall to 105°, a fact that is unfavorable for drying a dihydrate there. This hydrate does exist, but it is stable only up to 60°.

The anhydrous compound is obtained very easily, even at

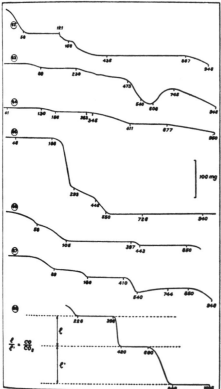

Figures 62 to 68. Thermolysis Curves

62. Cinchonine germanomolybdate
63. Thallic oxide by ferricyanide
64. Thallic oxide by electrolysis
65. Zirconium mandelate
66. Phosphomolybdic acid
67. Ammonium phosphomolybdate
68. Check of atomic weight of carbon

105°, if the temperature is maintained there long enough, but the best way is to dry the precipitate at 145°. Constant weight is rapidly established at this temperature and there is no danger of loss, even though the heating is prolonged for several hours. The result is not affected by using as much as 100% excess reagent. In every case, a temperature of 160° produces low results.

Cuprous Iodide Precipitate. If a cupric salt is treated with sulfur dioxide and then with potassium iodide, the resulting cuprous iodide precipitate gives the curve shown in Figure 60. The salt is stable up to 296°, and it then loses iodine rather rapidly and takes up oxygen. Above 482° there is a horizontal corresponding to cupric oxide, CuO. Consequently, it seems fair to ask whether the fusion temperature 628°, the boiling temperature 772°, and the allotropic transformation point 402°, which are given in the literature, are not fantasies, since the curve indicates that the iodide has ceased to exist at these three temperatures.

Precipitation of Antimony by Oxine. This precipitation has given rise to direct contradictions. In his classic text, Prodinger states that it is incomplete, whereas Pirtea (*95*) claims that it is quantitative, and that the yellow precipitate contains 21.97% antimony, which agrees with the formula $Sb(C_9H_6NO)_3$. Pirtea also indicates a pH precipitation zone from 6.0 to 7.5, which seems impossible of realization without contaminating the precipitate with antimony oxychloride. Figure 61 shows a good horizontal up to 111°, which, as might be expected, does not agree at all with the formula just given, but which, in the author's opinion, contains 54.03% antimony, and consequently the precipitate contains excess metal. The determination of antimony by means of oxine does not seem to be worth considering (*79*).

Determination of Germanium by Cinchonine. The author (*28*) has repeated the procedure of Davies and Morgan (*12*), who claimed that they obtained a precipitate with the composition $H_4[Ge(Mo_{12}O_{46})](C_{19}H_{22}ON_2)_4$. Figure 62 presents the curve corresponding to 2 mg. of metal, while another—not given here—corresponded to 9 mg., because these workers stated that low results are obtained above 5 mg. The author has found that the precipitates produced in this case do not give reproducible curves. Although a vague horizontal is obtained between 92° and 121°, it does not accord with the formula just given, nor to the more rational $H_4[Ge(Mo_2O_7)_6](C_{19}H_{22}ON_2)_4$. The horizontal which extends from 450° and 900° does not correspond to the usual mixture $GeO_2 + 12MoO_3$; it contains too much molybdenum. Therefore, this procedure is not recommended for determining germanium; instead, the use of 5,7-dibromo-oxine is advised.

Phosphomolybdic Acid. This material was prepared by treating ammonium phosphoduodecimolybdate with aqua regia. Since Finkener, it is usually given the formula $P_2O_5.24MoO_3.-3H_2O + 58H_2O$. The thermolysis curve (Figure 66) given by one preparation indicates that the residue between 510° and 850° consists entirely of the mixture $P_2O_5 + 24MoO_3$; the volatilization of the latter anhydride proceeds rapidly beyond 880°. From 120° to 355°, there is a horizontal which corresponds equally well to the ensemble $P_2O_5.24MoO_3.3H_2O$ or $H_3PO_4.-12MoO_3$, but beyond that temperature things go badly. The initial material never contains 58 H_2O of crystallization, but quantities which, from preparation to preparation, vary between 31 and 33.5 H_2O. Of course, the crystallization is always made from a quantity of acidulated water, such that there is an excess of water in comparison with the 58 H_2O included in the foregoing formula. The reasons for such wide disagreement are not known, but the author prefers the automatic graphic recording when dealing with these superabundant molecules of water.

As the acid has been discussed, its ammonium salt can hardly be ignored, first because it serves in the rapid determination of phosphorus, and secondly because, even in the modern texts on analysis, it gives rise to calculations which do not seem to be clear. For the precipitation, the directions given in Treadwell's text were followed, weighing in the form of $(NH_4)_3PO_4.12MoO_3$ after prolonged heating at 170°. According to Woy (*111*), gentle ignition produces the mixture $P_2O_5 + 24MoO_3$. The curve (Figure 67) was obtained from a moist precipitate of ammonium phosphomolybdate, which is usually written as $(NH_4)_3$-$PO_4.12MoO_3.2HNO_3.H_2O$. This loses nitric acid and water up to 180°. However, if the recording is made with a precipitate that has been air-dried, the loss in weight up to 180° corresponds to $2HNO_3$ and not to $2HNO_3 + H_2O$. The difference is not significant and has no influence on the determination of phosphorus. A strictly horizontal portion of the curve extends from 180° to 410°; it closely corresponds to the formula given by Treadwell.

Beginning at 410°, two molecules of this compound share the loss of six molecules of ammonia and three molecules of water up to 540°; however, all of the author's recordings show that the decomposition is more deep seated than this because the curves have a very clear bend, and then rise again. Consequently, there is probably a transient reduction of molybdenum, followed by a reoxidation by the air to produce the ensemble $P_2O_5.24MoO_3$. At 540°, the material in the crucible is green—i.e., it is a mixture of molybdenum blue and yellow phosphomolybdate. The ascent of the curve is very slow. Theoretically, the horizontal is reached only between 812° and 850°, but the admissible region of ignition may extend from 600° to 850° without much error with respect to phosphorus. Above this temperature, the molybdic anhydride sublimes rapidly. Below 600°, all the possible approximate formulas for the residue may be found.

UNEXPLAINED PHENOMENA

The various studies described above should not lead the reader to believe that the author has been able to interpret all the singularities of the curves. A few findings remain unexplained.

Example I. When thallium trioxide is precipitated by the ferricyanide method, it maintains a constant weight between 126° and 230° (*91*). Beyond that temperature, Figure 63 reveals a first loss of oxygen from 230° to 375°; the residual material has the composition $3Tl_2O_3.Tl_2O$. There is a more rapid fall from 408° to 596–600°, and a certain part of the thallous oxide of the preceding system disappears. The curve rises from 600° to 720°, and according to this ascending branch, $^3/_4$ of a Tl_2O molecule recovers its oxygen, with the result that pure thallic oxide is regained between 720° and 745°. A new loss of weight begins at 745°; it is due to a new dissociation and to volatilization. The residue at 946° consists of a mixture $Tl + Tl_2O_3$. If the chemistry of thallium were not filled with such strange happenings, we might be surprised to see a single oxide possessing two domains of existence. However, that is not all. If the thallic oxide produced by the electrolysis of a thallous salt in a sulfuric acid medium and in the presence of acetone, is collected at the anode, the material contains $0.8\ H_2O$. The hydroxides $Tl(OH)_3$ and $TlO(OH)$ are obtained here no more than in the preceding case. Between 156° and 233° the pyrolysis curve (Figure 64) has an almost horizontal section, which agrees fairly well with the composition Tl_2O_3. It then has a new horizontal between 411° and 677°, which corresponds to the double oxide $3Tl_2O_3.Tl_2O$ referred to above, but up to 950°, the limit of the experiment, there is neither a minimum nor maximum in the loss of weight, which is perfectly regular. The residue at 950° is again a mixture $Tl + Tl_2O_3$ in the process of evolution. Therefore, it appears that above 677° there are two different forms of the double oxide $3Tl_2O_3.Tl_2O$, and yet the Debye and Scherrer spectra are identical.

Example II. Kumins (*73*) suggested the excellent method in which zirconium is precipitated by means of mandelic acid. The corresponding curve has a good horizontal up to 188° (Figure 65), which fits the formula $Zr(C_6H_4.CHOH.COO)_4$ very well. After destruction of the organic matter, the zirconium seems to be partially reduced; in fact, beyond 570° the curve rises again in such manner that a constant weight of zirconia is obtained at least at 950°.

This is not an isolated case and occurs with all the precipitates produced by zirconium and arsenical compounds (arrhenal, atoxyl, propylarsinate, phenylarsinate, hydroxyphenylarsinate, etc.) but in a much more marked fashion. Consequently, the author decided to isolate and analyse the mixture appearing at the minimum of the curves. Even though the composition

agrees with $Zr_2O_3 + ZrO_2$, the spectra of the powders is always the same—that of pure zirconia. The matter is still undecided (*105*).

INSTRUCTIONAL EXPERIMENT

Check of Atomic Weight of Carbon. Figure 68 consists of the right side of Figure 8, which represented the decomposition of calcium oxalate. It is known that one molecule of carbon monoxide is evolved from 400° to 420° and that a molecule of carbon dioxide escapes from 660° to 840°. Therefore:

$$\frac{CO}{CO_2} = \frac{l}{l'} \text{ or } \frac{C + 16}{C + 32} = \frac{l}{l'}$$

Six runs by students gave the following values for C: 12.00, 12.06, 12.08, 12.00, 12.00, and 12.00.

It is likely that eventually the thermobalance will be used not to measure an atomic weight but to control the weighing temperatures and the purity of the reagents employed. A priori, the number of suitable instances appears to be large.

AUTOMATIC DETERMINATIONS (*37, 45, 46, 48, 56*)

The classic determination, developed by Feigl (*60*), of copper by benzoinoxime has been selected as a model. As shown by Figure 71, the complex $CuC_{14}H_{11}O_2N$, after losing some wash

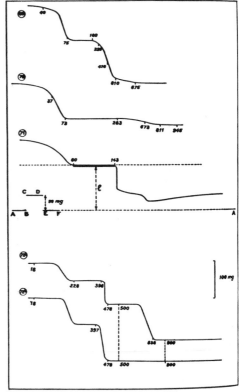

Figures 69 to 72. Thermolysis Curves

69. Filter paper
70. Asbestos
71. Copper benzoinoxime complex (principle of automatic determination)
72, *a.* Calcium oxalate
72, *b.* Magnesium oxalate

liquid, is stable from 60° to 143°. The ordinate of the corresponding horizontal is noted and compared with that of the straight line of the calibration.

The filtration must not be through paper, because the latter changes the material during the incineration, and furthermore, as shown by Figure 69, the paper which burns loses weight up to 675° (*38*). Asbestos in Gooch crucibles can be used up to 283° (Figure 70), where it begins to lose weight (*38*). Fritted glass serves up to 520° (gaining slightly in weight), but the corresponding crucibles are usually too heavy for the ring of the thermobalance, and they are hard to wash. Instead, simple glass crucibles may be used, with perforated bottoms and fitted with a mat of fat-free glass wool. Each crucible weighs only 3 grams, and the mat is discarded after each determination. Quartz or porcelain crucibles are used above 510°, but for rapid automatic gravimetric procedures, precipitates have been selected which come down, wash, and dry instantly, and which produce their horizontals of constant weight below 200°.

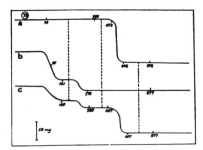

Figure 73.   Analysis of Silver-Copper Alloys
a.   Thermolysis curves of silver nitrate
b.   Copper nitrate
c.   Mixture of silver nitrate and cupric nitrate

Figure 71 illustrates the method of operation.

The straight line *AB* is inscribed on the paper for 5 minutes. The excess weight of 50 mg. is placed on the platform of the silica rod; the spot traces the straight line *CD* during another 5 minutes. The surcharge is removed, the empty filtering crucible being kept always on the ring (these operations are obligatory only after a regulation, cleaning, or changing to a new brand of paper). The lamp is extinguished; the crucible is removed, the precipitate, brought down in advance, is filtered into the crucible with the aid of suction (finishing whenever possible with alcohol and ether). The crucible is put into the furnace, which is set at a temperature within the limits of the horizontal, and a record is made only of a portion of the horizontal (shown in heavy print in Figure 71). Finally a reading is made of the length, *l*, which is compared at once with the distance between *AB* and *CD*.

In certain cases (cadmium, uranium, etc.) the whole operation consumes not more than 17 minutes.

Another practical case involves a mixture of calcium and magnesium compounds. It is well known that the calcium oxalate precipitate carries down an unknown amount of magnesium oxalate when there is at least ten times as much magnesium as calcium present.

The pyrolysis curves of the two oxalates are shown in Figure 72*a* and *b*. The former (CaC_2O_4) was given in Figure 8; the other curve is much simpler because it does not have a horizontal pertaining to magnesium carbonate. In other words, when anhydrous magnesium oxalate, which is stable from 233° to 397°, decomposes, it loses carbon monoxide and carbon dioxide simultaneously and almost instantaneously. Consequently, at 500°, for instance, there is a mixture of $CaCO_3 + MgO$, whereas at 900° the mixture consists of $CaO + MgO$. Therefore, two first degree equations can be set up on the basis of the two weights (*90*).

The final example of an entirely automatic determination is based on the difference in the stabilities of silver nitrate and copper nitrate (*92*).

Figure 73, *a* and *b*, represents the pyrolysis of silver nitrate and copper nitrate, respectively. *c* is a resultant on which is found a mixture of $AgNO_3 + CuO$ at 400°, and of $CuO + Ag$ at 700°. The difference in the heights of these two horizontals permits the determination, with an accuracy of $^1/_{300}$, of the quantity of silver and copper contained in their binary alloys and without a preliminary separation.

It is precisely this technique which will be developed in the years to come, provided that we know the precipitates which furnish horizontals and the temperatures between which they exist.

CONCLUSION

There is now available to laboratories and plant works an analytical tool, which is simple, sturdy, not likely to get out of order, practically without maintenance cost, and manufactured commercially, which yields results that are tangible, permanent, and of high scientific value. Although the Chevenard thermobalance was initially conceived for the study of the oxidation of alloys, it is nevertheless true that docimasy, during the past three years, has been indebted to this instrument for furnishing valuable information about the behavior of compounds when heated; the temperatures to which a precipitate should be brought to acquire a correct weight; and a simple procedure for studies in series, permitting in fact as many as 35 samples to be run through the same determination in a day, and whose results can be interpreted immediately. Furthermore, the curves provide a permanent record of each determination.

BIBLIOGRAPHY

(1) Allen, N., and Furman, N. H., *J. Am. Chem. Soc.*, 54, 4625 (1932).
(2) Atack, F. W., *Analyst*, 38, 316 (1913).
(3) Carnot, A., *Compt. rend.*, 108, 741 (1889); 109, 109 (1889); *Ann. chim. anal.*, 22, 121 (1917); *Bull. soc. chim. France*, 21, 211 (1917).
(4) Champ, P., Fauconnier, P., and Duval, C., *Anal. Chim. Acta*, 5, 277 (1951).
(5) Chernikhov, Y. A., and Uspenkaya, T. A., *Zavodskaya Lab.*, 9, 276 (1940).
(6) Chevenard, P., Waché, X., and De la Tullaye, R., *Bull. soc. chim. France*, 10, 41 (1944).
(7) Claassen, A., and Visser, J., *Rec. trav. chim.*, 65, 211 (1946).
(8) Clercq, M. De, and Duval, C., *Anal. Chim. Acta*, 5, 282 (1951).
(9) *Ibid.*, in press (article on tungsten).
(10) Congden, L. A., and Chen, T. H., *Chem. News*, 128, 132 (1924).
(11) Currah, J. E., McBryde, W. A. E., Cruikshank, A. J., and Beamish, F. E., IND. ENG. CHEM., ANAL. ED., 18, 120 (1946).
(12) Davies, G. R., and Morgan, G., *Analyst*, 63, 388 (1938).
(13) Dupuis, T., *Compt. rend.*, 228, 841 (1949).
(14) *Ibid.*, 230, 957 (1950); *Mikrochemie*, 35, 476 (1950).
(15) Dupuis, T., Besson, J., and Duval, C., *Anal. Chim. Acta*, 3, 599 (1949).
(16) Dupuis, T., and Duval, C., *Ibid.*, 3, 183 (1949).
(17) *Ibid.*, p. 186.
(18) *Ibid.*, p. 190.
(19) *Ibid.*, p. 201.
(20) *Ibid.*, p. 324.
(21) *Ibid.*, p. 330.
(22) *Ibid.*, p. 345.
(23) *Ibid.*, p. 438.
(24) *Ibid.*, p. 589.
(25) *Ibid.*, 4, 50 (1950).
(26) *Ibid.*, p. 173.
(27) *Ibid.*, p. 180.
(28) *Ibid.*, p. 186.
(29) *Ibid.*, p. 201.
(30) *Ibid.*, p. 262.
(31) *Ibid.*, p. 615.
(32) *Ibid.*, p. 623.
(33) Dupuis, T., and Duval, C., *Compt. rend.*, 227, 772 (1948).

(34) *Ibid.*, **228**, 401 (1949).
(35) *Ibid.*, **229**, 51 (1949).
(36) Duval, C., *Anal. Chim. Acta*, 2, 92 (1948).
(37) *Ibid.*, p. 432.
(38) *Ibid.*, 3, 163 (1949).
(39) *Ibid.*, p. 335.
(40) *Ibid.*, p. 338.
(41) *Ibid.*, 4, 55 (1950).
(42) *Ibid.*, p. 160.
(43) *Ibid.*, p. 190.
(44) Duval, C., *Chim. anal.*, 31, 177 (1949).
(45) Duval, C., *Compt. rend.*, 224, 1824 (1947).
(46) *Ibid.*, 226, 1276 (1948).
(47) *Ibid.*, 227, 679 (1948).
(48) Duval, C., Conference held at Centre de Perfectionnement technique, Nov. 22, 1948; *Chim. anal.*, 31, 173, 204 (1949).
(49) Duval, C., Conference held at Iᵉʳ Congrès International de Microchimie, Gras, July 6, 1950; *Mikrochemie*, 36, 425–65 (1951).
(50) *Ibid.*, 35, 242 (1950).
(51) Duval, C., Troisième Rapport de la Commission des réactifs nouveaux, Paris, Librairie Istra, 1948.
(52) Duval, C., and Dat Xuong, Ng., *Anal. Chim. Acta*, 5, 160 (1951).
(53) *Ibid.*, in press (article on mercury).
(54) Duval, T., and Duval, C., *Anal. Chim. Acta*, 2, 103 (1948).
(55) *Ibid.*, p. 207.
(56) *Ibid.*, p. 223.
(57) Duval, R., and Duval, C., *Ibid.*, 5, 71 (1951).
(58) *Ibid.*, p. 84.
(59) Duval, C., and Morette, A., *Compt. rend.*, 230, 545 (1950); *Anal. Chim. Acta*, 4, 490 (1950).
(60) Feigl, F., *Ber.*, 56, 2083 (1923).
(61) Friedrich, K., *Metallurgie*, 6, 175 (1909).
(62) Garrido, J., *Anales fís. y quím. (Madrid)*, 43, 1195 (1947).
(63) Gentry, C. H. R., and Sherrington, L. G., *Analyst*, 70, 419 (1945).
(64) Girard, J., *Chim. anal.* 4, 382 (1899).
(65) Hecht, F., and Donau, J., "Anorganische Mikrogewichts-analyse," p. 205, Vienna, Librairie Springer, 1940.
(66) Herrmann-Gurfinkel, M., *Bull. soc. chim. Belg.*, 48, 94 (1939).
(67) Jamieson, G. S., and Wrenshall, R., *J. Ind. Eng. Chem.*, 6, 203 (1914).
(68) Jilek, A., and Rysanek, A., *Collection Czechoslov. Chem. Communs.*, 10, 518 (1938).
(69) Jolibois, P., and Lefèvre, H., *Compt. rend.*, 176, 1317 (1923).
(70) Kolthoff, I. M., and Bendix, G. H., IND. ENG. CHEM., ANAL. ED., 11, 94 (1939).
(71) Kolthoff, I. M., and Meene, G. H. P. van der, *Z. anal. Chem.*, 72, 337 (1927).
(72) Krustinsons, J., *Ibid.*, 125, 98 (1943).
(73) Kumins, A., ANAL. CHEM., 19, 376 (1947).
(74) Marin, Y., diplôme d'études supérieures, Paris, Nov. 28, 1949.

(75) Marin, Y., and Duval, C., *Anal. Chim. Acta*, 4, 393 (1950).
(76) Mawrow, W., and Muthmann, F., *Z. anorg. Chem.*, 13, 209 (1897).
(77) Merritt, L. L., and Walker, J. K., IND. ENG. CHEM., ANAL. ED., 16, 387 (1944).
(78) Moeller, T., and Frits, N. D., ANAL. CHEM., 20, 1055 (1948).
(79) Morandat, J., and Duval, C., *Anal. Chim. Acta*, 4, 498 (1950).
(80) Müller, W., *Ber.*, 35, 1587 (1902).
(81) Nieuwenburg, C. J. Van, and Hoek, T. Van der, *Mikrochemie*, 18, 175 (1935).
(82) Panchout, S., and Duval, C., *Anal. Chim. Acta*, 5, 170 (1951).
(83) Parks, W. G., and Prebluda, H. N., *J. Am. Chem. Soc.*, 57, 1676 (1935).
(84) Pauling, L., *Ibid.*, 55, 1895, 3052 (1933).
(85) Pavelka, F., and Zucchelli, A., *Mikrochemie*, 31, 69 (1943).
(86) Peltier, S., diplôme d'études supérieures, Paris, June 11, 1947.
(87) Peltier, S., and Duval, C., *Anal. Chim. Acta*, 1, 346 (1947).
(88) *Ibid.*, p. 351.
(89) *Ibid.*, pp. 355, 348, 362.
(90) *Ibid.*, p. 408.
(91) *Ibid.*, 2, 211 (1948).
(92) Peltier, S., and Duval, C., *Compt. rend.*, 226, 1727 (1948).
(93) Pirtea, T. I., *Z. anal. Chem.*, 118, 26 (1939).
(94) Rammelsberg, C., *Pogg. Ann.*, 44, 577 (1838); *Ber.*, 1, 70 (1868).
(95) Rây, H. N., *J. Indian Chem. Soc.*, 17, 586 (1940).
(96) Robinson, P. L., and Scott, W. E., *Z. anal. Chem.*, 88, 417 (1932).
(97) Rogers, L. B., and Caley, E. R., IND. ENG. CHEM., ANAL. ED., 15, 209 (1943).
(98) Solodovnikov, P. P., *Trans. Kirov. Inst. Chem. Tech. Kazan*, No. 8, 57–60 (1940).
(99) Soule, B. A., *J. Am. Chem. Soc.*, 47, 981 (1925).
(100) Spacu, G., and Dick, J., *Z. anal. Chem.*, 78, 241 (1929).
(101) Spacu, G., and Dima, L., *Ibid.*, 120, 317 (1940).
(102) Spacu, G., and Macarovici, C. G., *Ibid.*, 102, 350 (1935).
(103) Spacu, G., and Spacu, P., *Ibid.*, 96, 30 (1934).
(104) Spakowski, A. E., and Freiser, H., ANAL. CHEM., 21, 984 (1949).
(105) Stachtchenko, J., and Duval, C., *Anal. Chim. Acta*, in press (article on zirconium).
(106) Vanino, L., and Guyot, O., *Arch. Pharm.*, 264, 98 (1926).
(107) Vejdilek, Z., and Vorisek, J., *Chem. Obzor*, 20, 138 (1945).
(108) Voter, R. C., Banks, C. V., and Diehl, H., ANAL. CHEM., 20, 459 (1948).
(109) Wenger, P. E., Cimerman, C., and Corbaz, A., *Mikrochim. Acta*, 2, 314 (1938).
(110) Willard, H. H., and Hall, D., *J. Am. Chem. Soc.*, 44, 2219 (1922).
(111) Woy, R., *Chem. Ztg.*, 21, 441 (1897).

RECEIVED December 13, 1950.

29

Reprinted from *J. Phys. Chem.*, **62**, 394–397 (Apr. 1958)

THE APPLICATION OF THERMOANALYTICAL TECHNIQUES TO REACTION KINETICS.[1] THE THERMOGRAVIMETRIC EVALUATION OF THE KINETICS OF THE DECOMPOSITION OF CALCIUM OXALATE MONOHYDRATE

By Eli S. Freeman[2] and Benjamin Carroll

Chemistry Department, Rutgers University, Newark 2, N. J., and The Pyrotechnics Chemical Research Laboratory, Picatinny Arsenal, Dover, N. J.

Received July 26, 1957

The application of thermoanalytical techniques to the investigation of rate processes is discussed. Equations have been derived for non-reversing reactions, which may be used to calculate energy of activation and order of reaction from thermogravimetric and volumetric curves. An equation, recently presented in the literature, for evaluating these parameters by the technique of differential thermal analysis has also been considered, so as to eliminate the trial and error procedure. The thermal decomposition of calcium oxalate monohydrate, which involves dehydration, decomposition of calcium oxalate and calcium carbonate, is used to illustrate the applicability of the derived relationships.

Introduction

Thermoanalytical methods, such as thermogravimetry, thermovolumetry and differential thermal analysis, are being employed increasingly in the investigation of chemical reactions in the liquid and solid states at elevated temperatures. These techniques involve the continuous measurement of a change in a physical property such as, weight, volume, heat capacity, etc., as sample temperature is increased, usually at a predetermined rate. In this article, equations are derived for non-reversing reactions so that rate dependent parameters such as energy of activation and order of reaction may be calculated from a single experimental curve. For this purpose a relationship between specific rate and temperature is assumed

$$k = Ze^{-E*/RT}$$

A general derivation is presented and applied to thermogravimetry. For the method of differential thermal analysis, the derivation of Borchardt and Daniels[3] has been expanded upon, so that the trial and error procedure now required for evaluating order of reaction and activation energy, may be replaced by a graphical or analytical solution.

It should be kept in mind that the treatment may be applied to the measurement of any physical property which is unaffected by sample temperature. The advantages of evaluating reaction kinetics by a continuous increase in sample temperature are that considerably less experimental data are required than in the isothermal method, and the kinetics can be probed over an entire temperature range in a continuous manner without any gaps. In addition, where a sample undergoes considerable reaction in being raised to the temperature of interest, the results obtained by an isothermal method of investigation are often questionable.

Theory and Derivation

Consider a reaction, in the liquid or solid states, where one of the products B is volatile, all other substances being in the condensed state.

$$aA = bB(g) + cC$$

The rate expression for the disappearance of reactant A from the mixture is

$$-\frac{dX}{dt} = kX^x \tag{1}$$

where

X = concn., mole fraction or amount of reactant, A
k = specific rate
x = order of reaction with respect to A

It is assumed that the specific rate may be expressed as

$$k = Ze^{-E*/RT} \tag{2}$$

Solving for k in (1) and substituting (2) for k gives

$$Ze^{-E*/RT} = \frac{-(dX/dt)}{X^x} \tag{3}$$

where

Z = frequency factor
$E*$ = energy of activation
R = gas constant
T = absolute temperature

The logarithmic form of equation 3 is differentiated with respect to, dX/dt, X and T, resulting in equation 4.

$$\frac{E* dt}{RT^2} = d \ln (-dX/dt) - x \, d \ln X \tag{4}$$

Integrating the above relationship gives

$$\frac{-E*}{R} = \Delta \left(\frac{1}{T} \right) = \Delta \ln \left(\frac{-dX}{dt} \right) - x\Delta \ln X \tag{5}$$

Dividing (4) and (5) by d ln X and Δ ln X, respectively, one obtains equations 6 and 7.

(1) This paper has been presented in part at the North Jersey Meeting in Miniature of the A.C.S. in Jan., 1957, and before the Division of Physical and Inorganic Chemistry at the National Meeting of the A.C.S. in April, 1957.

(2) Pyrotechnics Chemical Research Laboratory, Bldg. 1512, Picatinny Arsenal, Dover, New Jersey.

(3) H. J. Borchardt and F. D. Daniels, *J. Am. Chem. Soc.*, **79**, 41 (1957).

$$\frac{E^* \, dT}{RT^2 \, d \ln X} = \frac{d \ln (-dX/dt)}{d \ln X} - x \qquad (6)$$

$$\frac{-\dfrac{E^*}{R} \Delta \left(\dfrac{1}{T}\right)}{\Delta \ln X} = \frac{\Delta \ln (-dX/dt)}{\Delta \ln X} - x \qquad (7)$$

From (6) and (7) it is apparent that plots of

$$\frac{dT}{T^2 \log X} \quad vs. \quad \frac{d \log (-dX/dt)}{d \log X}$$

and

$$\frac{\Delta (1/T)}{\Delta \log X} \quad vs. \quad \frac{\Delta \log (-dX/dt)}{\Delta \log X}$$

should result in straight lines with slopes of $+$ or $-E^*/2.3R$ and intercepts of $-x$.

Let us consider the cases where X refers to mole fraction of A, molar concentration and amount of reactant.

1. Mole fraction of A, $X = n_a/M = N_A$

where: n_a = no. of moles of A at time t
M = total no. of moles in reaction mixture

(a) Total number of moles is constant during reaction. Substituting for X in (3) results in the relationships

$$\ln k = \ln M^{x-1} + \ln (-dn_a/dt) - x \ln n_a \qquad (8)$$

and

$$\frac{-(E^*/R)\Delta(1/T)}{\Delta \ln n_a} = -x + \frac{\Delta \ln (-dn_a/dt)}{\Delta \ln n_a} \qquad (9)$$

Equation 9 also may be written in differential form as equation 4.

(b) Total number of moles is not constant. For this case

$$\ln k = (x-2) \ln M - x \ln n_a + \ln \left(n_a \frac{dM}{dt} - M \frac{dn_a}{dt}\right) \qquad (10)$$

$$\frac{\frac{E^*}{RT^2} \, dT}{d (\ln M - \ln n_a)} = x + \frac{d \ln \left(n_a \dfrac{dM}{dt} - M \dfrac{dn_a}{dt}\right) - 2 \, d \ln M}{d (\ln M - \ln n_a)} \qquad (11)$$

and

$$\frac{-\dfrac{E^*}{R} \Delta \left(\dfrac{1}{T}\right)}{\Delta (\ln M - \ln n_a)} = x + \frac{\Delta \ln \left(n_a \dfrac{dM}{dt} - M \dfrac{dn_a}{dt}\right) - 2 \, \Delta \ln M}{\Delta (\ln M - \ln n_a)} \qquad (12)$$

2. Molar concn. $X = n_a/V$

where V = volume of reaction mixture

The equations which result are identical to the case of mole fraction with the exception that V replaces M.

3. $X = n_a$
For this case

$$\ln k = -x \ln n_a + \ln (-dn_a/dt) \qquad (13)$$

The final equation is identical to (9).

The above relationships may be applied simply to measurements of weight or volume changes by the appropriate substitutions for M and n_a. After evaluating x and E^* the frequency factors may be calculated by combining equations 2 and 1.

Let us consider the case of differential thermal analysis.

In a recent article[3] an equation was derived from which order of reaction and energy of activation was determined using differential thermal analysis. The expression given was

$$k = \left(\frac{KAV}{n_0}\right)^{x-1} \frac{C_p \dfrac{d\Delta T}{dt} + K\Delta T}{(K(A-a) - C_p \Delta T)^x} \qquad (14)$$

where

K = heat transfer coefficient
A = area under curve
ΔT = differential temp. at a particular time
$d\Delta T/dt$ = rate of change of differential temp. at the point where ΔT is measured
V = volume of solution
n_0 = initial number of moles of reactants
C_p = total heat capacity of reactant solution or liquid
a = area under curve up to time where ΔT and $d\Delta T/dt$ is taken
x = order of reaction with respect to one component

The method used by the authors[3] to determine x and E^* is as follows. A value of x is chosen and used to calculate k over an entire temperature range using equation 14. A graph of log k vs. T^{-1} was then plotted. If a linear relationship was obtained, it was assumed that the value of x was valid and the energy of activation could then be calculated from the slope of the line.

A method of evaluating x and E^* which eliminates this trial and error procedure becomes apparent if equation 2 is substituted in (14). The resulting expression written in logarithmic form is

$$\ln Z - \frac{E^*}{RT} = (x-1) \ln \frac{KAV}{n_0} - x \ln (K(A - a) - C_p \Delta T) + \ln \left(C_p \frac{d\Delta T}{dt} + K\Delta T\right) \qquad (15)$$

Differentiating and integrating (15) gives equations 16 and 17, respectively.

$$\frac{\frac{E^* \, dT}{RT^2}}{d \ln (K(A-a) - C_p \Delta T)} = -x + \frac{d \ln \left(C_p \dfrac{d\Delta T}{dt} + K\Delta T\right)}{d \ln (K(A-a) - C_p \Delta T)} \qquad (16)$$

$$\frac{-\dfrac{E^*}{R} \Delta \left(\dfrac{1}{T}\right)}{\Delta \ln (K(A-a) - C_p \Delta t)} = -x + \frac{\Delta \ln \left(C_p \dfrac{d\Delta T}{dt} + K\Delta T\right)}{\Delta \ln (K(A-a) - C_p \Delta T)} \qquad (17)$$

From the above it is clear that plots of

$$\frac{dT}{T^2} \over d \log (K(A-a) - C_p \Delta T) \quad vs. \quad \frac{d \log \left(\dfrac{d\Delta T}{dt} + K\Delta T\right)}{d \log (K(A-a) - C_p \Delta T)}$$

and

$$\frac{\Delta \left(\dfrac{1}{T}\right)}{\Delta \log (K(A-a) - C_p \Delta T)} \quad vs. \quad \frac{\Delta \log \left(\dfrac{d\Delta T}{dt} + K\Delta T\right)}{\Delta \log (K(A-a) - C_p \Delta T)}$$

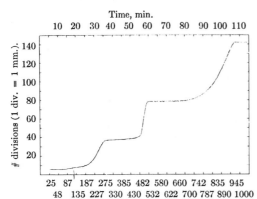

Fig. 1.—Thermogram for decomposition of calcium oxalate monohydrate, weight (mg.) vs. temp. and time: 2 mm. on X-axis = 1 min.; 1 mm. of Y-axis = 2.05 mg.

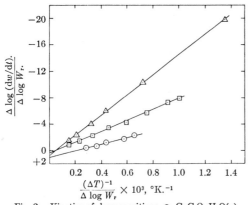

Fig. 2.—Kinetics of decomposition: O, $CaC_2O_4 \cdot H_2O(s) = CaC_2O_4(s) + H_2O$ (g); $\triangle$, $CaC_2O_4(s) = CaCO_3(s) + CO(g)$; $\square$, $CaCO_3(s) = CaO(s) + CO_2(g)$.

should result in straight lines with intercepts at $-x$ and slopes of $+$ or $-E^*/2.3R$ for any unique physical or chemical reaction. It should be noted, however, that (14) and subsequent equations are valid only where the volume of the reaction mixture does not change appreciably.

Experimental

The Testut magnetic transmission type continuously recording thermobalance[4] employed for the thermogravimetric studies, was supplied by the Testut Company of Paris, France.[5] A weight of 423 mg. of calcium oxalate monohydrate was placed in #000 Coors glazed porcelain crucible and heated in air, from room temperature to 1000° at a rate of 10°/min. A Marshall Product's furnace was used for this purpose. A chromel–alumel thermocouple was located directly beneath the sample. The increase in furnace temperature was regulated by a Guardsman Indicating Pyrometric Stepless Program Controller which was purchased from the West Instrument Company. A continuous record of weight change as a function of time and temperature was obtained as shown in Fig. 1. The calcium oxalate monohydrate was of C.P. grade and purchased from the Fisher Scientific Company.

(4) An article describing and evaluating this balance will be submitted for publication in the Journal of Analytical Chemistry. Also see Abst. Paper #12, Div. of Analytical Chem. Natl. Meeting, A.C.S., April 1957.

(5) Testut Company, 9, Rue Brown Sequard, Paris XV, France.

The rate of change in sample weight was obtained by going, horizontally, two divisions to the right and two divisions to the left of a particular point on the curve and extending two vertical lines in opposite directions at these positions to the curve. The vertical distance between the points on the curve was taken as the measure of the rate of change in weight. Since two horizontal divisions is equal to 1 min., the rate of change in weight is given as change in the number of divisions for every 2 min. Substantially the same results were obtained by drawing tangents to the curve. The sample was weighed before and after heating, on an analytical balance, and was within 2% of the theoretical weight loss.

Results and Discussion

Figure 1 is a continuous tracing of change in sample weight as a function of time and temperature where $CaC_2O_4 \cdot H_2O$ is heated from 25 to 1000°. Three distinct reactions are apparent, dehydration, decomposition of CaC_2O_4 to $CaCO_3$ and CO, and decomposition of $CaCO_3$ to CaO and CO_2. The weight losses are within 2% of the theoretical values. Since the reactions under consideration involve solid state decomposition it is assumed that the rate expression may be given in terms of amount of reactant where equations 9 and 13 apply.

The following relationships may be used to relate number of moles of reactant to weight

$$-\frac{dn_a}{dt} = -\frac{n_0}{w_c}\frac{dw}{dt} \qquad (18)$$

and

$$W_r = w_c - w \qquad (19)$$

where

n_0 = initial number of moles of A
w_c = weight loss at completion of reaction
w = total weight loss up to time, t

combining (18) and (19) with (9), equation 20 is obtained, which is used to evaluate the reaction kinetics

$$\frac{-E^*}{2.3R}\frac{\Delta\left(\frac{1}{T}\right)}{\Delta \log W_r} = -x + \frac{\Delta \log dw/dt}{\Delta \log W_r} \qquad (20)$$

Figure 2 is a graph of

$$\frac{\Delta \log dw/dt}{\Delta \log W_r} \quad vs. \quad \frac{\Delta(T^{-1})}{\Delta \log W_r}$$

for the three reactions. For the purpose of this plot, dw and W_r can be determined directly from the thermogram in terms of number of divisions. The values of order of reaction and energy of activation for dehydration, decomposition of CaC_2O_4 and $CaCO_3$ were found to be 1.0, 0.7, 0.4 and 22, 74 and 39 kcal./mole, respectively. See Table I.

Table I compares these results with those reported in the literature. The referenced rate studies were conducted under vacuum. There appears to be general agreement among investigators that the kinetics of decomposition is dependent on particle size, which may partially account for the lack of better agreement shown in Table I. For calcium carbonate the order of reaction falls between zero and unity with an energy of activation in the neighborhood of 42 kcal./mole, for decomposition under vacuum. This is in reasonable agreement with the values determined

TABLE I

COMPARISON OF RESULTS WITH REPORTED VALUES

Reaction	Order of reaction Exp.	Lit.	Energy of activation (kcal./mole) Exp.	Lit.
$CaC_2O_4 \cdot H_2O = CaC_2O_4 + H_2O$	1.0		22	
$CaC_2O_4 = CaCO_3 + CO$	0.7[6]		74	
$CaCO_3 = CaO + CO_2$	0.4	0.20–0.53[7]	39	
		0–1[8]		49[8]
		0.3[9]		35–42[9]
		1[10]		95[10]
		1[11]		41–44[11]
		0.3[12]		37–39[12]
				48[13]

(6) E. Moles and C. D. Vallamil, *Anal. soc. espan. fis. quím.*, **24**, 465 (1926).

(7) P. Vallet and A. Richer, *Compt. rend.*, **238**, 1020 (1954).

(8) G. F. Huttig and H. Kappal, *Angew. Chem.*, **53**, 57 (1940).

(9) H. T. S. Britton, S. J. Gregg and G. W. Winsor, *Trans. Faraday Soc.*, **48**, 63 (1952).

(10) Maskill and W. E. S. Turner, *J. Soc. Glass Tech.*, **16**, 80 (1932).

(11) G. Slonim, *Z. Elektrochem.*, **36**, 439 (1930).

(12) J. Splichel, St. Skramovsky and J. Goll, *Collection Czechoslov. Chem. Commun.*, **9**, 302 (1937).

(13) J. Zawadzki and S. Bretsznajder, *Z. Elektrochem.*, **41**, 215 (1935).

in this investigation, in air, 0.4 and 39 kcal./mole, respectively.

The reported order of reaction for the decomposition of calcium oxalate is unity. The order of reaction determined in this study is 0.7. No previous work has been found in the literature dealing with the kinetics of the dehydration of calcium oxalate monohydrate.

Acknowledgment.—The authors wish to express their appreciation to Mr. Jean Carton and the Testut Company of Paris, France for providing the thermobalance for this research.

The value of 0.3 was derived from the author's equation[9]

$$(w/w_0)^n = kt + a$$

where

$$n = 0.7$$

The above equation was differentiated with respect to (w/w_0) and t. This gives

$$\frac{- \, d \, (w/w_0)}{dt} = nk \left(\frac{w}{w_0}\right)^{1-n}$$

where

$$1 - n = \text{order of reaction}$$

ERRATA

Page 394, equation (4), the first element should read $\dfrac{E^* dT}{RT^2} \cdots$

Page 394, equation (5), should read $\dfrac{-E^*}{R}\left[\Delta\left(\dfrac{1}{T}\right)\right] = \cdots$

Page 395, equation (17), should read $\dfrac{\dfrac{-E^*}{R}\Delta\left(\dfrac{1}{T}\right)}{\Delta\ln\left(K(A-a)-Cp\,\Delta T\right)} = \cdots$

30

Reprinted from *Z. Anorg. Allg. Chem.*, **298**(3–4), 202–211 (1959)

Eine vielseitig verwendbare Thermowaage hoher Genauigkeit[1]

Von Horst Peters und Hans-Georg Wiedemann

Inhaltsübersicht

Es wird über eine Thermowaage berichtet, die in beliebiger, nicht aggressiver Atmosphäre und im Vakuum arbeitet und Temperaturen bis 1200° C erreichen läßt. Ferner wird der Einfluß der Gasatmosphäre auf den Nullpunkt und der temperaturabhängige Auftrieb von Substanz, Tiegel und Tiegelträger untersucht. Unter Berücksichtigung dieser Faktoren ist im zugänglichen Temperaturbereich mit der beschriebenen Waage bei einer maximalen Belastbarkeit von etwa 10 g eine Wägegenauigkeit von ± 0,08 mg zu erreichen.

Summary

A thermobalance is described, which works in any non-corrosive atmosphere and in vacuum up to 1200° C. Besides the influence of the atmosphere on the zero position and the buoyancy of the substance, the crucible, and the crucible bearer, which strongly depends on the temperature has been examined. Regarding these factors the described balance permits an weighing accuracy of 0.08 mg up to 1200° C, with a maximum load-carrying capacity of 10 g.

Im Verlauf einer Untersuchung der Reduktion von Oxyden durch eine Gasphase versuchten wir, für diesen speziellen Zweck eine Thermowaage zu verwenden, da sie es gestattet, an einer einzigen Probe den Ablauf der Reaktion, die mit einer Gewichtsänderung der Substanz verknüpft ist, in Abhängigkeit von der Temperatur und der Zusammensetzung der reduzierenden Atmosphäre zu verfolgen. Wir entwickelten eine vielseitig verwendbare, sehr genau arbeitende Thermowaage. Hier wird berichtet über ihren dem Problem angepaßten Aufbau und einige wichtige Effekte, die stets bei thermogravimetrischen Arbeiten auftreten, aber vielfach noch nicht genügend beachtet werden. In einer anschließenden Veröffentlichung wird gezeigt werden, daß sich unter sorgfältiger Berücksichtigung aller möglicherweise störenden Faktoren bei thermogravimetrischen Zersetzungen die Genauigkeit guter analytischer Waagen erreichen läßt.

---

[1] Teil der Diplomarbeit H. G. Wiedemann, Rostock 1955.

Allgemeiner Aufbau der Waage

Der allgemeine Aufbau der Waage ergab sich aus den durch das zu untersuchende Problem gestellten Anforderungen. Da die mit der Reaktion verbundenen relativen Gewichtsänderungen der Substanz klein sind, ist eine hohe Empfindlichkeit bei gleichzeitig größerer Belastbarkeit erforderlich. Diese Bedingung erfüllen Federwaagen, die verhältnismäßig einfach herzustellen sind, nicht, so daß man unter diesen Umständen eine Balkenwaage verwenden muß. Leider haben diese außer ihrem komplizierteren Aufbau den Nachteil, bei längerer Versuchsdauer ihren Nullpunkt und in geringerem Maße auch ihre Empfindlichkeit zu verändern. Diese Störung ist nach unseren Erfahrungen auf

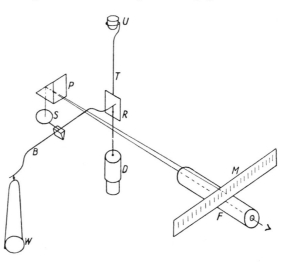

Abb. 1. Schematische Darstellung der Waage; B Waagebalken; S Spiegel; M Skale; F Fernrohr; U Tiegel; T Tiegelträger; R Aluminiumrahmen; D Dämpfungsglocke; W Waagschale; P Prisma

den Einfluß des Metallzeigers zurückzuführen und wird völlig vermieden, wenn man ihn entfernt und den Ausschlag der Waage nach Abb. 1 mittels eines am Waagebalken befestigten Spiegels mit Skala und Fernrohr beobachtet. Auf eine automatische Kompensation der Gewichtsänderungen oder Registrierung des Ausschlags wurde zunächst wegen des damit verbundenen Aufwands verzichtet. Von besonderer Bedeutung für das einwandfreie Funktionieren einer Thermowaage ist, wie schon Chevenard[2]) feststellte, die Anordnung des Ofens, in dem die Substanz erhitzt werden soll.

Befindet er sich unterhalb der Waage, so tritt, da der Ofen zur Durchführung des Drahtes, an dem die Probe hängt, nach oben offen sein muß, eine starke Konvektion der heißen Gase auf. Die dadurch auf den im Ofen befindlichen Tiegel einwirkenden, rasch schwankenden Kräfte verursachen eine Unruhe der Waage, die nur durch eine starke Dämpfung beseitigt werden kann. Viele der bisher beschriebenen Thermowaagen be-

---

[2]) P. Chevenard, X. Wache u. R. de la Tullaye, Metaux et Corrosion 18, 121 (1943).

14*

sitzen aus diesem Grunde eine Öldämpfung[3]), die aber eine Verringerung der Wäge-genauigkeit zur Folge hat. Ein weiterer Nachteil dieser Ofenanordnung liegt darin, daß die Möglichkeit einer einseitigen Erwärmung des Waagebalkens besteht, die sich in einer langsamen Nullpunktsverschiebung und einer Änderung des Waagearmverhält-nisses während des Betriebes bemerkbar macht.

Bei der von uns konstruierten Waage[4]) befindet sich der nach unten offene Metallblock-Ofen oberhalb des Waagegehäuses, wodurch allerdings ein besonderes Gestell, auf dem der Tiegel mit der Substanz ruht, not-wendig wird. Die Konvektion im Ofenraum ist bei dieser Anordnung so klein, daß eine einfache Luftdämpfung, wie sie bei Analysenwaagen üblich ist, völlig ausreicht. Die schematische Abb. 1 zeigt auf der rechten Seite des Waagebalkens B das Gehänge für den nach oben in den Ofen ragenden dünnen Tiegelträger T aus Quarzglas (1,5 mm Durch-messer), der oben einen Ring zur Aufnahme des Tiegels U besitzt.

Das Gehänge besteht aus einem leichtem, rechteckigen Aluminiumrahmen R, der unten an einem Metallstab die Dämpfungsglocke D aus Aluminium trägt. Im Innern dieser Glocke befindet sich ein kleines Stabilisierungsgewicht, das die Anordnung, die in das normale Steingehänge der Waage eingehängt ist, senkrecht hält. Etwa auftretende Pendelschwingungen der Einrichtung werden ebenfalls durch die Glocke D wirksam gedämpft.

Da die Reduktionsversuche in einem $CO-CO_2$-Gemisch genau be-kannter Zusammensetzung durchgeführt werden müssen, ist ein völlig gasdichter Verschluß der Apparatur nach außen erforderlich. Hierzu wurde die eigentliche Waage nebst Waagschale W und Dämpfungsein-richtung in einem dickwandigen, evakuierbaren Messinggehäuse, das an der Vorderseite mit einer Plexiglasscheibe verschlossen ist, unter-gebracht, wodurch gleichzeitig für eine gleichmäßige Temperatur-verteilung gesorgt ist.

Beschreibung der Waage

Da thermogravimetrische Untersuchungen sich im allgemeinen über mehrere Stunden erstrecken, wurde bei der Konstruktion auf große Stabilität Wert gelegt. Eine schwere, mit drei Schrauben (13, Abb. 2) horizontierbare Grundplatte (12) trägt drei Säulen aus Eisenrohr, von denen die beiden längeren in ihrem oberen Teil mit Schienen versehen sind, zwischen denen der Ofen (17) auf drei Rollen sicher auf und ab laufen kann. Sein Gewicht wird von zwei Bleigewichten (15), die von dünnen Drähten geführt werden, nahezu kom-pensiert. Der Ofen besteht aus einer starkwandigen, oben geschlossenen Edelstahlhaube, die eine bifilare, bis 1200° C verwendbare Heizwicklung trägt, die zur Isolation in eine keramische Masse eingebettet wurde. Durch die bifilare Wicklung werden magnetische Einwirkungen auf die Waage oder die Substanz vermieden. Es wurde ferner dafür Sorge getragen, daß die Temperatur im Ofen von unten nach oben etwas ansteigt, wodurch

---

[3]) K. Honda, Sci. Rep. Tôhoku Imp. Univ. **4**, 7 (1915); M. Guichard, Bull. Soc. chim. France **37**, 62, 251, 381 (1925).

[4]) H. Peters u. H. G. Wiedemann, Chem. Techn. **8**, 689 (1956).

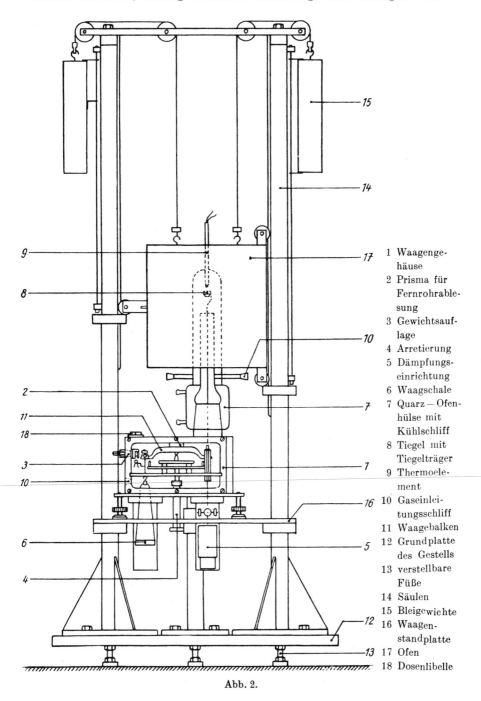

1 Waagenge-
   häuse
2 Prisma für
   Fernrohrable-
   sung
3 Gewichtsauf-
   lage
4 Arretierung
5 Dämpfungs-
   einrichtung
6 Waagschale
7 Quarz — Ofen-
   hülse mit
   Kühlschliff
8 Tiegel mit
   Tiegelträger
9 Thermoele-
   ment
10 Gaseinlei-
   tungsschliff
11 Waagebalken
12 Grundplatte
   des Gestells
13 verstellbare
   Füße
14 Säulen
15 Bleigewichte
16 Waagen-
   standplatte
17 Ofen
18 Dosenlibelle

Abb. 2.

241

Konvektion verhindert wird. Auf einer dünneren Aluminiumplatte (16), die auf drei an den Säulen festgeklemmten Ringen ruht, steht in dem dickwandigen Messinggehäuse (1), das mittels der Dosenlibelle (18) und dreier Stellschrauben waagerecht ausgerichtet werden kann, die eigentliche Waage. An der Unterseite des Gehäuses befinden sich zwei durch die Platte (16) nach unten hindurchtretende abschraubbare Aluminiumhülsen, von denen die linke die Waagschale (6) und die rechte die Dämpfungsglocke (5) beherbergt. Diese Hülsen waren zunächst aus Plexiglas gefertigt worden, das sich jedoch wegen der bei jeder Berührung auftretenden, sehr störenden elektrostatischen Aufladung als ungeeignetes Material erwies. Über dem rechten Gehänge der Waage, das bereits oben beschrieben wurde, ist ein weiter Metallschliff in das Gehäuse eingelötet, durch den der Tiegelträger nach oben in den Ofen ragt. Innerhalb dieses Schliffes befindet sich ein Satz blanker Reflexionsbleche aus Nickel, die die Wärmestrahlung des glühenden Ofens vom Waageninneren fernhalten. Da die Waage gasdicht abgeschlosen sein muß, ist über Tiegel (8) und Tiegelträger eine weite Quarzhülse gestülpt, die Schliffe für den durchzuleitenden Gasstrom besitzt. Ein am unteren Ende angeschmolzener Kühlschliff (7) wird während des Betriebs der Waage von einem HÖPPLER-Thermostaten temperiert. Es hat sich im Verlauf der Untersuchungen, wie später ausführlich behandelt werden soll, als unumgänglich erwiesen, die Waage während der Messung auf konstanter Temperatur zu halten. In einem Einzug am oberen halbrunden Ende der Quarzhülse steckt das Thermoelement (9), dessen Lötstelle so in die unmittelbare Umgebung der untersuchten Substanz kommt. Die eigentliche Waage mit der Arretierung, die von unten durch den mit einem Schliff abgedichteten Zapfen (4) betätigt wird, ist auf eine Platte montiert, die von vorn in das Gehäuse (1) eingeschoben wird. Diese Teile sowie die Steingehänge stammen von einer alten BUNGE-Analysenwaage. Da sich der Schwerpunkt des Waagebalkens durch den bereits oben begründeten Fortfall des Zeigers so weit nach oben verlagerte, daß durch Verstellen der Justiergewichte ein stabiles Gleichgewicht nicht mehr zu erreichen war, mußte das die Oberkante des Balkens bildende, jetzt überflüssige Reiterlineal entfernt werden. Um die Empfindlichkeit der Waage voll ausnützen zu können, wurde noch eine von außen zu betätigende Gewichtsauflage (3) angebracht. Der dazu erforderliche Mechanismus besteht aus zwei ineinander sitzenden konischen Schliffen, die gasdicht in die linke Gehäusewand eingesetzt sind. Der innere Schliff verschiebt beim Drehen mittels eines Zahnrades eine kleine Zahnstange, die an ihrem vorderen Ende einen Finger trägt, mit dem es möglich ist, von einem im Gehäuse befindlichen Gestell Reiter auf einen am linken Gehänge der Waage angebrachten Bügel zu setzen. Die Zahnstange läuft in einer Führung, die am äußeren der beiden Schliffe befestigt ist, so daß durch Drehen dieses Schliffes die Richtung der Zahnstange geändert werden kann. Für den Finger resultiert damit eine zweidimensionale Beweglichkeit. Mit vier Reitern von 10, 20, 40 und 80 mg können unter Ausnutzung einer Empfindlichkeit von 10 mg für den Vollausschlag vom linken zum rechten Ende der Skala Gewichtsänderungen bis 160 mg ausgeglichen werden. Außer der Gewichtsauflage befindet sich in der linken Gehäusewand noch eine Schlifföffnung (10), durch die die Waage evakuiert (erreichbares Vakuum 10^{-3} Torr) oder ein leichter Gasstrom geschickt werden kann, der bei der Zersetzung der untersuchten Substanz eventuell auftretende aggressive Gase oder kondensierbare Stoffe von der eigentlichen Waage fernhält.

Einfluß von Auftriebseffekten auf die Wägung

Bei einer normalen analytischen Waage ist die Nullage nur geringen, unregelmäßigen Schwankungen unterworfen. Anders bei einer Thermowaage, deren Gehäuse evakuiert, oder mit Gasen wechselnden Druckes oder Zusammensetzung gefüllt werden kann. Ihr Aufbau ist, abgesehen von Differential-Thermowaagen, stets unsymmetrisch, und die Teile der Waage, die auf dem linken und dem rechten Stein des Balkens ruhen, haben unterschiedliche Volumina bei gleichem Gewicht. Eine Änderung der Gasdichte, wie sie durch eine Änderung der Zusammensetzung oder der Temperatur der Atmosphäre hervorgerufen wird, hat damit eine unterschiedliche Änderung des Auftriebs dieser Waagenteile und eine Verschiebung des Nullpunktes zur Folge. Die hier beschriebene Waage zeigte, wenn die Luft im Gehäuse durch Wasserstoff ersetzt wurde, einen Ausschlag, der einer einseitigen Mehrbelastung von etwa 12 mg entsprach. Wie die Abb. 3 zeigt, ist die Verschiebung der Gasdichte proportional. Es ist zweckmäßig, diesen Effekt von vornherein beim Einbringen der Substanz und der Einregulierung des Ausgangspunktes an Hand einer ein für allemal zu bestimmenden Eichkurve zu berück-

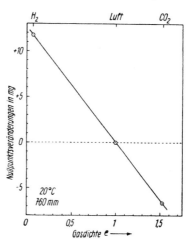

Abb. 3. Abhängigkeit der Nullage von der Gasdichte (Dichte der Luft unter Normalbedingungen = 1)

sichtigen, da es sonst vorkommen kann, daß nach dem Einleiten des Gases, in dem die Untersuchung vorgenommen werden soll, der Ausgangspunkt auf einer ungünstigen Stelle oder gar außerhalb der Skala liegt.

Abb. 3 läßt auch erkennen, welche Bedeutung die Temperaturkonstanz des Waagengehäuses während des Versuches für die Nullage hat.

Eine Dichteänderung der Gasphase um $1/_{300}$, — dies entspricht bei 27° C einer Temperaturänderung um 1°, — bewirkte bei der beschriebenen Waage eine Verschiebung des Nullpunkts um 0,04 mg. Die Temperatur des Messingsgehäuses wurde daher stets kontrolliert und eventuell durch Änderung der Temperatur des durch den Kühlschliff (7) (Abb. 2) strömenden Wassers konstant gehalten. Außerdem wurde, um die bei hohen Ofentemperaturen sehr störende Wärmestrahlung unschädlich zu machen, ein in Abb. 2 nicht gezeichnetes, spiegelndes Nickelblech zwischen Gehäuse und Ofen angebracht. Ohne diese Vorsichtsmaßregeln erhöht sich die Temperatur der Waage während eines mehrstündigen Versuchs leicht um 5 bis 10 Grad, was einer Nullpunktsverschiebung von 2 bis 4 mg entspricht. Hierdurch wird aber die Wägegenauigkeit, die sonst $\pm$ 0,08 mg beträgt, wesentlich erniedrigt.

Wie eine einfache Rechnung zeigt, ist die Gasdichte ϱ nicht nur für die Nullage, sondern auch für die Empfindlichkeit der Waage maßgebend. Dies folgt daraus, daß im allgemeinen der Volumschwerpunkt und der Massenschwerpunkt des Waagebalkens nicht mit dem Drehpunkt der Mittelschneide zusammenfallen.

Ist E die Empfindlichkeit der Waage in g^{-1}, r die Länge eines Waagarmes in cm, V_B das Volumen des Balkens, V_W das Volumen der Teile, die auf den äußeren Steinen ruhen, d der vertikale Abstand des Volumschwerpunktes des Balkens von der Mittelschneide und a der Abstand der Mittelschneide von der Verbindungsgeraden, der äußeren Steine, so ergibt sich:

$$\frac{dE}{d\varrho} = \frac{E^2}{r}(V_B\, d + V_W\, a).$$

Abb. 4 zeigt den an der Waage beobachteten Effekt, der so groß ist, daß es sich empfiehlt, bei Beginn jedes Versuches die Empfindlichkeit der Waage in der benutzten Atmosphäre gesondert zu bestimmen.

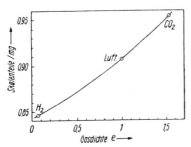

Abb. 4. Änderung der Empfindlichkeit der Waage mit der Gasdichte

Nach der angegebenen Gleichung sollte allerdings die Änderung von E mit ϱ um mehr als eine Zehnerpotenz geringer sein, wenn die bekannten Werte für E, r, V_B, V_W und sinnvolle Schätzwerte für a und d eingesetzt werden. Bisher gelang es nicht, diesen Widerspruch zu klären.

Normalerweise wird bei analytischen Wägungen die Reduktion des Ergebnisses auf den luftleeren Raum[5]) nicht vorgenommen. Die Fehler, die durch diese Vernachlässigung auftreten, können aber auch bei den in der analytischen Chemie üblichen Bestimmungen von Gewichtsverhältnissen größer als 1 Promille sein. Bei thermogravimetrischen Arbeiten dagegen muß auf jeden Fall der Gasauftrieb der Substanz und der Teile der Waage, die sich auf höherer, von der des Waagegehäuses abweichender Temperatur befinden, berücksichtigt werden. Mit zunehmender Temperatur nimmt die Dichte der Gasphase stark ab. Bei 300° C ist die Dichte und damit der Auftrieb nur noch halb so groß wie bei 25° C. In Luft ergibt sich hieraus pro cm³ eine scheinbare Gewichtsänderung von 0,6 mg.

Bei Annahme des idealen Gasgesetzes ist die Änderung des Auftriebs für einen Körper der Masse M und der Dichte d beim Erhitzen von der Temperatur τ (Zimmer- bzw. Waagentemperatur) auf die Ofentemperatur T:

$$\Delta A = M\,\varrho_\tau \frac{1}{d}\left[1 - \frac{\tau}{T}\right].$$

⁵) A. v. Lüpke, Angew. Chem. 68, 515 (1956).

Hier bedeutet ϱ_τ die während des Versuchs konstante Dichte der Gasatmosphäre im Waagegehäuse bei der Temperatur τ.

In Abb. 5 ist das für einen kleinen Quarztiegel von 1,8553 g errechnete ΔA (Luft) gegen die Temperatur aufgetragen.

Für Tiegel und Substanz ist eine rein rechnerische Ermittlung von ΔA möglich, da sie sich ja während des Versuchs auf konstanter Temperatur befinden, nicht aber für den Tiegelträger. Um für ihn ΔA zu erhalten, müßte das Temperaturgefälle zwischen Ofen und Waagegehäuse, das auch noch von der Wärmeleitfähigkeit des Gases abhängig ist, genau bekannt sein.

In Abb. 6 ist der experimentell gefundene Auftrieb für einen Tiegel und den Tiegelträger in verschiedenen Gasen gezeichnet. Wenn das Temperaturgefälle längs des Tiegelträgers in allen drei Fällen (Wasserstoff, Luft, Kohlendioxyd) gleich wäre, sollten bei jeder Temperatur die Abstände der Kurven von der Nullinie sich wie die Molgewichte bzw. die Dichten ϱ_τ der Gase verhalten. Während dies für Luft und CO_2 auch zutrifft, treten besonders bei höheren Temperaturen in Wasserstoff starke Abweichungen auf.

Es ist daher zweckmäßiger, für den Tiegelträger die Änderung des Auftriebs im benutzten Gas durch einen Blindversuch zu bestimmen. Die unterste Kurve in Abb. 5 zeigt das Ergebnis eines Blindversuchs für einen Tiegelträger aus Quarzglas von 1,5 mm Durchmesser und 32 cm Länge. Durch Addition der Werte für Tiegelträger und Tiegel erhält man die zweitoberste Kurve, die mit der obersten, die direkt durch einen Blindversuch mit Tiegelträger und eingesetztem leeren Tiegel erhalten wurde, übereinstimmt. Bemerkenswert ist in der Abbildung der rasche Anstieg aller Kurven bei niedrigen Temperaturen, der wieder die Bedeutung einer konstanten Waagentemperatur erkennen läßt. Eine kleine

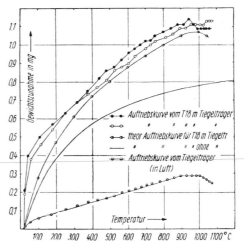

Abb. 5. Änderung des Auftriebs eines Tiegels und eines Tiegelträgers mit der Temperatur (in Luft)

Änderung von τ hat eine verhältnismäßig große Änderung von ΔA zur Folge. Das am rechten Ende der obersten Kurve (ΔA für Tiegel und Tiegelträger) erkennbare Absinken ist auf eine gegen Ende des Versuchs aufgetretene geringe Erhöhung der Waagentemperatur durch die Strahlung des Ofens zurückzuführen.

Nach diesen Ausführungen sind thermogravimetrische Zersetzungs-kurven, die durch langsames Erhitzen der Substanz und Registrieren

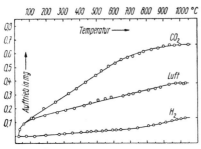

der gleichzeitigen Gewichtsver-änderung erhalten werden, in folgender Weise auszuwerten:

Abb. 6. Änderung des Auftriebs eines Tie-gels mit Tiegelträger mit der Temperatur in verschiedenen Gasen

Zunächst ist von der erhalte-nen Kurve der zu den entsprechen-den Temperaturen gehörende ΔA-Wert für den Tiegel und den Tie-gelträger abzuziehen. Aus der so korrigierten Kurve ist jetzt zu be-stimmen, ob und welche definier-ten Zersetzungsprodukte entstan-den sind. Mit den für diese Sub-stanzen bekannten, oder, falls diese fehlen, geschätzten Dichten ist deren Auftrieb

$$A = \frac{M}{d} \cdot \varrho_\tau \frac{\tau}{T}$$

zu berechnen und ebenfalls zu subtrahieren.

Bei der Anwendung dieses Korrekturverfahrens sind die benutzten Gewichtsstücke nicht mit ihrem Massenwert m, sondern mit ihrem „Luft"-Gewicht m⁺ einzusetzen.

$$m^+ = m \left[1 - \frac{\varrho_\tau}{d} \right].$$

Zur Ermittlung des Auftriebs der Substanz und der Zersetzungsprodukte genügt es, in die oben angegebene Gleichung für M an Stelle der Massen die „Luft"-Gewichte einzu-setzen.

Wird der Einfluß des Auftriebs bei thermogravimetrischen Unter-suchungen nicht berücksichtigt, so sind, wie sich aus der Literatur nach-weisen läßt, Fehldeutungen der Ergebnisse möglich. Zum Beispiel ist aus der durch die Abnahme des Auftriebs A mit zunehmender Tem-peratur vorgetäuschten Gewichtszunahme mehrfach auf eine Aufnahme von Sauerstoff aus der Luft durch in Wirklichkeit völlig inerte Sub-stanzen geschlossen worden. Diese Irrtümer treten dann besonders leicht auf, wenn Tiegel und Tiegelträger verhältnismäßig groß und schwer sind. Die hier besprochenen Auftriebseffekte sind nicht nur für Waagen vom hier beschriebenen Typ von Bedeutung, sondern auch für alle anderen Thermowaagen. Ist bei einer Waage der Ofenraum mit einem anderen Gas als Luft zu bespülen, so tritt stets eine Nullpunkts-verschiebung auf, so auch bei der bekannten, von DUVAL benutzten CHEVENARDschen Waage.

Die Genauigkeit der hier beschriebenen Waage ist sehr hoch; sie ist jedoch nur unter Berücksichtigung der oben behandelten Faktoren voll auszunutzen. Im Betrieb ergab sich eine Konstanz des Nullpunkts über mehrere Tage hinweg von $\pm$ 0,03 mg. Der mittlere Fehler einer Einzelwägung ist, wie in längeren Meßreihen festgestellt wurde, etwa $\pm$ 0,05 mg. Bei thermogravimetrischen Untersuchungen wird der Fehler durch die oben besprochenen Korrekturen etwa größer, bleibt aber in der Regel unter $\pm$ 0,08 mg.

In der anschließenden Veröffentlichung wird als Beispiel für die Genauigkeit der Waage die thermische Zersetzung des Calciumcarbonats und des Calciumoxalats und an Hand der erhaltenen Zersetzungskurven die Auswertung mitgeteilt werden.

Rostock, Institut für Physikalische Chemie der Universität.

Bei der Redaktion eingegangen am 31. Juli 1958.

31

Reprinted from *Anal. Chem.*, **32**(5), 271R–289R (1960)

Review of Fundamental Developments in Analysis

Automatic and Recording Balances

Saul Gordon and Clement Campbell

Picatinny Arsenal, Dover, N. J.

MODERN TRENDS in laboratory research investigations have been towards the automatic and continuous recording of data which could previously be obtained only by tedious and time-consuming manual measurements involving point by point readings and subsequent plotting of the results. In recent years, commercially available instrumentation has been very widely applied to electroanalytical techniques (potentiometry, amperometry, polarography, conductometry, coulometry), absorption spectrophotometry, chromatography, gas analysis (thermal conductivity, combustion analyzers), the measurement of physical properties such as density, viscosity, resistivity, and of conditions such as temperature, pressure, and relative humidity. However, despite these rapid and profound advances in laboratory automation, the fundamental operation of precision weighing has been relatively neglected until recent years (*94*). Although a number of automatic and recording balances were developed and used by individual research investigators, most laboratory investigations involving a study of weight changes as a function of time, temperature, or any other variable, were accomplished by means of point by point measurements. Despite these many balances reported in the literature, it was not until the Chevenard thermobalance was marketed in France about 1950 that a recording laboratory balance became commercially available. With the publication of Duval's monograph (*55*) describing thermogravimetric studies of inorganic analytical precipitates, using the Chevenard thermobalance, the application of automatic recording balances was brought to the attention of chemists in many fields for whom such equipment could be of inestimable value.

The past 2 years have seen increased activity among instrument manufac-

turers both here and abroad, witnessed by the present choice of 18 commercially available recording balances: six domestically manufactured, one each in England and Switzerland, three in France and seven in Western Germany. In addition to the use of an automatic recording balance for thermogravimetric analysis and the evaluation of precipitates proposed for gravimetric analysis, there are a great many other applications for which a recording balance may be used (Table I) in fields such as metallurgy, textiles, ceramics, paint and coal technologies, biochemistry, biological sciences, mineralogy, and most of the other fields of physicochemical endeavor. This review describes the types of automatic recording balances, the principles of their operation, and the areas in which they have been applied. The authors have tried to include descriptions of all the balances that are automatically recording, or obviously lend themselves to such operation. In addition, there is a comprehensive survey of the commercially available recording balances manufactured both here and abroad.

PRINCIPLES OF OPERATION

The primary requirements of a suitable automatic and continuously recording balance are essentially those for an analytical balance—accuracy, sensitivity, reproducibility, capacity, rugged construction, and insensitivity to ambient temperature changes. In addition, the recording balance should have an adequate adjustable range of weight change, a high degree of electronic and mechanical stability, be able to respond rapidly to changes in weight, be relatively unaffected by vibration, and be of sufficiently simple construction to minimize the initial cost and need for maintenance. From a practical point of view, the balance should

Table I. Applications of Automatic and Recording Balances

Corrosion of metals
Thermal decomposition of inorganic and organic compounds
Solid state reactions
Roasting and calcination of minerals
Thermochemical reactions of ceramics and cermets
Pyrolysis of coals, petroleum, and wood
Determination of moisture, volatiles, and ash
Absorption, adsorption, and desorption properties of materials
Rates of evaporation (drying curves) and sublimation; latent heats
Dehydration and hygroscopicity studies
Effect of load stress on fibers
Sedimentation of aerosols and powdered materials
Rates of crystallization and diffusion
Studies of permeability and osmotic pressure changes
Changes in specific gravity or gas density
Specific surface as determined by weight of adsorbed gas
Metabolic rates of small plants and animals
Variations in weights of small laboratory animals
Automatic thermogravimetric analysis
Effect of radiation on various substances
Thermal and oxidative degradation of polymers
Growth of living organisms in controlled environments
Weight of effluent from chromatographic or other separatory system
Development of gravimetric analytical procedures

be simple to operate and versatile in that it can be used for varied applications by the addition of auxiliary equipment, such as sample holders, heating and cooling devices, and vacuum or controlled atmosphere jackets.

The recording balances reported in the literature or commercially available may be grouped into two types on a basis of their mode of operation. They are the null-point and the deflection types of instruments, schematically illustrated in Figures 1 and 2.

[*Editors' Note:* Figures 12, 13, 16–20, and 22–26 have been omitted because the original halftones were unavailable.]

The automatic null-point balance, Figure 1, incorporates a sensing element which detects a deviation of the balance beam from its null position; horizontal for beam balances and vertical for the electromagnetic suspension type. Through appropriate electronic or mechanical servo-linkage, a restoring force—electrical or mechanical weight loading—is then applied to the beam, returning it to the null point. This restoring force, proportional to the change in weight, is recorded either directly or through an electromechanical transducer. Deflection balances, Figure 2, involve the conversion of balance beam deflections about the fulcrum into photographically recorded traces; recorded electrical signals generated by appropriate displacement measuring transducers; or electromechanically drawn curves. Related types of deflection balances are: the helical spring wherein changes in weight are manifested in contractions or elongations which may be automatically recorded by suitable transducers; the cantilevered beam, constructed so that one end is fixed and the other end, from which the sample is suspended, is free to undergo deflection; the suspension of a sample by an appropriately mounted strain gage that stretches and contracts in proportion to the weight changes; and the attachment of a beam to a taut wire which serves as the fulcrum and is rigidly fixed at one or both ends so that deflections are proportional to the changes in weight and the torsional characteristics of the wire.

A number of electronic and electromechanical principles may be employed for detecting deviations of the balance beam from its horizontal or vertical rest point in the null-point recording balance. These methods are summarized in Table II. Photoelectric and photomechanical detectors make use of the varying intensities of a light source impinging upon a phototube as a result of balance beam fluctuations about the null position. This is accomplished by a light source, intermediate shutter or mirror, and either single or dual phototubes. The displacement of the shutter attached to the balance beam intercepts the light beam so as to either increase or decrease the intensity of light acting upon the phototube due to the change in weight. The resulting change in the magnitude of the current from the phototube is used to electronically or electromechanically restore the balance to its null point. It is also possible to use a mirror mounted on the balance beam to reflect the light upon a dual phototube, so that fluctuations of the beam result in a deflection of the light to one photocathode or the other as changes in weight occur. Other

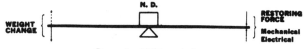

Figure 1. Null-type balance

Restoring force proportional to change in weight. Null detector, N.D., may be optical (electronic), mechanical, electromechanical, electrical, or radiation detector

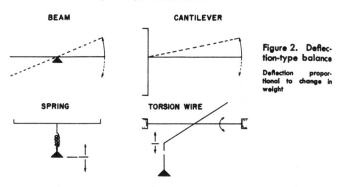

Figure 2. Deflection-type balance

Deflection proportional to change in weight

methods for electronic null detection include the change in: capacity of a condenser as a function of balance beam displacement; inductive coupling between plates and/or coils; magnetic coupling; the output of a differential transformer as a function of armature displacement; nuclear radiation flux; or output of a strain gage circuit. In each of these cases, one element of the transducing system is attached to the balance beam in close proximity to the other element which is maintained in a fixed position. Mechanical methods, although less satisfactory, are also feasible for this purpose. These involve "feelers" which operate servomotors through appropriate relays, or are modifications of the older type mechanical galvanometric stripchart recording potentiometers.

Having detected the departure of the balance beam from its rest point, any one of several methods summarized in Table III may be used to restore the position of the balance to its null point. The appropriate restoring force can be applied by either electronic or electromechanical means. Electronic restoring devices may be solenoid coils operating upon a magnetic armature attached to the balance beam and freely suspended within the coil. A current of sufficient magnitude is applied to the coil so that the balance beam remains at its null point. This current can be obtained either directly from the null-detecting circuit or through the operation of a servo-driven potentiometer. The restoring force can also be supplied by the repulsion of a coil fixed in position below an opposing short-circuited coil, which is attached to the balance beam; the necessary current may be supplied in the same

manner as for the afore-mentioned solenoid. An additional modification of a restoring force coil involves energizing a coil attached to the beam and located in the field of a permanent magnet. A completely mechanical form of weight loading uses a servo-mechanically driven chain on a conventional type of chainomatic balance. To restore the balance beam after a gain or loss in weight, it is necessary to use either two motors or a reversing motor with the appropriate electronic circuits for operation by the null detector. Completely mechanical linkage between a chain, balance beam, and an older type of mechanically recording potentiometer can also be utilized for automatically restoring the balance to its rest point. The restoring force can also be applied by the electrolytic dissolution or deposition of an electrode suspended from the balance beam.

Recording the changes in weight, methods for which are shown in Table IV, is best accomplished by measuring the current or voltage applied to the null-point restoring device, such as the solenoid or the inductive coil. Alternatively, precision potentiometers may be coupled to the servomotor-operated chain drive, or to the mechanical device used to energize the restoring force coil. This precision potentiometer is made a part of a Wheatstone bridge circuit or energized by a battery with suitable limiting resistors and then connected to a recording instrument. A balance consisting of a beam, chain, and old mechanical recorder can be modified to produce a record directly as a result of the mechanical linkages involved.

The classical method of recording with deflection type balances is that of

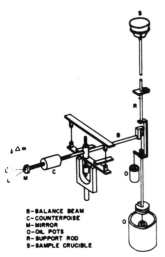

B - BALANCE BEAM
C - COUNTERPOISE
M - MIRROR
O - OIL POTS
R - SUPPORT ROD
S - SAMPLE CRUCIBLE

Figure 3. Chevenard photographically recording deflection thermobalance

measuring the balance beam deflections photographically by means of a light beam reflected from a mirror mounted on the balance beam; relatively small deflections are magnified by the use of a sufficiently long optical path. This technique involves recording the deflections directly on a photographic paper wrapped around a motor-driven drum. The deflections may be measured electronically by a shutter attached to the balance beam and operating so as to intercept a beam of light impinging upon a phototube; the light intensity is a measure of the balance beam displacement, and therefore of the change in weight. A capacity-sensitive device, inductive-sensing element, radiation detector, or linear variable differential transformer can also be used to measure beam deflections electronically. The first three methods require that a condenser element, induction coil, or radioactive material be suspended from the beam; last uses an armature freely suspended from the beam into the differential transformer coil. Strain gages connecting the beam to the base of the balance have also been used. Appropriate circuits are employed with these devices to convert the changes in electrical properties to signals which may then be recorded.

The selection of the type of automatic recording balance to be purchased or constructed depends largely upon the application and a consideration of the physical and chemical parameters involved. Among these are the load, range of weight change, accuracy, sensitivity, speed of response, environmental conditions of atmosphere, tem-

Table II. Transducers for Null-Point Detection and for Measuring Changes in Weight with Deflection-Type Balances

OPTICAL

Light source-mirror-photographic paper
Light source-shutter-photocell

ELECTRONIC

Capacitance bridge
Mutual inductance: coil-plate, coil-coil
Differential transformer or variable permeance transducer
Radiation detector (Geiger tube)
Strain gage

MECHANICAL

Pen electromechanically linked to balance beam or coulometer

Table III. Methods of Applying a Restoring Force to Null-Type Balances

MECHANICAL

Addition or removal of discrete weights; or beam-rider positioning
Incremental or continuous application of torsional or helical spring force
Incremental or continuous chainomatic operation
Incremental addition or withdrawal of liquid (buoyancy)
Incremental increase or decrease of pressure (hydraulic)

ELECTROMAGNETIC INTERACTION

Coil-armature
Coil-magnet
Coil-coil

ELECTROCHEMICAL

Coulometric dissolution or deposition of metal at electrode suspended from balance beam or coulometer

perature and pressure, type of recording system, and constructional complexity of the apparatus. Although a null-type balance has an inherently larger load-to-range ratio than a deflection balance, their accuracy, sensitivity, and speed of response are comparable as they are primarily a function of the electronic or electromechanical servo-systems required for automatic graphic recording. This limitation is inherent in any of the electronic recorders which have a practical accuracy no better than ±0.2%, even though other components of the balance may be capable of greater accuracy. With automatic weight loading and multirange recording over several full-scale excursions, a greater accuracy can be achieved. However, the null balances are also more complex because of the null detector and restoring-force system which are not required in deflection-type instruments. The choice of recording technique must be made among the photographic, mechanical-pen linkage, and electronic graphic methods. With the photographic techniques, which are confined to the deflection

Table IV. Recording Techniques for Automatic Balances

MECHANICAL

Pen linked to potentiometer slider
Pen linked to chain-restoring drum
Pen or electric arcing-point on end of beam
Pen(s) linked to servo-driven photoelectric beam-deflection follower

PHOTOGRAPHIC

Light source-mirror-photographic paper
Drum: time base
Flatbed: temperature base-mirror galvanometer

ELECTRONIC

Current generated in a transducing circuit—e.g., photocell, differential transformer, variable permeance transducer, strain gage, bridge, radiation detector, capacitor, inductor
Current passing through the coil of an electromagnet

balances, extremely accurate recordings can be obtained at the expense of the inconvenient handling and processing involved with light-sensitive papers. Electronic recording is more versatile and convenient than a mechanically linked system because of the many transducers that can be used to obtain an electrical signal proportional to the change in weight as measured by either a null or deflection balance. Furthermore, the continuously recorded analog data from the primary curve can be simultaneously translated into other useful forms such as derivatives, integrals, logarithms, or any other desired function, many of which lend themselves to the digital operations associated with automatic computation and automated processes.

AUTOMATIC AND RECORDING BALANCES

Deflection Balances. Most of the automatic and recording balances reported in the literature were designed for specific studies and then used only in the laboratories of the respective investigators. The earliest automatically recording balance was constructed by Kuhlmann (90) in 1910. His apparatus, using the deflection principle, consisted of an analytical balance with a mirror mounted on the center of the beam. A light focused on this mirror was reflected onto a piece of photographic paper wrapped around a clock-motor-driven drum. The relative deflection on the paper was proportional to the change in weight. When the light beam reached the edge of the photographic paper, a photoelectric relay actuated a mechanism which arrested the balance while a weight was dropped onto the pan. Upon release the balance beam returned to its original position to begin a new deflection. Abderhalden (1, 2) described a similar completely automatic photographically recording deflection balance for studies of animal metabolism which included a means of extending the range by the automatic electromechanical addition and subtraction of weights. Another single range general purpose deflection balance of this type was developed by Kohler (*89*).

The first automatically recording thermobalance was reported by Dubois (*52*), who made use of the afore-mentioned photographic recording technique. Samples under investigation were suspended into a furnace from the balance beam. Jouin (*83*), Zagoraki (*165*), and Vytasil *et al.* (*159*) constructed similar equipment to study mineralogical specimens, the pyrolysis of analytical precipitates, and the drying of enamels, respectively. Chevenard, Waché, and de la Tullaye (*38*) developed a recording thermobalance in which the balance beam was suspended by wires and the light source—mirror—photographic drum recording technique was used. This instrument was the prototype for the present commercially available Chevenard thermobalance (*5*) shown schematically in Figure 3 and described later in this review. Longchambon (*97*) applied the photographic principle to a unique method of recording changes in weight directly as a function of temperature for studies of the dehydration of minerals. This deflection balance had a mirror mounted on the beam so that the vertical components of light deflections on the photographic paper were proportional to the change in weight. A galvanometer mirror was also placed in the light path so that the horizontal component of the light deflection was proportional to the temperature of the furnace into which the sample was suspended. Similar recording thermobalances were later constructed by Orosco (*111*) for studying the dehydration and decomposition of mineralogical samples and by Spinedi (*159*) for investigating the oxidation of metals and the decomposition of metallic oxides.

Tryhorn and Wyatt (*155*) utilized the hydrostatic principle for automatically recording weight changes in connection with their studies of adsorption and in measurements of diffusion coefficients. Their recording hydrostatic balance, made from an ordinary analytical balance, was a deflection-type instrument with a rod suspended from one end of the balance beam and partially immersed in a nonvolatile liquid. The deflection of the beam from its equilibrium position was a function of the change in weight, and the linear range was determined by the radius of the rod and the density of the liquid. A mirror, mechanically linked to the balance beam, was used to de-

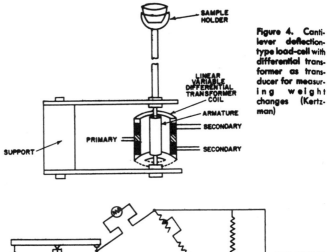

Figure 4. Cantilever deflection-type load-cell with differential transformer as transducer for measuring weight changes (Kertzman)

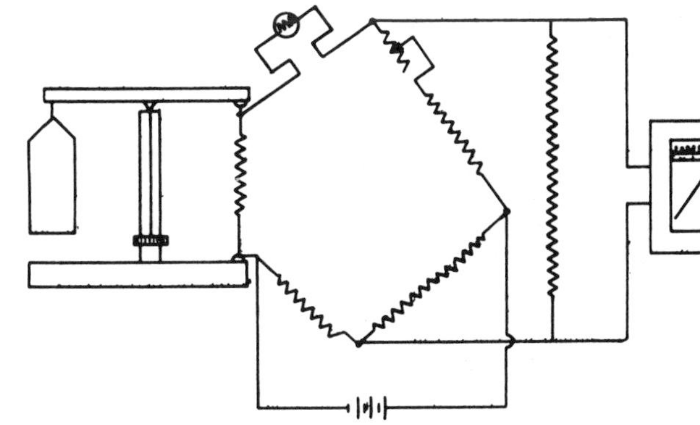

Figure 5. Strain gage-type deflection balance (Bartlett and Williams)

flect a beam of light onto a drum camera and thereby record the change in weight as a function of balance beam displacement. This type of hydrostatic balance was adapted by Campbell and Gordon (*33*) for graphic electronic recording by using a linear variable differential transformer to measure the balance beam deflections. Another automatic recording balance based on the hydrostatic principle was developed by Dervichian (*50*) for recording variations in surface tension as a function of the surface and of time. This deflection balance had a microscope cover glass suspended from one end of the beam into the liquid whose surface tension was to be measured, and a rod suspended from the other end into an oil bath which determined the range and acted as a damper. A mirror fixed to the beam reflected a light spot onto photographic paper for automatic recording. By using a torsion wire balance with magnetic damping, Andersson and the Stenhagens (*10*) improved upon the Dervichian-type surface tension recording balance. Spinedi (*138*) studied the anodic aqueous dissolution of metals by means of a photographi-

ically recording deflection balance having a mirror mounted on the beam pointer to deflect the beam of light. Brefort (*27*) in his studies of the dehydration of salts employed a photographically recording thermobalance which made use of two optical systems. The change in current of a photocell, upon which was focused a light intercepted by a shutter attached to the deflection balance beam, was proportional to the beam deflections. This current was recorded by means of a mirror galvanometer reflecting a beam of light onto a drum-wound photographic paper.

A simple torsion-type sedimentation balance for particle size analysis was devised by Bostock (*24*) for reading directly the beam deflections on a frosted glass scale in a manner also used by Peters and Wiedemann (*115*) in their manual beam deflection thermobalance. These balances are of interest because of the ease with which they could be converted to automatic recording instruments. An electrical setup which relates deflection of the analytical balance beam to electrical

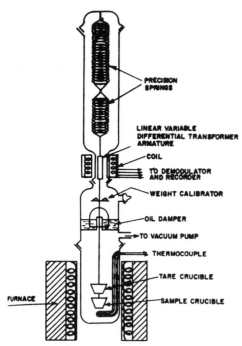

Figure 6. Spring-type deflection balance using differential transformer as transducer (Campbell and Gordon)

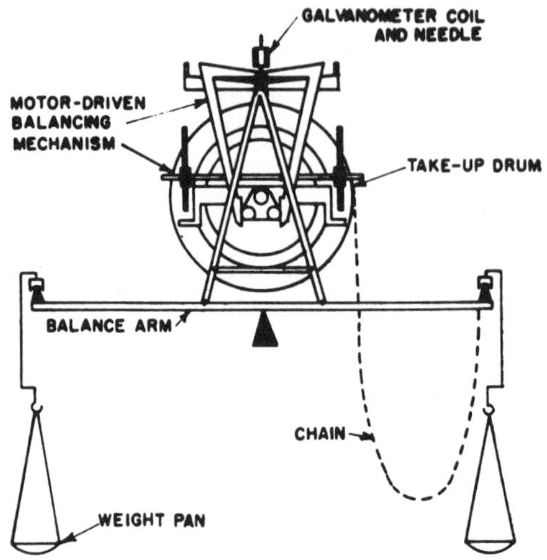

Figure 7. Null-type recording balance constructed from mechanical galvanometric recording potentiometer (Ewald)

current, has been described by Deryagin *et al.* (*51*).

Binnington and Geddes (*18*) reported an automatic recording mechanical balance of the deflection type which they used to conduct drying studies of grain products. A record of the beam displacement was obtained by a timed spark emitted from a beam-mounted stylus to the metallic drum of a variable-speed kymograph through a chart paper wrapped around the drum. At the point of maximum beam deflection, a wire attached to the beam contacted a pool of mercury and actuated a device which sequentially dropped steel balls onto the balance pan, thus restoring the beam to its starting position. Rogers and Earle (*124*), in their studies on macaroni drying, adapted a solution balance for automatic recording using an incremental weight loading system and recording technique similar to that of Binnington and Geddes.

Ahmad (*4*) devised an automatic recording deflection balance for measuring the moisture content of hygroscopic materials. The range of beam deflection was controlled by a spring, with the pen connected to the beam by a system of levers. For a study of the evaporation of solvents, Couleru (*45*) built a deflection thermobalance with

an off-center-pivoted beam. The sample was suspended into a furnace from the center of the long arm, at the end of which was a pen used to record changes in weight of several grams on a rotating cylinder.

Linear variable differential transformers have been recently employed as transducers for several deflection-type automatic recording balances. By suspending the transformer armature from one end of the balance beam and recording the weight as a function of beam deflections, Peterson (*116*) was able to convert an ordinary analytical balance to a recording microbalance. Campbell and Gordon (*35*) used this transducer and the hydrostatic principle of Tryhorn and Wyatt (*155*) for a simple conversion of analytical balances to the automatic recording of weight changes from micrograms to grams. They (*69*) also modified a photographically recording Chevenard thermobalance for graphic pen-and-ink recording by using the differential transformer to transduce the balance beam deflections. Šatava (*130a*) used a differential transformer to obtain meter readings proportional to the deflections of a thermobalance beam. Projection-type single pan constant load balances have also been modified for automatic recording over a 100-mg.

range by means of this type of transducer as reported by both DeLong (*49*) and Williamson (*164*).

For absorption analyses Claesson (*40*) used an instrument which automatically recorded weight as a function of refractive index. A current proportional to the change in refractive index was obtained by means of a photocell arrangement and fed into a mirror-galvanometer. The balance mechanism was the cantilever type with a mirror mounted on the sample end of the beam to deflect a light beam a distance proportional to the change in weight. By properly positioning the mirrors on the balance and galvanometer, a direct plot of weight *vs.* refractive index was obtained on photographic paper in a manner similar to that used by Longchambon (*97*). A cantilevered load cell with a differential transformer for measuring deflections, illustrated in Figure 4, has been proposed by Kertzman (*85*) as an automatic recording balance.

Blau and Carlin (*90*) described the application of radioactivity to the detection and measurement of balance beam displacement by coupling a piece of radioactive foil to the balance arm near a double ionisation chamber. A nonrecording version of this microbalance was subsequently exhibited by Arthur S. LaPine & Co. (*92*). Feuer (*65*) developed a similar sensitive radioactive electronic detector that was used in a commercially available Seeder-Kohlbusch microbalance to obtain a reported sensitivity of at least 1 γ per ml.

A strain gage was used to weigh eluriant solutions in a precision re-

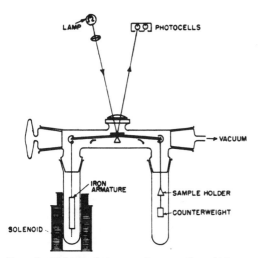

Figure 8. Null-type electromagnetic vacuum thermobalance with photoelectric null detector (I. Eyraud)

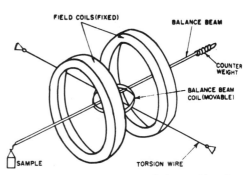

Figure 9. Null-type torsion microbalance with coil-coil electromagnet null detector and restoring force (Vieweg and Gast)

cording refractometer developed by Trenner, Warren, and Jones (*154*) for chromatographic analysis. The output of the strain gage bridge circuit, proportional to changes in weight, was automatically plotted on a potentiometric recorder. This principle is also being used by the Phoenix Precision Instrument Co. in its recording differential refractometer (*118*) with which the refractive index of a liquid can be plotted directly as a function of the mass of sample passing through the instrument. A strain gage wire was also used by Bartlett and Williams (*14*) in their recording thermobalance for studying the oxidation of metals, shown schematically in Figure 5. One end of the beam of an analytical balance is fastened to the base with the strain gage wire so that as the sample suspended from the other end undergoes an increase in weight, the resistance of the wire increases. This strain wire is part of a Wheatstone bridge circuit from which a signal proportional to the weight of the sample is fed into a potentiometric recorder.

By attaching a mirror to a tungsten spring balance so that a beam of light was deflected onto a photographic paper wrapped around a rotating drum, Isvekov (*80*) automatically recorded changes in weight of a sample suspended into a furnace. Van Nordstrand (*157*) suspended the armature of a linear variable differential transformer in line with the sample from a quartz spring deflection balance. This apparatus was used for automatic determinations of the specific surface of powdered materials by the B.E.T. method, wherein the change in weight is a function of the amount of gas adsorbed by the

sample at low temperatures. Vacuum thermobalances using quartz or metal springs and differential transformers have been reported by Hooley (*78*) and by Campbell and Gordon (*34*), illustrated in Figure 6. Klevens (*88*) has also used this kind of spring balance for adsorption and desorption studies. A vacuum thermobalance of the spring type, the Thermo-Grav, is being manufactured by the American Instrument Co. (*8*) and is described later. A simple recording metal spring balance for particle size determinations by sedimentation was constructed by Rabatin and Gale (*123*). Changes in weight were converted into electrical signals for graphic recording by a light-shutter-photocell system. Stephenson, Smith, and Trantham (*143*) also developed a spring-type recording vacuum balance for low temperature sublimation studies. The spring in this apparatus was made of copper and served as one lead of a copper-constantan thermocouple embedded in the sample, the other lead running along the wall of the balance chamber sample. Recording was achieved by projecting the image of a cross-hair on the spring onto a piece of moving photographic film.

Okeanov (*110*) constructed a device for recording the rates of vaporization of liquids which consisted of a spring balance carrying a sample-containing disk, the displacement of the disk during evaporation being transmitted to a recording stylus by a system of relays.

Null Balances. The largest number of automatic and recording balances described in the literature have been of the null type, Figure 1, which basically consists of the balance

beam assembly, a means of detecting deviations from the equilibrium point, and a means of applying a force of the proper magnitude and direction to restore the beam to its null point. The simplest approach to automatic operation using the null principle is to simulate the manual operation of an analytical balance by mechanically adding weights incrementally to the balance pan to maintain equilibrium.

Oden (*108*) utilized this principle in an automatic incremental-weight recording analytical balance which he used for sedimentation studies of the particle size distribution of soils and clays. When beam deflections equivalent to predetermined weight changes were realized, electrical contacts were established between small electrodes mounted on the beam and fixed mercury pools, at which time small weights were electromechanically dropped onto the pan and timing marks were simultaneously recorded. Improvements in the operation of the balance and recording system were made and described subsequently (*107, 109*). Suito and Arakawa (*145*) constructed a somewhat similar electromechanical sedimentation balance. Starodubtsev (*142*), in his sedimentation balance, connected one end of the balance beam to a motor-driven cam by means of a sensitive spring. As the beam deflects, electrical connection is made between a contact mounted on the beam and one just above it, causing the motor to turn the cam and incrementally restore the balance to equilibrium. A pen is linked to the cam so that as the cam is turned the pen is moved vertically on a motor-driven recording drum.

The deficiency inherent in this method is that the changes in weight are compensated for incrementally rather than continuously. Therefore, as in the development of analytical balances, advantage was taken of the chainomatic principle to vary continuously the gravitational force required to maintain the balance beam at its null point. The application of this improvement for recording balances was first reported by Müller and Garman (104). They described an electromechanical chainomatic balance in which the detector consisted of a light source, shutter, and phototube arranged to sense a deflection from the null point resulting from an increase or decrease in weight. The signal generated was amplified and applied to a reversing motor through a phase-discriminating circuit employing thyraton switching tubes. The motor rotated in the direction required to drive a chain-restoring mechanism and bring the balance beam back to the null point. This weight was recorded by picking off potentials from a battery-energized slide-wire potentiometer which was mechanically coupled to the chain-restoring mechanism. The position of the slide-wire contactor at any time produced a voltage which was equivalent to the instantaneous weight of the sample. An automatic recording null-type thermobalance developed by Hyatt, Cutler, and Wadsworth (79) for ceramic investigations consisted of a similar servomotor-driven chainomatic balance with a light source-mirror-dual photocell null detector and a precision potentiometer geared to the chain drive.

Ewald (59) converted a standard analytical balance and mechanical galvanometer potentiometric recorder to an automatically recording balance shown schematically in Figure 7. As the null point deviations were mechanically linked to the galvanometer of the recorder, the normal restoring action of the recorder was used to drive a chain attached to the balance beam. This restored the balance to its rest point and simultaneously recorded the change in weight. The chain was selected and mounted in such a way as to obtain any desired range of recorded weight changes. Thall (152) combined an analytical balance and mechanical potentiometric recorder to accomplish automatic recording in much the same way as Ewald.

In order to record automatically large changes in weight during drying operations, Muller and Peck (105) devised a null balance with a reversible motor-driven, weight-restoring chain operated by an electromechanical beam-contacting null detector. The recording chart was connected to the chain drum, and the pen was moved along the time axis by a modified clock motor. A similar balance was employed by Besser and Piret (17) to study the drying of materials by dielectric heating.

Jones and Tinklepaugh (81) modified a chainomatic analytical balance so that a radio-frequency capacitance

bridge, one part of which was mounted on the balance beam, was used to detect deviations from the null position of the beam. The high-frequency voltage generated by this pickup unit was proportional to the position of the beam, and a portion of this signal was used to drive the restoring chain servomechanically. The recording of weight was accomplished by means of a battery-energized Wheatstone bridge circuit, one leg of which was a potentiometer-connected slide-wire mounted along the rear of the chain scale on the balance. Auxiliary weight changers 100-mg. weights were mounted behind each pan to add automatically whenever the 100-mg. chain-limited range was exceeded.

By attaching a shutter to the balance pointer of a chainomatic balance, Lohmann (96) made a light source-shutter-phototube null detector which, at equilibrium, allowed half of the maximum illumination to impinge upon the tube. Changes in the photoelectric current due to balance beam deflections caused a servomotor to drive the chain and restore the equilibrium position. Simultaneously the motor drove the slider of a potentiometer connected in a bridge network. The output of this circuit, proportional to the change in weight, was fed into a recorder. This balance also included a range extender consisting of a mechanism for adding stainless steel balls to the pan. Allen and Wright (7) and Kalinin and Kuznetsov (83) constructed similar nonrecording chainomatic balances. The latters' instrument utilized a transmitting potentiometer, geared to the automatic chain balancing mechanism, for manually reading changes in weight. A photoelectric null detector assembly which could be applied to ordinary analytical balances was reported by Rulfs (125).

An automatic recording chainomatic balance for measuring the strength of fibers was devised by Sinclair. The prototype (135) was a manually operated instrument but this was later automated (136). The fiber being tested is suspended from one end of the balance beam, and from the other end, the armature of a linear variable differential transformer which is used as the null detector. As tension is applied to the fiber, the balance is maintained in equilibrium by a servomotor driving the chain. A pen attached to the vertical indicator carrying the chain records the force applied to the fiber on a motor-driven drum. A similar apparatus using a photoelectric null detector was reported by Lasater et al. (93). Garn (68) also used a linear variable differential transformer as a null detector with a servomotor-driven chain as the restoring force in this general purpose balance. The

motor driving the chain also turns a potentiometer matched with the potentiometric slide-wire in an electronic strip-chart recorder to provide graphic recording. One of the commercially available thermobalances, the Testut (151), described later in this paper, makes use of this motor-driven chain principle.

The ideal restoring force would be one capable of continuous operation with infinite resolution. This can be most nearly achieved by means of the interaction between an electromagnetic coil and another coil, a magnet, plate, or armature, one of which is fixed and the other attached to one end of the balance beam. Most of the null balances that have been reported make use of this principle in any one of its many modifications.

The forerunner of modern electronic analytical balances was reported by Ångström (11) in 1895. This apparatus consisted of a null-type balance with a magnet suspended from one end of the beam into a solenoid coil. As the sample under study underwent a change in weight, the current through the coil was manually adjusted to maintain the balance beam at its null point. This current proportional to the weight of the sample was then read with a mirror galvanometer. Urbain (156) used the same principle for a thermobalance in an investigation of the rates of reaction of gas-producing compounds. A similar apparatus, enclosed in glass for use under vacuum and in controlled atmospheres, has been constructed (102). This method, in conjunction with modern servomechanisms and recording techniques, has been used by more recent investigators to develop completely automatic and recording balances. To investigate the high temperature evaporation of oxides, Blewett (23) utilized a manually operated electromagnetic null balance with a small magnet attached to the end of the balance beam which was surrounded by a Helmholtz coil. Observations with a cathetometer were used for null-point detection. This principle was previously employed by Stock (144) for a gas density balance. A very sensitive automatic electromagnetic weighing and gas-density balance of this type was developed by Simons, Scheirer, and Ritter (134).

Brown et al. (31) combined a standard analytical balance with a mechanical recording potentiometer. The current produced by the dual photocell null-point detector, when the balance is displaced, varies the current supplied to the damping coils so as to restore the beam to its rest point by the action of a stationary coil upon the beam-suspended magnet. As the linear drum potentiometer which controlled the current supplying the restoring coil was attached to the shaft of the mechanical

254

recorder pen mechanism, the weight at any time was indicated by the position of the pen. Controls were provided to obtain desired damping and adjustable weight ranges for this instrument. Mauer (98, 99) developed a balance similar to that of Brown et al. This balance is now available from the Niagara Electron Laboratories (106) and is described in the section on commercially available balances. The quick acting null-type balance constructed by Smith and Stevens (137) for repetitive weighings utilizes a capacitance-sensitive displacement detector, one plate of which is attached to the balance beam. Signals proportional to the beam deflection are amplified and fed to a restoring coil mounted on the opposite end of the balance beam. Electromagnetic interaction of the solenoid with a permanent magnet fixed to the base serves to bring the balance beam to the equilibrium point. The restoring force currents are measured on a meter and could be readily recorded.

Groot and Troutner (72) also made use of the electromagnetic principle in their null-type recording thermobalance equipped with a light source-shutter-photocell null detector. The current required to maintain the null position was recorded as a function of time equivalent to the linear heating rate of the furnace. Another thermobalance of this type was developed by Błażek (21, 22) that can also be used to obtain differential thermal analysis curves and gas analysis of decomposition products. It makes use of a solenoid-iron core electromagnetic weight compensator with a photoelectric null detector having electromechanical feedback. Magnet-solenoid coil-type recording thermobalances have been reported (67, 140). Pope (121) reported an automatically recording balance which represented an improvement of Gregg and Wintle's sorption balance described below (71). His contributions involved replacement of the beam-mounted coil with a magnet and circuitry for electronic balancing and recording. A vacuum apparatus for studying the kinetics of oxidation and reduction processes was reported by Chizhikov and Gvelesiani (39). This system consisted of a suspended magnet-restoring coil null-type thermobalance with photoelectric null detector mounted in a bell jar equipped with suitable pressure controlling and gas collecting accessories.

Instead of a magnet, a rod of magnetic material can be suspended into the field of the electromagnet to provide the null-point restoring force. Svedberg and Rinde (146) used this variation in their automatic but nonrecording sedimentation balance for particle size distribution studies. A sliding contact on the current-controlling resistor was motor-driven by null-point-

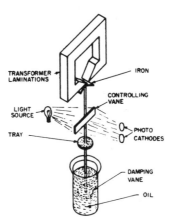

Figure 10. Electromagnetic levitation-type beamless null balance with photoelectric null detector (Clark)

detecting electromechanical contacts on the end of the beam, the current read on an ammeter being proportional to the change in weight.

Coutts et al. (46) improved upon the Oden (107–109) balance and the incremental-weight loading technique previously employed. A continuous record of the change in weight was obtained by means of a pen connected to the motor-driven slide-wire contactor used to control the current in the restoring coil, with a clock-motor-driven drum providing the time axis. The authors alluded to the commercial availability of this all-purpose balance, and thus it would represent the first fully automatic and recording analytical balance to be marketed. However, the apparent absence of further reference to this apparatus would seem to indicate that it achieved limited acceptance. A similar automatic recording sedimentation balance was reported by Bishop (19) using a magnet rather than a magnetic rod in the coil and a photoelectric null detector. A null-point balance for vacuum or controlled atmosphere adsorption studies, shown in Figure 8, was devised by I. Eyraud (64) which has a photoelectric null detector and this same type of electromagnetic restoring force. An improved version of this apparatus, designed by C. Eyraud (60–63), is now being manufactured by A.R.A.M. (12) in France and is described in the section on commercially available recording balances. Another null-type balance of somewhat similar design also using a light-shutter-photocell null detector has been constructed by Scholten, Smit, and Wijnen (131). As in this instrument the center of gravity of the beam is raised to the fulcrum and the change in weight is proportional to the current through a coil surrounding the

beam-suspended magnet, a range of sensitivities can be easily obtained electrically and there is no need to maintain a horizontal null point.

Kinjyo and Iwata (87) reported a relatively simple null thermobalance using a vacuum thermocouple, in series with an electromagnetic restoring force coil and a battery, as a transducer for recording weight changes as a function of the manually regulated restoring force currents.

A manually operated magnetically controlled quartz fiber microbalance was constructed by Edwards and Baldwin (55). This instrument was operated as a null balance with a small magnet attached to the beam and suspended into a coil. Microscopic observations of the deviations from equilibrium were used to inform the operator when the solenoid current should be varied to restore the beam to its null position. The voltage drop across a precision resistor in series with the coil was proportional to the change in weight over a range of about 1 mg. Csanderna and Honig (48) developed a similar automatic quartz-beam microbalance which they employed for gas density measurements. A magnetized wire was used instead of a magnet. An automatic recording version of this type microbalance was reported by Cochran (45) who used a variable permeance transducer as a null detector in the servosystem which applied a restoring force through a compensation solenoid surrounding a magnet suspended from one end of the beam.

Cannon (36) developed an automatic recording high-pressure null-type sorption balance utilizing a novel photoresistive cell as the null detector which senses the position of a beam of light, and through a servo network nulls the arm of a field coil-magnet electromagnetic counterpoise balance system enclosed in a metal shell.

A projection-type single pan constant load balance was converted to automatic recording over a large range of weight changes by Teetzel et al. (148). They used a differential transformer as a null detector which in turn controlled the electromagnetic restoring force.

Another form of electromagnetic interaction was applied to a manually operated electromagnetic weight-loader for analytical balances, by Hodsman and Brooke (77). It depended upon the application of known variable forces to the swinging beam of the balance by using the repulsion between a metal plate placed on the beam and an adjacent rigidly mounted alternating current-operated coil. The force applied to the beam was proportional to the square of the current through the coil. An optical lever, used to indicate angular position, served as the null detector.

For following the rates of absorption of vapors by silica gel, Anderson (9),

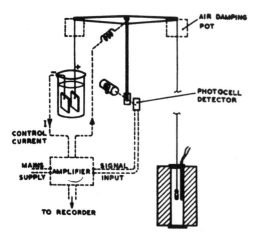

Figure 11. Null-type differential thermobalance with coulometric restoring force and photoelectric null detector (Waters)

developed a manually operated electromagnetic vacuum balance, the operation of which was based upon the interaction between the fields of series-connected coils mounted above and below a third coil which is wired in opposition and suspended from one end of the balance beam. Vieweg and Gast (158) described a recording null-type torsion microbalance, illustrated in Figure 9, for measuring diffusion through plastic membranes, which depends upon a coil-coil inductive interaction. This instrument has been marketed as the Sartorius Elmic, now the Electrona (128), a vacuum balance described under commercially available balances. A recording null balance for following weight changes in thin films of treated fats and oils has been reported by Kaufmann (84). The sensing element is a moving coil in the field of a permanent magnet and a high-frequency coil, the latter being used to maintain the null point.

An automatic recording null balance having a glass tubing beam was constructed by Gregg and Wintle (70, 71) for sorption studies in controlled atmospheres. The null detector consisted of a mirror mounted on the beam above the fulcrum and two photocells. Deflection of a light beam onto one or the other of these photocells operated a reversing motor which controlled the amount of current passing through a fixed restoring coil surrounding an inner coil suspended from one end of the balance beam. Their interaction served to maintain the balance at its null position.

In the recording balance reported by Caule and McCully (57), a deflection of the beam was detected by the change in mutual inductance between two coils,

one of which was a short-circuited coil suspended from the balance beam. The attraction between these coils served as the restoring force, a portion of the current supplied to the fixed coil being rectified and recorded as a measure of the weight. The force in milligrams was a linear function of the square of the output voltage.

The D'Arsonval galvanometer principle was used by Brockdorff and Kirsch (30) to construct a null torsion balance in which the beam corresponded to the needle of a galvanometer. The sample was attached to the beam, and the current supplied to the coil to maintain the equilibrium position was recorded as the change in weight. Pouradier and Dubois (122) and Padday (112) employed the same technique in an automatic recording surface tensiometers. The Cahn electrobalance (32) is a manually operated commercial instrument of this general type, described in a later section.

Electromagnetic Levitation Balances. A unique electromagnetic levitation principle was applied by Clark (42) in the construction of a beamless null balance shown in Figure 10. It consisted of an oil-damped sample tray suspended in mid-air by an electromagnet through which sufficient current was passed to maintain the null position as detected by a shutter intercepting the light beam incident upon two photocells. A similar nonrecording automatic levitation balance was described by Beams et al. (15) which involved attaching the sample to a ferromagnetic body magnetically sus-

pended in the field of a solenoid. The equilibrium position was maintained by a servomechanical null detector consisting of a light beam and multiplier phototube which regulated the current in the solenoid.

TORSION BALANCES

Sarakhov (126) developed a unique nonrecording null-point torsion balance for adsorption studies. An electromagnetic coil-magnet interaction was used to restore the beam in accordance with the demands of the photoelectric null detector. A manually operated balance of similar construction was reported by Szabo and Király (147) for use in adsorption and desorption studies, the measurement of gas density, and thermogravimetry. Picon (119) also constructed an interesting torsion balance, based upon the null principle, for measuring surface tension. The null detector consisted of a contactor on the beam used to energize a solenoid whose armature is mechanically coupled to a movable rider on the balance which is incrementally positioned to maintain equilibrium. A pen connected to the same rider-positioning shaft is used to record the change in weight on a drum. Using an optical system consisting of a light source, lenses, and mirrors with one reflector mounted on the torsion wire, Guastalla (73) was able to measure surface tensions manually on a calibrated ground glass scale.

A torsion wire null balance was reported by Waters (160–162) for the thermogravimetric study of coals. The restoring force necessary to maintain the beam at its null point was achieved through a photocell null detector by a special type of servomotor rotating the torsion head. Wendlandt (*165*) converted a torsion-wire–type balance into an automatic recording thermobalance, using a light source-mirror-dual photocell arrangement to maintain the balance in equilibrium by means of a reversible synchronous motor connected to the torsion wire. The servomotor also rotates a recording drum proportionally to the change in weight, the pen being driven at a constant rate by another motor. A comparable instrument developed by Cueilleron and Hartmanshenn (*47*) utilizes a double contact meter-relay to operate the reversible null-point restoring motor to which the recording pen is connected.

An automatic and recording balance was applied to the determination of particle size and distribution of powdered materials by an air sedimentation technique reported by Payne (*114*) This balance, the basis for the Micromerograph developed in the Research Laboratories of the Sharples Corp., had a torsion wire-supported balance arm and a null detector which was operated by the interaction of four fixed coils and shielding vanes attached to the balance arm. A static-type motor, activated by the detector, was used to restore the null position of the balance, with the necessary current supplied to this motor serving as a measure of weight changes. This instrument was later modified by Eadie and Payne (*54*) to incorporate a linear variable differential transformer for null detection with an electromagnetic restoring system, and is now available as the Sharples Universal recording balance (*132*) described below. Other torsion balances previously described include those of Vieweg and Gast (*158*), Bostock (*24*), Andersson and the Stenhagens (*10*), the Elmic and the Electrona (*128*).

MISCELLANEOUS BALANCES

A novel principle of hydrostatic compensation for a manually operated null-type balance was introduced by Guichard (*74, 75*) in a thermobalance for kinetic dehydration studies. It consisted of a rod attached to one end of the beam and dipping into a liquid contained in a U-tube connected to a buret. The balance beam was maintained at its null position by adding or removing a known volume of liquid which was proportional to the change in weight. With modern electronic technology, this principle could be readily adapted to automatic operation and recording.

Haller and Calcamuggio (*76*) devised a manually operated pressure-controlled hydraulic balance for the measurement of extremely small weight,

Table V. Commercially Avail-

Manufacturer	Principle of Balance	Remarks
Wm. Ainsworth & Sons 2151 Lawrence St. Denver 2, Colo.	DEFLECTION: Linear variable permanence transducer in bridge circuit with slide-wire of recorder measures deflection of balance beam	200-gram capacity: One chart-width equivalent to 100 mg. Range-extending weights are automatically added or subtracted at 100-mg. increments. Weight switching allows total range of 4 grams. Models available for either ambient or vacuum and controlled atmosphere operation
American Instruments Co. 8030 Georgia Ave. Silver Spring, Md. (Thermo-Grav thermobalance)	DEFLECTION: Extension of precision spring measured by linear variable differential transformer	15-gram capacity, 200-mg. range. Load and range can be varied with interchangeable springs. Recording as a function of either temperature or time. Vacuum or controlled atmosphere operation. Dual furnaces and temperature programmer supplied
A.R.A.M. 12 Quai Rambaud Lyon, France (Eyraud recording balance)	NULL: Electromagnetic coil-armature restoring force; shutter-light source-photocell used as null detector	100-gram capacity, 100-mg. range. Parabolic recording scale. System can be evacuated for low pressure studies. Can be supplied with a furnace
C. W. Brabender Instruments, Inc. 50 E. Wesley St. South Hackensack, N. J.	DEFLECTION: Stylus attached to end of balance beam is cyclically pressed against pressure-sensitive recorder chart paper	Thermobalance for ambient and controlled atmospheres with 1-gram weight loss range measured on a 17.5-cm. strip chart recorder; supplied with 1200° C. furnace and pyrometric programming controller having interchangeable templates for various heating cycles
Cahn Instrument Co. 14511 Paramount Blvd. Paramount, Calif. (RM electrobalance)	NULL: Electromagnetic D'Arsonval movement restoring force; shutter-light-source-phototube used as null detector	35-mg. and 175-mg. total load capacity for counterweighted 0–20 and 0–100 mg. weight change ranges; D.C. signal output for potentiometric recording; 0.1% accuracy; 0–100% variable suppression for expanded scale operation. Can be used in vacuum or humid chamber atmospheres. Manually operated model available
E-H Research Labs. 2161 Shattuck Ave. Berkeley, Calif.	NULL: Electromagnetic beam-frame coil pair interaction; radio frequency coupling null detector	200-gram capacity; 30- to 3000-mg. range; meter or digital readout; potentiometric recording; electronic taring; 1% accuracy for routine weight measurements
Fisher Scientific Co. 711 Forbes Ave. Pittsburgh 19, Pa. (Recording Balance accessory)	NULL: Electromagnetic coil-magnet restoring force; differential transformer used as null detector	Device for converting any analytical balance to automatic electronic recording with slight or no modification, for use with any millivolt recorder
Firma Linseis Germany	NULL: Servomotor-driven chain restoring force; photoelectric null detector	Sedimentation balance; thermobalance for ambient and controlled atmosphere operation; strip chart recorder with full scale range of 0.25 to 4.0 grams; sample suspended from balance beam in vertically mounted furnace; 1550° C. furnace with temperature programming controller
E. Mettler* Pelikanstrasse 19 Zurich, Switzerland	DEFLECTION: Servomotor-driven photoelectric balance beam follower mechanically coupled to a multi-stylus recorder for sequential scale expansion to 1000 units in five 200 unit incremental spans	Two models of the single pan balances with large optical scale ranges adapted to recording system for maximum range of 1020 mg. or 1020 grams; 160- and 800-gram loads, respectively. Additional stylus for continuously recording complete change in weight as single curve

able Recording Balances

Manufacturer	Principle of Balance	Remarks
Gebrüder Netzsch[b] Germany	NULL: Servomotor-driven chain restoring; photoelectric null detector	Thermobalance for ambient and controlled atmosphere with 1500° C. furnace and programming temperature controller for 1° to 10° C. per min.; 100-gram load and 200-mg. recording range
Niagara Electron Labs. Andover, N. Y. (Mauer recording balance)	NULL: Equilibrium maintained by electromagnetic coil acting upon magnet suspended from end of beam; phototube-mirror-light source null detector	Utilizes a conventional type analytical balance; available as custom-made apparatus with various accessory equipment. 100-mg. full-scale range with range extender to several grams
Sartorius-Werke, A.G.[b] Göttingen, Germany (De Keyser differential thermobalance)	DEFLECTION: Differential weight change of twin samples in identical furnaces recorded photographically	2.5-gram sample weights; 5-mg. per 180-mm. differential weight sensitivity; supplied with furnaces and programmers
Sartorius-Werke, A.G.[b] Göttingen, Germany Electrona (formerly Elmic)	NULL: Torsion beam-mounted coil maintained at null position by compensating current of fixed coil	Recording microbalance; 0.1- to 20-mg. ranges; 1000-mg. load; two models available, one for vacuum or controlled atmosphere operation
Sartorius-Werke, A.G.[b] Göttingen, Germany Selecta (formerly Recordal)	DEFLECTION: Beam displacement converted to electrical signal by light source-shutter photocell system	Recording attachment for Selecta analytical balance; 200-gram capacity, 100-mg. range
Sartorius-Werke, A.G.[b] Göttingen, Germany (Sedibal)	DEFLECTION: Light source-photocell null detector. Motor twists a torsion wire to restore beam to null position incrementally.	Automatic-recording sedimentation balance. Weight range 500 mg., particle size range 1 to 60 microns
Sharples Corp. Research Laboratories Bridgeport, Pa.	NULL: Torsion beam suspension positioned by current through restoring coil in field of permanent magnet; inductive position-sensing null detector	25-gram capacity; 25- to 250-mg. standard ranges; 1 mg. to 1 gram on special order; bell jar supplied for vacuum operation. Strip-chart recording
Société A.D.A.M.E.L.[c] Paris, France (Chevenard thermobalance)	DEFLECTION: Wire-supported beam balance. Deflections recorded photographically by beam-mirror reflections of fixed light source, or mechanically recorded on pen and ink recorder using stylet interchangeable with mirror	10-gram capacity; 50- to 400-mg. range. Furnished with furnace and temperature programmer. Photographic, mechanical, or electronic recording. Electronic recording attachment, Microverter, available from American distributor
Stanton Instruments, Ltd.[d] London, England	DEFLECTION: Capacitance-sensitive follower; incremental servo-operated electric weight loading	Two models: 1-gram weight-loaded range at 1-mg. sensitivity, full-scale chart range 100 mg., 0.1 gram weight-loaded range at 0.1-mg. sensitivity, 10-mg. full-scale chart range; 100-gram load; available with or without furnace and temperature programmer
Testing Equipment Co. Murray Hill, N. J. (Load cells)	DEFLECTION: Double-cantilevered beam; deflections measured by linear variable differential transformers	Proper selection of load cell cantilever material and construction provides ranges from milligrams to pounds
Testing Equipment Co. Murray Hill, N. J. (Balance conversion kit)	DEFLECTION: Deflection of standard analytical balance measured by linear variable differential transformers	Transducer attachment for converting any analytical balance to a recording microbalance
Testut Co. 9 Rue Brown Sequard Paris XV, France	NULL: Servomotor-driven chain loading; mechanical beam-contacting null detector	200-gram capacity; 400-mg. range; other ranges available; with or without furnace and controls. Can be supplied with magnetic balance-beam-follower for use in closed systems

[a] Mettler Instrument Corp., Hightstown, N. J.
[b] C. A. Brinkmann & Co., Inc., 115 Cutter Mill Rd., Great Neck, L. I., N. Y.
[c] Cooke, Troughton & Simms Co., Inc., 110 Pleasant St., Malden 48, Mass.
[d] Burrell Corp., Fifth Ave., Pittsburgh, Pa.

volume, and density changes which they used for studies of the underwater corrosion of silicate glasses. The null indicator consisted of a pressure-sensitive glass float carrying the sample and operating on the principle of the Cartesian driver. With appropriate transduction it, too, could be made automatically recording.

A remotely operated semimicrobalance for the weighing operations involved in performing radiochemical and chemical analyses on highly radioactive materials was constructed by Moses, King, and Wendelyn (103). Although not designed as a recording balance, the use of solenoids for weight loading and selsyn systems for chainomatic operation lends itself to such further development.

DIFFERENTIAL THERMOBALANCES

A modified approach to thermogravimetry that has recently received a great deal of attention, with respect to the development and use of novel automatic recording balances, is that of differential thermogravimetric analysis. This technique has proved valuable because the curves obtained are the derivative of primary weight-change plots and, therefore, closely resemble the complementary differential thermal analysis curves with which they are compared in many thermoanalytical investigations. As derivative plots, the curves obtained consist of sequential or overlapping bands corresponding to the rate of change in the weight of the sample as a function of time or temperature. Several instruments have been expressly devised for this purpose, one of which, the Sartorius-De Keyser differential thermobalance (127), is available commercially.

It was the prototype of this apparatus (86) for automatically recording differential changes in weight as a function of time at a linear heating rate, that introduced this new concept in thermogravimetry. A description of its operation is presented in a later section of this review.

Another differential thermobalance was developed by Erdey, Paulik, and Paulik (57, 58). Their apparatus consisted of a permanent magnet attached to one end of the beam of an analytical balance and suspended inside a symmetrically placed coil. Changes in weight of the sample cause a motion of the permanent magnet, thus inducing a current in the coil proportional to the rate of weight change. By galvanometrically measuring this current, the derivative of the change in weight may be obtained and plotted. A recording version of this apparatus called a "Derivatograph" has since been developed for simultaneously conducting differential thermal analysis, thermo-

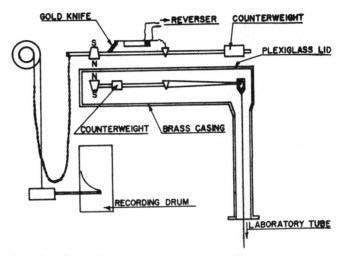

Figure 14. Testut null-type vacuum thermobalance with servo-operated chain for restoring force and electromechanical null detector

gravimetry, and derivative thermogravimetry (*113*).

A relatively complex electromechanical apparatus for obtaining both primary and derivative thermogravimetric curves with an analytical balance was constructed by Lambert (*91*) for ceramic studies. The null detector was a capacitance-sensing device and the restoring force was supplied by coupling the balance beam to the needle of the galvanometer of an ammeter. The current necessary to maintain the balance at a dynamic equilibrium point is recorded as the change in weight while the derivative of these signals is obtained at discrete intervals using a dual-potentiometer mechanism which generates a signal proportional to the change in weight during the interval.

A very unique recording differential thermobalance for thermogravimetric investigations of coal has been described by Waters (*160–162*). This balance, illustrated in Figure 11, operates on a null principle with the counterbalancing force being applied at a rate proportionally controlled by an electromotive force equivalent to the rate of loss in weight. An analytical balance was adapted so that the sample was counter-balanced by the suspended electrode of a silver coulometer. Changes in the sample weight were compensated by corresponding changes in weight of the silver electrode due to a control current passing through the cell. The necessary voltage and current

controls were effected by a photocell null detector and magnetic amplifier with an output rectifier. The voltage drop across a precision resistor in the output circuit enabled a portion of the controlling voltage or current to be recorded as a measure of the differential weight change. Integration of this function by electromechanical means also enabled an accurate record of the cumulative weight change to be made simultaneously on a multipoint recorder.

An electronic differentiator for continuously recording the derivative of any primary curves plotted on null-balance recorders has been reported by Campbell, Gordon, and Smith (*35*). This device consists of a battery-energized transmitting potentiometer mounted on the same shaft as the balancing slide-wire in the null-type servo-operated strip-chart potentiometric or Wheatstone bridge recorder. The output of the transmitter, which varies in direct proportion to the primary signal being recorded as a function of time, is differentiated by means of a simple R-C circuit and then automatically plotted on either a time-base or x-y function recorder. This method is particularly suited for differential thermogravimetric studies with thermobalances that involve converting the change in weight to electrical signals that are then recorded in the above manner.

COMMERCIALLY AVAILABLE AUTOMATIC RECORDING BALANCES

The commercially available automatic recording balances to be described in this section range from simple "do-it-yourself" conversion kits to completely

packaged apparatus for vacuum thermogravimetry or particle size determinations by sedimentation. These instruments, manufactured both here and abroad, are summarized in Table V.

The Chevenard thermobalance, reported in 1944 (*38*) and subsequently manufactured by the Société A.D.A.M.E.L. in France (*3*), was the first recording thermobalance to become commercially available. This instrument, consisting of a wire-suspended balance beam, is of the deflection type with change in weight ranges of 50 to 400 mg. and a sample plus crucible or specimen holder load of 10 grams. The change in weight may be recorded photographically by the deflection of a beam of light from a mirror mounted at one end of the balance beam onto a photosensitive paper wrapped around a synchronously rotating drum. An interchangeable mechanically recording unit, Figure 12, is also available for use with this balance, the mirror being replaced by a stylet which is mechanically linked to the pen-and-ink recorder. Included in the apparatus are the furnace and temperature programmer. Aspects of the performance of the photographic recording balance were reported by Claisse, East, and Abesque (*41*) and of the pen-recording balance by Simons, Newkirk, and Aliferis (*133*). A method for easily converting the photographic recording balance to electronic pen-and-ink recording by using a linear variable differential transformer has been reported by Gordon and Campbell (*69*), and is now available as accessory equipment for this balance (*44*).

Another recording balance manufactured in France is the Testut (*151*)

Figure 15. Stanton recording deflection thermobalance

instrument, Figure 13, which is available as a general purpose balance, a thermobalance with furnace, a heavy duty industrial scale, or a gas-density balance. The operation is based upon the null-point restoring force principle, wherein deviations from the rest point are detected by an electrical contractor which actuates a reversible servomotor-driven chain to increase or decrease the load on the beam by winding or unwinding a chain. The beam is maintained at its equilibrium point dynamically by sinusoidal incremental loading and unloading of the beam due to the reversing motor being intermittently energized by the single contact on the beam. The curve is recorded by mechanical linkage of the chain mechanism to the pen-and-ink recorder, the range of weight change being a function of the chainweight. A typical laboratory balance has a load capacity of 100 grams with a weight change range of 300 mg. recorded on a 7-inch chart mounted on a motor-driven drum.

A modification of this balance, shown schematically in Figure 14, can also be used for operation under vacuum or in controlled atmospheres. It consists of a second beam contained in a non-magnetic chamber and magnetically coupled to the primary balance beam by means of mutually repelling magnets attached to the ends of each beam.

The Eyraud balance, which utilises a combination of the deflection and null principles, is manufactured by Atelier Reparation Appareils Mesures (A.R.A.M.) in France (*62*). The complete balance assembly can be evacuated for low pressure applications, and the sample suspended from one end of the balance beam can be heated, or cooled, as desired. An iron rod and a shutter are suspended in line from one end of the balance beam with the shutter controlling the amount of light impinging upon a photoelectric cell. The current produced by the photocell is amplified and passed through a coil surrounding the iron rod. A gain or loss in weight of the sample produces a change in the illumination on the photocell, resulting in an increase or decrease in the current through the coil, and a corresponding variation in the force exerted by the magnetic field on the iron rod. This adjustment of the magnetic force compensates for the change in weight of the sample, a stable equilibrium is obtained, and the change in weight can be recorded by measuring the current through the coil. As the magnetic force is proportional to the square of the current in the coil the recording of the weight change is parabolic. The load capacity of this balance is 100 grams and the range of weight change is 100 mg. For applications as a thermobalance, furnaces and temperature programming equipment are available.

The Stanton (*141*) recording balance or thermobalance, Figure 15, is a deflection-type instrument in which a capacitance-follower plate located over the balance beam is used to measure electromechanically and record the change in weight for a predetermined beam movement. At the end of this deflection, a servomechanism brings electric weight-loading into operation, so that changes in weight up to 1 gram at 1-mg. sensitivity or 0.1 gram at 0.1-mg. sensitivity may be followed and recorded automatically. Dual recorders and a temperature programmer are provided with the thermobalance to obtain the change in weight and furnace temperature simultaneously as a function of time on curvilinear chart paper.

A regular analytical-type balance operating on the deflection principle with a special beam and double air dampers has been developed for use as an automatic recording balance by Wm. Ainsworth and Sons, Inc. (*5*). The Type BR balance with AU recorder has a capacity of 200 grams and records a full-scale change in weight of 110 mg. This can be extended to a total range of 4 grams by means of automatically switched weights which are added or subtracted in 100-mg. increments. A variable permeance transducer is installed in the balance with its armature attached to the beam. As two halves of the transducer form a bridge circuit with the slide-wire in the recorder, any unbalance in the bridge circuit gives a signal which is amplified to operate the recorder and so rebalance the bridge. This instrument is readily adaptable to various applications for which it can be used in conjunction with auxiliary equipment. A version of this instrument (*6*) has been adapted for use in vacuum and controlled atmospheres. For this apparatus, the balance mechanism is mounted on a base plate and covered by a bell jar. Controls for manual and automatic operation of the balance are mounted on the frame which supports the base plate, as shown in Figure 16.

The American Instrument Co. (8) has developed an automatic recording vacuum thermobalance, the Thermo-Grav, that can be used to record automatically changes in weight as a function of either temperature—programmed at a selected heating rate—or of time—at a constant temperature—in vacuum or controlled atmospheres. It is a spring balance, Figure 17, the deflections of which are measured and recorded by means of a linear variable differential transformer. Although the standard capacity and range of weight change are 10 grams and 200 mg., respectively, other values can be obtained by changing the spring. The spring, suspended in a glass housing, is provided with an oil damper and weight calibrator that can be operated under the experimental conditions at any time. Dual furnaces allow consecutive determinations without delay due to "cool down" time, and temperature controls permit either six linear heating rates, ranging from 3° to 18° C. per minute or constant temperature settings. For the latter, four time bases ranging from 50 to 400 minutes can be obtained. The change in weight can be recorded as a function of the temperature with full scale deflections of 0° to 200°, 500°, or 1000° C., as well as expanded scale ranges of 200° or 500° starting at any of the selected 100° intervals over the range of 0° to 800° C. A vacuum pump and pressure or controlled-atmosphere controls are also provided.

The Brinkmann Recorbal (23) is a Sartorius analytical balance modified for operation as a deflection-type recording instrument. On the end of the beam is mounted a shutter which alters the amount of light incident to a phototube as the beam undergoes deflections due to changes in weight. The output of the phototube, proportional to the change in weight, is then recorded. This instrument has been superceded by an automatic electro-optical attachment (129) which may be used with all models of the Sartorius Selecta analytical balances. The capacity and ranges are those of the balance with which the recording attachment is used.

An electrical microgram recording balance, the Sartorius Electrona (128) formerly the Elmic is based on the instrument developed by Vieweg and Gast (158). This instrument consists of a quartz-rod beam to one end of which is attached a small coil situated so that as the beam suspended from a torsion wire swings, the coil moves in the magnetic field of an outer coil system. The voltage generated is used to detect displacements of the beam and through suitable amplification re-establish the null point of the beam. This compensating current is directly proportional to the change in weight and may be used to indicate and record its value. The maximum load is 1000

mg., and the three ranges available are 1, 2.5, and 5 mg., with 0.001-mg. accuracy. There are two models of this recording balance available—one for ambient operation and the other for use in vacuum or controlled atmospheres (Figure 18).

An automatic instrument for the determination of particle size distribution of powdered materials is the Sartorius Sedibal (130), Figure 19, a sedimentation balance which records the increase in weight of a sediment deposited on the pan from a liquid dispersion as a function of time. The grain-size distribution between 1 and 60 microns can be determined from the curve obtained. Deposition of the suspended particles on the pan in the sedimentation vessel causes the balance beam to deflect, thus allowing a light beam to energize a photocell and operate a stepping motor which twists a torsion wire to a preset angle. The elastic force transmits a torsional moment to a small beam fixed to the torsion wire, and, in turn, to the main balance beam by a mechanical coupling which turns it back to its null position. This automatically compensates the moment developed by the load on the pan, and is repeated until the suspended particles are completely settled. As the recording pen moves 0.08 cm. on the chart in synchronism with each action of the stepping motor, and the paper moves at a constant action speed at right angles to the pen movement, a stepped curve is obtained for the weight as a function of time. A maximum deposit of 500 mg. can be measured with steps for every 2 mg. of sediment. Recorder chart speeds of 60 to 600 mm. per hour are provided.

The Sharples Universal recording balance (132), Figure 20, is a null-type balance having a torsion tube suspension mounted on a seismic mass in an enclosed steel and glass housing that may be evacuated or used for controlled-atmosphere studies. An inductive-position-sensing element is used to detect deflections of the balance beam about the torsion suspension. The detector delivers a signal to the electronic restoring unit which supplies current to a

restoring force coil located in the field of a permanent magnet and connected to one end of the balance beam. The current in this coil serves to restore the beam to its null point and is recorded as the instantaneous value of the change in weight. Load capacity is 25 grams and the standard model has adjustable ranges of 25, 50, 100, and 250 mg.

The Mauer recording balance (98, 99), Figure 21, being manufactured by the Niagara Electron Laboratories (106), is a null-type instrument using a conventional analytical balance. A completely electronic method is used for restoring the balance to its null point by means of the electromagnetic interaction between a fixed coil and a magnet which is suspended into it from one end of the beam. The null detection system consists of a light source, beam-mounted mirror, and a dual phototube which controls the current through the restoring coil. This current, which is a linear function of the change in weight, is recorded by measuring the voltage drop across a precision resistor in series with the solenoid. The standard balance can be used for recording a basic full-scale change in weight of 100 mg. An incremental weight loading device has been added to increase the range of weight change to several grams.

A direct meter-reading or electronic recording automatic analytical balance, Figure 22, manufactured by the E-H Research Laboratories (56), consists of a conventional twin-pan analytical beam balance that has been modified for null operation by means of beam and frame coils which are mounted at both ends of the balance beam and energized by a radio-frequency (R. F.) oscillator. The R. F. signals generated by increased coupling between the coils, proportional to the change in weight on one pan of the balance as the beam deflects, are amplified, rectified, and fed back to the concurrently direct current energized coils so as to increase the electromagnetic interaction of the coils and restore the beam to its equilibrium point. In this way, the coils serve as both the null point detector and the means for applying the restoring force, the amount of direct current required

Figure 21. Niagara Electron Laboratories Mauer recording null thermobalance

to maintain the null position of the beam being read on a meter or recorded potentiometrically. The balance has a 1-second measurement time; automatic selection of 5 weight ranges from 30 to 3000 mg.; extension of this range to 200 grams and an ultimate accuracy of 0.2%.

The Sartorius-De Keyser differential thermobalance (*127*), Figure 23, employs two identical samples of equal weight suspended into identical furnaces from opposite ends of a balance beam. These furnaces are heated at the same linear rate but one is kept approximately 4° cooler than the other. As the reactions involving changes in weight in the two samples occur consecutively there is a weight differential, the magnitude of which is proportional to the heating rate. Beam deflections corresponding to these weight differentials are photographically recorded by using a mirror attached to the beam to deflect a light onto the photosensitive paper wrapped around a drum. The resulting curve approximates the derivative of the weight change plotted as a function of temperature, the area under this curve being the total change in weight of the sample. Ranges of weight change of 5, 10, and 20 mg. can be obtained for sample weights within the capacity of an analytical balance.

The Mettler Co. has adapted its single pan balances that have a large optical scale range, for use as automatic recording deflection balances (*100, 101*), Figure 24. For this purpose, it developed a new type of recording system to obtain an effective recording span of 1000 mm. with a chart width of 200 mm. for a maximum change in weight of 1020 mg. or 1020 grams. Light reflected from a mirror which is attached to the balance beam impinges upon a pair of matched photocells that are mounted on a servomotor-driven carriage. As the balance beam deflects in response to weight changes, an unequal amount of light strikes the two photocells and generates a signal which is amplified and fed to the reversible servomotor. This action moves the photocells in the direction necessary to illuminate the cells equally and thereby follow changes in the balance beam position. These displacements of the photocell carriage are mechanically coupled to a unique strip-chart recording system utilizing five current-carrying recording styli mounted on a continuous-loop steel tape. These styli sequentially make their full-scale excursions across the chart synchronously with the changes in weight, each stylus recording one fifth of the maximum optical scale range of the balance. A sixth stylus simultaneously records the complete change in weight as a continuous curve on the same chart through a 1:10 mechanical reduction.

The Brabender thermobalance (*26*), Figure 25, developed in the chemical laboratories of the Steinkohlenbergbauverein Essen, consists of a deflection balance with built-in strip-chart recorder, a controlling and programming pyrometer having interchangeable plastic templates for obtaining the desired temperature cycles, and a 200°–1200° C. furnace for ambient or controlled atmosphere studies. A stylus attached to the end of the balance beam is automatically pressed against the pressure-sensitive chart paper at regular intervals every few seconds, thus recording the change in weight as a function of time.

The Netzsch recording thermobalance (*29*), Figure 26, consists of a short-beam symmetrical knife-edge balance system with a photoelectric null detector that drives a servomotor-operated chain to maintain the balance in equilibrium and produce a signal proportional to the change in weight. A full scale range of 100 mg. is recorded on a 5-inch chart, simultaneously with temperature, up to 1500° C. for loads up to 100 grams. Linear heating rates of 1° to 10° C. per minute can be obtained.

The Linseis thermobalance (*95*) also uses a photoelectric null detector to operate a servomotor-driven chain along with the pen of a strip-chart recorder with full-scale ranges of 0.25 to 4.0 grams. An unusual feature is the way in which the sample is supported by a horizontally mounted furnace. A modified form of this system is available as a recording sedimentation balance.

A manually operated null-type electronic microbalance has been developed by the Cahn Instrument Co. (*32*). In this system the pan, calibrating mass, and sample are successively balanced by passing an electric current through a precision torque motor. As the beam is always returned to the same balance position the electromagnetic interaction is in the form of a current which may be measured with high accuracy. The range of sample weights may be varied from 0 to 5 mg. to 0 to 100 mg., with a precision of 0.20% and accuracy of about 0.05% of full scale on each range.

A recording model of this null balance, the RM automatic Electrobalance, is available which utilizes a shutter mounted on the balance beam in conjunction with a phototube-light source as the null detector which operates through a servo-amplifier to restore the beam to its equilibrium point. The output signal, proportional to the change in weight, can be plotted on conventional D.C. recording potentiometers; and at rapid rates because of the fast servo-system employed. It is also designed for use in small evacuated or high humidity chambers. Total load capacities are 35 and 175 mg.

Miniature cantilever-type load cells including differential transformers for accurately measuring change in weight are available from the Testing Equipment Sales Co. (*149*). This company also supplies a "Do-It-Yourself" differential transformer kit for converting any balance into an automatically recording unit (*150*).

Another analytical balance accessory which has been developed by the Fisher Scientific Co. (*66*) provides the means for recording changes in weight automatically with most two-pan balances. It consists of a differential transformer whose armature is suspended from one end of the balance beam to serve as a null detector and a magnet hanging from the other end of the beam which passes through an electromagnetic restoring coil. The alternating current signal from the null detector is amplified, rectified, and applied to the coil, thus restoring the original equilibrium point of the balance beam. As the current necessary to maintain the null position is proportional to the change in weight, it can be automatically and continuously recorded.

The Arnold O. Beckman gas density balance (*16*) consists of a horizontal quartz fiber supporting a lightweight dumbbell, one ball of which is punctured so as not to experience buoyancy effects. As the other ball tends to rise or dip as the density of the gas increases or decreases, the dumbbell experiences a rotational force about the quartz fiber, the magnitude of which is proportional to the density of the sample gas. A conductive coating on the suspension, which is held in the electrostatic field of a pair of electrodes maintained at a fixed potential, allows the establishment of an electrostatic force, the magnitude of which is proportional to the density of the gas. An optical null detector with a mirror on the suspension is employed to maintain electronically the beam at its null point, the magnitude of the signal being a function of the gas density and suitable for recording.

Although a majority of these commercially available automatic recording balances are thermobalances, almost all of them can be used for other purposes such as those mentioned in Table I. The range of applicability varies with their construction, operating characteristics, availability of auxiliary equipment, and the ingenuity of those planning to use the equipment.

Some of the principles for developing recording balances are being applied to the automation of industrial weighing operations (*13, 155*). The Philip Morris "Autovar" (*117*) is a deflection balance used to evaluate the uniformity of individual cigarettes electronically on a statistical basis and provide the data required for manufacturing plant quality control. An automatic moisture tester was developed by Pillsbury Mills (*120*) to determine accurately the moisture content of flour by means of a recording deflection balance. A commercial model of this instrument is manufactured by the Brabender Corp. (*25*). These instruments are representative of the ways in which the principles of automated gravimetry are being extended from the laboratory to the production line.

LITERATURE CITED

(1) Abderhalden, Emil, *Fermentforschung* 1, 155–64 (1914).
(2) Abderhalden, Emil, *Skand. Arch. Physiol.* 29, 75–83 (1913).
(3) A.D.A.M.E.L., Paris, France, Bull., "Chevenard Thermobalance."
(4) Ahmad, Nasir, *Proc. Natl. Inst. Sci. India* 7, 79–87 (1941).
(5) Ainsworth, William, & Sons, Inc., Bull. **158**.
(6) *Ibid.*, Bull. **459**.
(7) Allen, P. W., Wright, R. D., *J. Sci. Instr.* 29, 235 (1952).
(8) American Instrument Co., Bull. **2304**.
(9) Anderson, J. S., *Trans. Faraday Soc.* 11, 69–75 (1915).
(10) Andersson, K. J. I., Ställberg-Stenhagen, S., Stenhagen, E., *The Svedberg* (Mem. Vol.) 1944, 11–32.
(11) Angström, K., *Öfversigt Kongl. Vetenskaps-Akad. Förh.* 1895, 643–655.
(12) A.R.A.M. (Atelier Reparation Appareils Measures), Lyon, France, Bull., "Eyraud Recording Balance."
(13) Avion Div., ACF Industries, Inc., Paramus, N. J., Bull., "Electronic Weighing Device."
(14) Bartlett, E. S., Williams, D. N., *Rev. Sci. Instr.* 28, 919–21 (1957).
(15) Beams, J. W., Hulburt, C. W., Lotz, W. E., Jr., Montague, R. M., Jr., *Ibid.*, 26, 1181 (1955).
(16) Beckman, Arnold O., Inc., South Pasadena, Calif., Bull., "Gas Density Balance."
(17) Besser, E. D., Piret, E. L., *Chem. Eng. Progr.* 51, 405–10 (1955).
(18) Binnington, D. S., Geddes, W. F., IND. ENG. CHEM., ANAL. ED. 8, 76–79 (1936).
(19) Bishop, D. L., *J. Research Natl. Bur. Standards* 12, 173–84 (1934).
(20) Blau, M., Carlin, J. R., *Electronics* 21, 78–82 (1948).
(21) Blažek, A., *Hutnické listy* 12, 1096–102 (1957).
(22) Blažek, A., *Silikáty* 1, 158–63 (1957).
(23) Blewett, J. P., *Rev. Sci. Instr.* 10, 231–33 (1939).
(24) Bostock, W., *J. Sci. Instr.* 29, 209–11 (1952).
(25) Brabender Corp., Rochelle Park, N. J., Bull., "Automatic Moisture Recorder."
(26) Brabender, C. W., Instruments Inc., South Hackensack, N. J., Bull., "Thermobalance."
(27) Brefort, J., *Bull. soc. chim.* 1949, 524–8.
(28) Brinkmann, C. A., & Co. Great Neck, Long Island, N. Y., Bull., "Recorbal."
(29) *Ibid.*, Bull., Netzsch Recording Thermobalance."
(30) Brockdorff, U. V., Kirsch, K., *Elektrotech. Z.* 71, 611 (1950).
(31) Brown, F. E., Loomis, T. C., Peabody, R. C., Woods, J. D., *Proc. Iowa Acad. Sci.* 59, 159–69 (1953).
(32) Cahn Instrument Co., Downey, Calif., Bull., "Cahn Electrobalance."
(33) Campbell, C., Gordon, S., ANAL. CHEM. 29, 298–301 (1957).
(34) Campbell, C., Gordon, S., Pittsburgh Conference on Analytical Chemistry and Applied Spectroscopy, March 1957.

(35) Campbell, C., Gordon, S., Smith, C. L., ANAL. CHEM. 31, 1188–91 (1959).
(36) Cannon, P., *Rev. Sci. Instr.* 29, 1115–17 (1958).
(37) Caule, E. J., McCully, G., *Can. J. Technol.* 33, 1–11 (1955).
(38) Chevenard, P., Waché, S., de la Tullaye, R., *Bull. soc. chim.* 11, 41–7 (1944).
(39) Chishikov, D. M., Gvelesiani, G. G., *Zavodskaya Lab.* 22, 499–501 (1956).
(40) Claesson, S., *The Svedberg* (Mem. Vol.) 1944 82–93.
(41) Claisse, F., East, F., Abesque, F., Quebec, Dept. Mines Prelim. Repts. No. **305**, 1954.
(42) Clark, J. W., *Rev. Sci. Instr.* 18, 915–18 (1947).
(43) Cochran, C. N., *Ibid.*, 29, 1135–8 (1958).
(44) Cooke, Troughton, & Simms, Inc., Malden, Mass., Bull., "Chevenard Thermobalance."
(45) Couleru, A., *Rev. prods. chim.* 40, 611–18 (1937).
(46) Coutts, J. R. H., Crowther, E. M., Keen, B. A., Oden, S., *Proc. Roy. Soc.* (London) 106A, 33–51 (1924).
(47) Cueilleron, Jean, Hartmanshenn, Olivier, *Bull. soc. chim. France* 1959 164–76.
(48) Czanderna, A. W., Honig, J. M., ANAL. CHEM. 29, 1006–10 (1957).
(49) DeLong, W. B., E. I. du Pont de Nemours & Co., Wilmington, Del., private communication, 1956.
(50) Dervichian, D. G., *J. phys. radium* 6, 221–5 (1935).
(51) Deryagin, B. V., Timofeev, K. K., Abrikosova, I. I., Sachkov, Y. N., *Trudy Komissii Anal. Khim., Akad, Nauk, S.S.S.R., Otdel. Khim. Nauk* 5, (8), 152–61 (1954) *Chem. Abstr.* 49, 11330.
(52) Dubois, P., Ph.D. thesis, University of Paris, 1935.
(53) Duval, C., "Inorganic Thermogravimetric Analysis," Elsevier, Houston, 1954.
(54) Eadie, F. S., Payne, R. E., *Iron Age* 172, 99–102 (1954).
(55) Edwards, F. C., Baldwin, R. R., ANAL. CHEM. 23, 357–61 (1951).
(56) E-H Research Laboratories, Berkeley, Calif., Bull., "Automatic Analytical Balance," p. 301.
(57) Erdey, L., Paulik, F., Paulik, J., *Acta Chim. Acad. Sci. Hung.* 10, 61–97 (1956).
(58) Erdey, L., Paulik, F., Paulik, J., *Nature* 174, 885–6 (1954).
(59) Ewald, P., IND. ENG. CHEM., ANAL. ED. 14, 66–7 (1942).
(60) Eyraud, C., *Compt. rend.* 238, 1511–12 (1954).
(61) Eyraud, C., *Technica* 177, 2–4 (1954).
(62) Eyraud, C., Eyraud, I., Catalogue Soc. Fr. Phys. 50th Exposition, 163 (1953).
(63) Eyraud, C., Eyraud, I., *Laboratories* 12, 13–20 (1955).
(64) Eyraud, I., *J. chim. phys.* 47, 104 (1950).
(65) Feuer, I., ANAL. CHEM. 20, 1231–37 (1948).
(66) Fisher Scientific Co., Pittsburgh, Pa., Bull. FS-231.
(67) Formanek, Z., Bauer, J., *Silikáty* 1, 164–70 (1957).
(68) Garn, P. D., ANAL. CHEM. 29, 839–41 (1957).
(69) Gordon, S., Campbell, C., *Ibid.*, 28, 124–26 (1956).
(70) Gregg, S. J., *J. Chem. Soc.* 1946, 561–2.
(71) Gregg, S. J., Wintle, M. F., *J. Sci. Instr.* 23, 259–64 (1946).
(72) Groot, C., Troutner, V. H., ANAL. CHEM. 29, 835–9 (1957).
(73) Guastalla, J., *Proc. 2nd Intern.*

Congr. Surface Activity, London, vol. 3 p. 143–52 (1957).
(74) Guichard, M. M., *Bull. soc. chim.* 37, 251–3 (1925).
(75) *Ibid.*, 39, 1113–15 (1926).
(76) Haller, W. K., Calcamuggio, G. L., *Rev. Sci. Instr.* 26, 1064–8 (1955).
(77) Hodsman, G. F., Brooke, E. R., *J. Sci. Instr.* 28, 348–51 (1951).
(78) Hooley, J. G., *Can. J. Chem.* 35, 374–80 (1957).
(79) Hyatt, E. P., Cutler, I. B., Wadsworth, M. E., *Am. Ceram. Soc. Bull.* 35, 180–1 (1956).
(80) Isvekov, I. V., *Trudy Krym. Filiala, Akad. Nauk, S.S.S.R.* 4, (1) 81–2 (1953).
(81) Jones, G. A., Tinklepaugh, J. R., U.S.A.F. Tech. Rept. No. **6448**, November 1950.
(82) Jouin, Y., *Chim & ind.* 58, 24–47 (1947).
(83) Kalinin, P. D., Kusnetsov, A. K., *Zhur. Fis. Khim.* 32, 1658–60 (1958).
(84) Kaufmann, H. P., *Fette u Seifen, Anstrichmittel* 58, 844–7 (1956).
(85) Kertsman, J., Division of Analytical Chemistry, 130th Meeting, ACS, Atlantic City, N. J., September 1956.
(86) De Keyser, W. L., *Nature* 172, 364–5 (1953).
(87) Kinjyo, K., Iwata, S., *J. Chem. Soc. Japan, Ind. Chem. Sect.* 74, 642–4 (1953).
(88) Klevens, H. B., Mellon Institute, Pittsburgh, Pa., private communication, 1957.
(89) Kohler, G., *Svensk. Kem. Tidskr.* 38, 130–4 (1926).
(90) Kuhlmann, W. H. F., *Der Mechaniker* 18, 146–7 (1910).
(91) Lambert, A., *Bull. soc. franc. ceram.* 28, 23–8 (1955).
(92) LaPine, Arthur S., & Co., *Chem. Eng. News* 26, 3269 (1948).
(93) Lassater, J. A., Nimer, E. L., Quinn, E. L., Davidson, A., A.S.T.I.A. Tech. Rept. AD 19923 (1952).
(94) Lewin, S. Z., *J. Chem. Educ.* 36, A67–A76 (1959).
(95) Linseis, Max, *Keram. Z.* 11, 54–5 (1959).
(96) Lohmann, I. W., *J. Sci. Instr.* 21, 999–1002 (1950).
(97) Longchambon, H., *Bull. soc. franc. minéral.* 59, 145–61 (1936).
(98) Mauer, F. A., *Instrumentation* 7, 36–7 (1955).
(99) Mauer, F. A., *Rev. Sci. Instr.* 25, 598–602 (1954).
(100) Mettler Instrument Corp., Hightstown, N. J., Bull. **R5**, 1959.
(101) Mettler, E., "Mettler News," Zurich, Switzerland, June 1959.
(102) Mikulinskii, A. S., Gel'd, P. V., *Zavodskaya Lab.* 9, 921–2 (1940).
(103) Moses, A. J., King, L. O., Wendelyn, J. M., *Nucleonics* 15, 120 (1957).
(104) Müller, R. H., Garman, R. L., IND. ENG. CHEM., ANAL. ED. 10, 436–40 (1938).
(105) Muller, N. W., Peck, R. E., *Ibid.*, 15, 46–8 (1943).
(106) Niagara Electron Laboratories, Andover, N. Y., Bull., "Mauer Recording Balance."
(107) Oden, S., *Bull. Geol. Inst. Univ. Upsala* 16, 15–64 (1918).
(108) Oden, S., *Proc. Royal Soc. Edinburgh* 219–36 (1916).
(109) Oden, S., *Trans. Faraday Soc.* 17, 327–48 (1922).
(110) Okeanov, K. V., *Trudy Bryansk. Lesokhoz. Inst.* No. 6, 181–2 (1953); *Referat. Zhur. Khim.* 1955, No. 739.
(111) Orosco, E., A Termobalança, Inst. nacl. tecnol. (Rio de Janeiro) (1940).
(112) Padday, J. F., *Proc. 2nd Intern. Congr. Surface Activity, London* Vol. 1, p. 1–6 (1957).

(113) Paulik, F., Paulik, J., Erdey, L., Z. anal. Chem. 160, 241–52 (1958).
(114) Payne, R. E., Sharples Corp., Bridgeport, Pa., Bull. No. 1244.
(115) Peters, Horst, Wiedemann, H. G., Z. anorg. u. allgem. Chem. 298, 202–11 (1959).
(116) Peterson, A. H., Instr. and Automation 28, 1104–6 (1955).
(117) Philip Morris Co., Richmond, Va.; Ind. Labs. 6, (July 1955).
(118) Phoenix Precision Instrument Co., Philadelphia, Pa. Bull. R-1000, "Differential Refractometers."
(119) Picon, M., Ann. pharm. franc. 6 34–92 (1948).
(120) Pillsbury Mills, Inc., Minneapolis, Minn., Chem. Processing 18, (July 1955).
(121) Pope, M. I., J. Sci. Instr. 34, 229–32 (1957).
(122) Pouradier, J., Dubois, A., Research London 2, 119–21, 1949; Suppl. (Surface Chemistry).
(123) Rabatin, J. G., Gale, R. H., ANAL. CHEM. 28, 1314–16 (1956).
(124) Rogers, M. C., Earle, P. L., Ind. Eng. Chem. 33, 642–7 (1941).
(125) Rulfs, C. L., ANAL. CHEM. 20, 262–4 (1948).
(126) Sarakhov, A. I., Doklady Akad. Nauk S.S.S.R. 86, 989–92 (1952).
(127) Sartorius-Werke A. G., Gottingen, Germany, Bull., "De Keyser Differential Thermobalance."
(128) Ibid., Bull., "Electrona Recording Microbalance."
(129) Ibid., Bull., "Recording Attachment for Selecta Balances."
(130) Ibid., Bull., "Sedibal Recording Sedimentation Balance."
(130a) Šatava, V., Silikáty 1, 188–90 (1957).
(131) Scholten, P. C., Smit, W. M., Wijnen, M. D., Rec. trav. chim. 77, 305–15 (1958).
(132) Sharples Corp., Bridgeport, Pa., Bull. 103, "Universal Recording Balance."
(133) Simons, E. L., Newkirk, A. E., Aliferis, I., ANAL. CHEM. 29, 48–54 (1957).
(134) Simons, J. H., Scheirer, C. L., Ritter, H. L., Rev. Sci. Instr. 24, 36–42 (1953).
(135) Sinclair, D., J. Appl. Phys. 21, 380–6 (1950).
(136) Sinclair, D., Rev. Sci. Instr. 27, 34–6 (1956).
(137) Smith, B. O., Stevens, J. W., J. Sci. Instr. 36, 206–9 (1959).
(138) Spinedi, P., Chime ind. (Milan) 33, 777–82 (1951).
(139) Spinedi, P., Ricerca sci. 23, 2009–14 (1953).
(140) Splitek, R., Hutnické listy 8 697–705 (1958).
(141) Stanton Instruments, Ltd., London, England, Bull., "Recording Balances."
(142) Starodubtsev, S., Zavodskaya Lab. 13, 1096–9 (1947).
(143) Stephenson, J. L., Smith, G. W., Trantham, H. V., Rev. Sci. Instr. 28, 380–1 (1957).
(144) Stock, A., Z. physik. Chem. 139, 47–52 (1928).
(145) Suito, E., Arakawa, M., Bull. Inst. Chem. Research, Kyoto Univ. 22, 7–17 (1950).
(146) Svedberg, T., Rinde, H., J. Am. Chem. Soc. 45, 943–54 (1923).
(147) Szabo, Zoltan, Király, Deszo, Magyar Kém. Folyóirat 63, 158–65 (1957).
(148) Teetsel, F. M., Monroe, M. A., Williamson, J. A., Abbot, A. E., Stoneking, D. J., U.S.A.E.C. Bull., NLCO-713, 1958.
(149) Testing Equipment Sales Co. (Tesco), Murray Hill, N. J., Bull. No. 25, "Tesco Miniature Low Capacity Adjustable or Fixed Load Cells."
(150) Ibid., Bull. No. 26 "Do-It-Yourself Differential Transformer Analytical Balance Amplifier Kit."
(151) Testut Co., 9 Rue Brown Sequard, Paris XV, France, Bull., "Automatic Recording Balances."
(152) Thall, E., Can. Mining Met. Bull. 145, 663–70 (1946).
(153) Torsion Balance Co., Clifton, N. J.
(154) Trenner, N. R., Warren, C. W., Jones, S. L., ANAL. CHEM. 25, 1685–91 (1953).
(155) Tryhorn, F. G., Wyatt, W. F., Trans. Faraday Soc. 23, 238–42 (1927).
(156) Urbain, M. G., Compt. rend. 6, 347–9 (1912).
(157) Van Nordstrand, R. A., U. S. Patent 2,692,497 (Oct. 26, 1954).
(158) Vieweg, R., Gast, T., Kunststoffe 34, 117–19 (1944).
(159) Vytasil, Vladimir, Vytasilova, Sona, Hejl, Vladimir, Silikáty 2, 285–9 (1958).
(160) Waters, P. L., Coke and Gas 20, 252–6 (1958).
(161) Waters, P. L., J. Sci. Instr. 35, 41–6 (1958).
(162) Waters, P. L., Nature 178, 324–6 (1956).
(163) Wendlandt, W. W., ANAL. CHEM. 30, 56–8 (1958).
(164) Williamson, J. A., Summary Tech. Rept., National Lead Co. NLCO-650, 142–4 (1956).
(165) Zagorski, Z., Przemysl Chem. 31 (8), 326–30 (1952).

32

Reprinted from *Anal. Chem.*, **32**(12), 1558–1563 (1960)

Thermogravimetric Measurements

ARTHUR E. NEWKIRK

General Electric Research Laboratory, Schenectady, N. Y.

▶ Interest in thermogravimetry has increased in recent years because of the commercial availability of automatic, continuously recording thermobalances which are rugged, reliable, and accurate. Although isothermal or static thermogravimetry is an old art, the new instruments facilitate studies by dynamic thermogravimetry in which the sample is heated at a uniform rate. Dynamic thermogravimetry is, however, subject to errors which are often overlooked. These include the effect of changing air buoyancy and convection, the measurement of temperature, and the effects of atmosphere, heating rate, and heat of reaction. The present limitations of dynamic thermogravimetry would seem to indicate that its chief value is for the determination of the approximate quantitative behavior of a system but not, by itself, for the determination of precise constants.

T HERMOGRAVIMETRY is the science and art of weighing substances while they are being heated. While modern thermogravimetry appears to offer the possibility of studying the complete thermal decomposition behavior of a substance with a simple, automatic apparatus which in a few hours plots the weight-temperature relation directly, it requires the use of a sensitive balance under conditions which sometimes seem designed to make it as difficult as possible to obtain the correct result. In addition to the obvious sources of error, further limitations are inherent in almost any reaction system, and particular systems often present additional unusual difficulties. Despite these problems, thermogravimetric results can be surprisingly good, and many of the apparent limitations of thermogravimetric measurements can be effectively overcome.

Modern thermobalances facilitate studies in which a sample is subjected to conditions of continuous increase in temperature, usually linear with time. This may be called dynamic thermogravimetry, and is relatively recent (1923) compared to isothermal or static thermogravimetry, which has a long history. Dynamic thermogravimetry is particularly attractive in the early stages of an investigation because it is a way to survey rapidly the complete thermal behavior of many systems. Its value was pointed out by Guichard (*11*), who recommended continuous weighing at linearly increasing temperatures for the study of dehydration reactions. Within a few years he recommended its extension as a first method to be used in the exploration of all phenomena which take place with a change in weight. As he pointed out, it gives the principal results very easily, and for the delicate cases it often indicates the zone of temperature in which it is necessary to multiply one's observations.

Dynamic thermogravimetric measurements are subject to errors and limitations which do not seem to be readily recognized. Those discussed in this paper apply most directly to thermobalances, such as the Chevenard, in which the sample is exposed to a gas at 1-atm. pressure. Vacuum thermobalances have their own peculiarities and will not be considered.

EFFECT OF CHANGING AIR BUOYANCY AND CONVECTION

First we will consider the effects of air buoyancy, convection, and related effects on an inert sample supported from below in the center of a cylindrical, capped tube furnace. This arrangement is favored for thermobalances

because it minimizes both heating of the balance mechanism, and its disturbance by thermal convection currents. Under these conditions an inert sample may show an apparent weight gain of up to 10 mg., and correction for this is the most important correction that may have to be applied to the recorded weight change. The several causes of this change are known, but the relative magnitudes due to each are not, and even the need for applying the correction has sometimes gone unrecognized. Duval, for example (*5*), did not mention the effect for reasons which have become clear only recently (*8*).

The three causes of the apparent weight gain, decreased air buoyancy, increased convection, and the effect of heat on the balance mechanism have been discussed in detail (*1, 12, 13*). The apparent weight gain increases with increasing volume of the load, thereby showing its dependency on decreased air buoyancy. The calculated magnitude of this effect is a function of the gas, being for hydrogen 0.1 mg., for air 1.4 mg., and for argon 1.9 mg., for a temperature change from 25° to 1000° C., and a load of one No. 4-0 porcelain crucible. The apparent weight gain varies with heating rate, thereby showing its dependency on convection. Recently in this laboratory the apparent weight gain has been observed to increase markedly if the balance mechanism was shielded to prevent interference from stray air currents. This shielding also reduced heat dissipation, thereby showing the effect of heat on the balance mechanism which in the Chevenard thermobalance may be attributed to decreased buoyancy of warm oil in the dash pots. Inasmuch as the relative magnitude of these effects depends on the design of

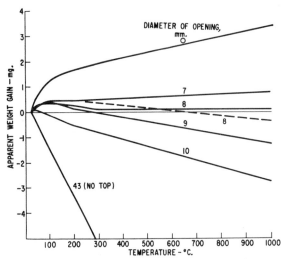

Figure 1. Effect of top opening on apparent weight gain

Chevenard thermobalance, heating rate 300°/hour
—— One porcelain crucible, Coors 230-000 (about 4 grams)
– – – Two crucibles

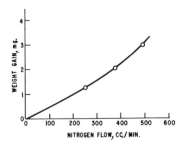

Figure 2. Effect of gas velocity on apparent weight gain of one porcelain crucible at room temperature

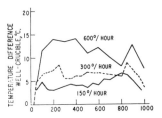

Figure 3. Effect of heating rate on crucible temperature

the balance and its environment, different apparent weight gain correction curves are obtained by different operators.

It has recently been suggested by Duval (8) that the cause of these differences is the extent to which the top cap of the furnace permits air to vent through the furnace tube, and he concludes that by proper choice of the size and position of the vents it is possible to obtain no apparent weight gain on heating. It is unlikely that this is correct in the general case. If the upward flow of air through a vented thermobalance furnace will by chance just compensate for the buoyant and turbulent effects on an empty crucible, it would be surprising that it should do so equally well for small and large loads in the same crucible. The apparent weight loss caused by the upflowing stream of air on the crucible and the apparent weight gain due to turbulence must be determined largely by the crucible size and shape, whereas the apparent weight gain due to decreased air buoyancy will depend not only on the crucible but on the volume of its contents. However, for small loads of relatively the same size and density, the differences would be negligible, and these were the conditions used by Duval in his study of analytical precipitates. He also advises using crucibles having the same form, and in particular the same large and small diameters.

Apparent weight change curves as a function of top opening are shown in Figure 1. These were obtained with a single empty crucible as described (18). Three characteristics are worth noting: Except when there is a very large opening, there is always an initial weight gain, even when, on further heating, there is an over-all weight loss; most of the curves can best be represented by a straight line in the temperature range 200° to 1000° C., and the lines shown are a least-squares fit to observations made at 100° intervals in this range; it is not possible to choose an opening which will give no apparent weight gain on heating over the whole temperature range, although the effect can be greatly reduced, and for many purposes rendered negligible. The effect of doubling the load from one to two crucibles with an 8-mm. opening is shown by the dashed line.

The choice of a top opening to minimize the apparent weight gain is also a function of the rate of heating. With the top closed, a marked difference is observed at heating rates of 150° and 300° per hour (18). With an 8-mm. top opening the differences are greatly reduced, but still apparent.

When samples are examined in the thermobalance using a gas tube to control the atmosphere, an additional apparent weight gain is caused by the gas flow, Figure 2. Its magnitude is affected by the molecular weight of the gas used.

Changes in air buoyancy and convection can cause apparent weight changes of sufficient magnitude so that for careful work correction curves should be obtained on inert samples under the same conditions used for the active samples. The magnitude of the apparent weight changes may be reduced considerably by proper venting.

MEASUREMENT OF TEMPERATURE

Thermobalances are regularly used for determining the drying ranges of precipitates and are becoming increasingly popular for the determination of kinetic constants of thermal decomposition reactions of technical importance; a warning about the significance of such measurements seems in order. Although some thermobalances have been described in which the sample temperature is measured directly, the more usual practice is to measure the temperature in the furnace near the sample. These measurements are subject to error in a manner which seems often to have been overlooked. The error is caused by the use of a heating program in which the temperature of the furnace is continually increased at a constant rate with the result that reaction temperatures so determined are usually higher than those determined by the more common process of heating a sample for a fixed length of time at each of a series of constantly increasing temperatures. Part of this difference is merely thermal lag, and part is due to the finite time required to cause a detectable weight change.

The effect of thermal lag is shown for three different heating rates in Figure 3.

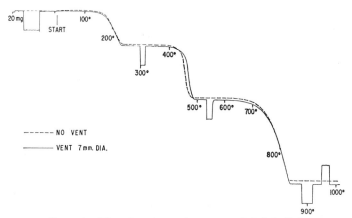

------ NO VENT

—— VENT 7 mm. DIA.

Figure 4. Effect of venting on thermogram of $CaC_2O_4 \cdot H_2O$

In these experiments a second thermocouple was located in a thin quartz sheath directly in the empty crucible, and the figure shows the extent to which the temperature of this thermocouple lagged behind the furnace temperature. The lag is significant, 3° to 14°, and is roughly proportional to the heating rate.

The effect of reaction time may be demonstrated by a very simple experiment. When a drop of water weighing 0.10 gram was heated in a porcelain crucible in a Chevenard thermobalance at a heating rate of 300° per hour, constant weight was attained at 122° C. It is obvious that the sample did not achieve this temperature, that it is a higher temperature than that necessary to dry the crucible (were longer drying times used at lower tempera-

tures), and that the observed value of the temperature will be influenced by the amount of water and the rate of heating. One would not think of specifying the higher temperature as a necessary lower limit for drying in this case, yet this conclusion is usually made about analytical precipitates and occasionally leads to confusion. A good example is the work on the use of zinc monosalicylaldoxime for the determination of zinc.

Duval (7) followed the procedure suggested by Flagg and Furman (9)

for the drying and decomposition of zinc monosalicylaldoxime, but, since his thermogram had no plateau for anhydrous zinc monosalicylaldoxime, he proposed abandoning this method. Interestingly enough, in neither of his papers (2, 7) does the thermogram indicated to be that of zinc monosalicylaldoxime correspond to the description given for it. A more detailed study by Rynasiewicz and Flagg (16) showed that zinc monosalicylaldoxime was stable from room temperature to 285° C., but that the plateau recorded by the thermobalance depended upon the initial water content. The thermolysis curve of a sample air-dried at room temperature to zero water content had a plateau from 25° to 285° C., and a partially dried sample with a water content of 50% gave a plateau from 135° to 290° C. Each of these samples was heated at a rate of 300° C. per hour, but held for 1 hour at a constant temperature below the decomposition temperature, 100° and 135°, respectively. A sample initially wet, 630% water, gave a plateau in the range 245° to 315° C. Inasmuch as Duval used a higher heating rate, on the order of 380° per hour, in this region, it is perhaps not surprising that he failed to find a plateau. The rejection of this precipitate as an analytical method for the determination of zinc on the grounds that it does not give a stable horizontal in the thermobalance at one particular

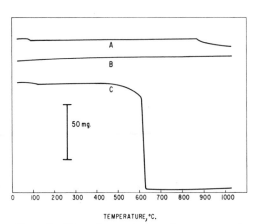

Figure 5. Effect of atmosphere on decomposition of Na_2CO_3; reaction $Na_2CO_3 + WO_3$ in CO_2

A. Sodium carbonate in air
B. Dried sodium carbonate in CO_2; gas flow 250 ml./min.
C. Sodium carbonate + tungsten oxide in CO_2; gas flow 250 ml./min. Weight change: calculated 0.0996 gram, observed 0.0995 gram

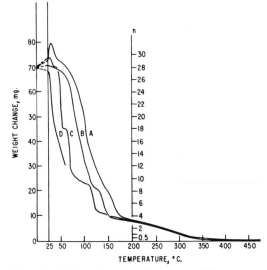

Figure 6. Dehydration of sodium (5 to 12) tungstate 28-hydrate

Atmosphere and heating rates
A. Humidified air, 300°/hr.
B. Humidified air, 150°/hr.
C. Humidified air, 10°/hr.
D. Room air, 10°/hr.
n = moles H_2O per $5Na_2O \cdot 12WO_3$; sample weight 0.5000 gram

267

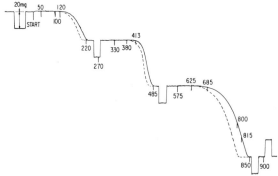

Figure 7. Effect of heating rate on the thermogram of $CaC_2O_4.-H_2O$

——— 300°/hr.
- - - -150°/hr. corrected for differences in scale and apparent weight gain

rate of heating is clearly unwarranted, although it may be reasonable to reject it for a program of automatic analysis of the kind developed by Duval. The results suggest that when the thermobalance is used to study the drying of bulky precipitates that contain considerable water, it would be well to use very slow rates of heating.

Another source of error, when a controlled atmosphere is used, is the effect of the gas stream in lowering the thermocouple temperature relative to the sample temperature. In the author's experience this difference amounts to a few degrees and is unimportant at high rates of heating where the true sample temperatures cannot be known to any better than this, since gradients of this order of magnitude exist in the sample itself. However, when the thermobalance is used as a constant temperature device, this effect must be taken into account.

Further examples of the difficulties which interfere with knowing the sample temperature will be given later in discussing the effect of rate of heating and heat of reaction, but the problem may be summarized as follows. Although the relation between the temperature in the thermobalance furnace (which may be measured precisely) and the temperature of the sample undergoing reaction (which is not measured) is indeterminate, certain qualitative generalizations may be made about the direction and magnitude of the difference since, for a given instrument, the geometry of the system and the procedure used are usually the same. The true temperature of a substance which does not change weight on heating will tend to be lower than the recorded temperature, the magnitude of the deviation depending on the mass of the sample, its heat capacity, thermal conductivity and degree of fineness, the rate of gas flow in the furnace, and the rate of heating. During an endothermic change in the sample, the deviation will of course be greater, while during

an exothermic change the true temperature may range from only slightly less than the recorded temperature to much greater depending on the amount of heat produced. However, there appears to be no simple way to estimate *a priori* the magnitude of the difference for a given sample.

The establishment of temperature ranges for drying (or the temperature range of stability of a compound) from thermobalance curves is, therefore, not simply a matter of reading the temperatures at the beginning and end of a level portion of a single thermogram. As a general rule this range will be approximately correct, but will have been shifted upward in temperature by the effect of the rate of heating. The statement of Duval (*6*) that when the pyrolysis curve of an analytical precipitate affords a horizontal, all dryings must be effected at a temperature situated between the two ends of the level, while correct for his apparatus and procedure, will not be correct in general. Fortunately, even though the temperature to be assigned to a point on a thermogravimetric curve may be indeterminate, the curves obtained by heating equal weights of the same material under the same conditions in separate runs are usually identical. A descriptive term to apply to such temperatures should be useful. The term "procedural decomposition temperature" or "p.d.t.," which was coined by Doyle in reporting on his extensive and thorough studies of the application of the thermobalance to polymer decompositions (*4*), is suggested.

EFFECT OF ATMOSPHERE

When a sample is dried or decomposed in a thermobalance in ambient air, the atmosphere near the sample is continuously modified by addition of gaseous decomposition products or the loss of part of the original gas by reaction with

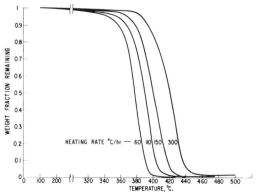

Figure 8. Pyrolysis of polystyrene in nitrogen at different heating rates

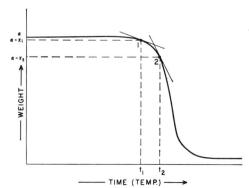

Figure 9. Determination of reaction rate and extent of reaction from a thermogram

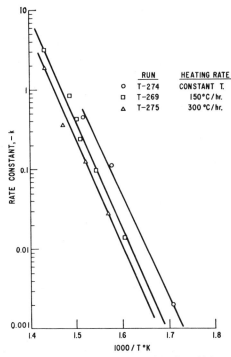

Figure 10. Rate of weight loss of Du Pont Mylar at three heating rates

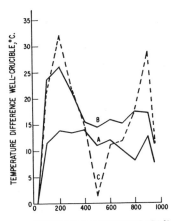

Figure 11. Temperature lag due to sample. Decomposition of CaC_2O_4.-H_2O

Heating rate 600°/hr.
A. Crucible only
B. Crucible + 0.2 gram $CaC_2O_4 \cdot H_2O$
C. Crucible + 0.6 gram of $CaC_2O_4 \cdot H_2O$

the sample. Even small changes in the composition of this atmosphere can affect the thermogram as shown by Figure 4. The dashed curve represents the decomposition of calcium oxalate monohydrate in a closed top furnace. By venting the top, the apparent weight gain correction is successfully reduced—on the plateaus the dashed curve lies above the solid curve and the difference increases with increase in temperature—but the course of the decomposition varies, as is particularly noticeable for the second step.

A second example is the decomposition of sodium carbonate Figure 5 (14). In air at a heating rate of 300° C. per hour, decomposition sets in above 850° C., curve A, but in CO_2 it does not occur even at 1050° C., curve B. By using a platinum crucible and a CO_2 atmosphere, sodium carbonate may be used as a thermogravimetric reagent without interference from the decomposition reaction. This compound has been used successfully in this laboratory to study the composition of complex polytungstates as indicated by curve C.

A third example illustrates the importance of both atmosphere control and heating rate. Sodium (5:12) tungstate 28-hydrate, when decomposed in ambient air at a heating rate as low as 10° per hour, gives no evidence of stable regions for either the 28-

hydrate or 18-hydrate, Figure 6, curve D. In humidified air good plateaus are obtained at this heating rate, curve C, but not at higher rates, curves B and A.

Further examples of the effect of atmosphere on the thermogram are given by Garn (10).

EFFECT OF HEATING RATE

The effect of heating rate on the thermogram obtained for a given sample of $CaC_2O_4 \cdot H_2O$ is illustrated more clearly in Figure 7. The two curves are plotted to the same temperature scale and corrected for differences in the apparent weight gain. For each of the three reactions—dehydration, loss of CO, and loss of CO_2—the temperature of apparent onset of decomposition is lower at the lower heating rate, and the decomposition is completed at a lower temperature, the lower the heating rate. By going to heating rates lower than 150° per hour, the thermogram would be shifted to lower temperatures. In this case there is independent evidence to show that the first and third plateaus, those for $CaC_2O_4 \cdot H_2O$ and $CaCO_3$, actually represent stable compounds in the lower parts of the indicated stability ranges, but careful work by Peters and Wiedemann (15) has shown that anhydrous calcium oxalate is de-

composed slowly at least as low as 300° C.

Similar results are obtained with polymer decompositions as shown for polystyrene by Figure 8. The effect of the different heating rates may be most readily appreciated by comparing the temperature at which a fixed amount of decomposition has occurred, and in this case for 10% decomposition the temperatures are 357° C. for a heating rate of 60° per hour, and 394° C. for a heating rate of 300° per hour, or a difference of 37° C. Sabatier (17) has observed a similar effect in the dehydration of mica. For samples heated at 12° to 30° per hour, the observed temperatures for a corresponding stage in weight loss ranged from 735° to 815°, a difference of 80° C.

The effect of heating rate is important if the thermogram is to be used for kinetic analysis. We can obtain from a thermogram for a series of temperatures both the sample weight remaining, $a - x_1$ and $a - x_2$, and the reaction rate, $\left(\dfrac{dx}{dt}\right)_1$ and $\left(\dfrac{dx}{dt}\right)_2$, shown in Figure 9 by the tangents to the curve at points 1 and 2 for the temperatures T_1 and T_2. In the case of Mylar the decomposition is first order, and the logarithm of the reaction rate constant k, $dx/dt = k(a - x)$, may be plotted against $1/T$ to yield a straight line. The results of measurements at constant temperature and at two heating rates are shown in Figure 10. The lines obtained are approximately parallel and, using the data obtained under isothermal conditions as a standard, we can calculate that the apparent temperature is decreased by 12° for a heating rate of 150° per hour, and by 19° for a heating rate of 300° per hour. The magnitude of these differences varies with the material and the form of the sample, but in any event, kinetic analyses made from dynamic thermograms should consider such effects. Doyle (*3*) has made extensive studies of polymer decompositions using dynamic thermogravimetry.

EFFECT OF HEAT ON REACTION

The heat of reaction will affect the difference between sample temperature and furnace temperature, causing the sample temperature to lead or lag the furnace temperature depending on whether the reaction is exothermic or endothermic. The method of differential thermal analysis is possible because this temperature difference exists, and it is common practice in DTA to use relatively high rates of heating, 600° per hour, to accentuate the differential temperature. Since such differential temperatures may range well above 10° at this heating rate, kinetic constants calculated from thermograms made at such high heating rates may be unavoidably and significantly in error. When the reaction is endothermic, the effect of temperature lag and differential temperature will be additive, but when the reaction is exothermic, the effects will tend to compensate each other. That the effect of the heat of reaction can be large is illustrated strikingly in the case of the oxidation of tungsten carbide in air (*13*). As shown by the thermograms, approximately the same rate of reaction was observed with the thermobalance furnace heated to 527° C. and held constant, as was observed when the furnace was heated continuously and uniformly.

The effect is perhaps more clearly shown in Figure 11 where the differential temperature between the furnace and sample was measured directly for the decomposition of calcium oxalate monohydrate at a high heating rate and for two different sample sizes. Curve *A* for the crucible alone shows a 10° to 14° lag in the range 100° to 1000° C. When a 0.2-gram sample is used, the endothermic loss of water results in a 25° lag at 200° C., the exothermic loss of CO brings the sample nearly back to the difference observed with the crucible alone, but the lag again increases during the endothermic loss of CO_2. With a 0.6-gram sample, these effects are accentuated and at one point the sample and furnace temperature are nearly the same.

ACKNOWLEDGMENT

The author thanks M. A. Bonanno and H. W. Middleton for assistance with the thermogravimetric measurements, and E. L. Simons for helpful discussions.

LITERATURE CITED

(1) Claisse, F., East, F., Abesque, F., "Use of the Thermobalance in Analytical Chemistry," Department of Mines, Province of Quebec, Canada, 1954.
(2) DeClerq, Monique, Duval, C., *Anal. Chim. Acta* 5, 282 (1951).
(3) Doyle, C. D., "Evaluation of Experimental Polymers—Aging Tests Conducted in the Thermobalance and Empirical Tests in Dry Nitrogen," Progress Rept. No. 6, Sept. 1 to Dec. 1, 1959; Contract AF 33(616)-5576, Project No. 7340.
(4) Doyle, C. D., Hill, J. A., "Evaluation of Experimental Polymers for Dielectric Applications," Second Quarterly Progress Rept., July 5 to Oct. 5, 1958, Contract AF 33(616)-5576, Project No. 8-(8-7371).
(5) Duval, C., "Inorganic Thermogravimetric Analysis," p. 26, Elsevier, New York, 1953.
(6) *Ibid.*, p. 34.
(7) *Ibid.*, p. 283.
(8) Duval, C., *Mikrochim. Acta* 1958, 705.
(9) Flagg, J. F., Furman, N. H., IND. ENG. CHEM., ANAL. ED. 12, 663 (1940).
(10) Garn, P. D., Kessler, J. E., ANAL. CHEM. 32, 1563 (1960).
(11) Guichard, Marcel, *Bull. soc. chim. France* 37, 62 (1925).
(12) Mielenz, R. C., Schieltz, N. C., King, M. E., "Clays and Clay Minerals," Natl. Acad. Sci.-Natl. Research Council Publ. 327, Washington, D. C., 1954, p. 285.
(13) Newkirk, A. E., *J. Am. Chem. Soc.* 77, 4521 (1955).
(14) Newkirk, A. E., Aliferis, I., ANAL. CHEM. 30, 982 (1958).
(15) Peters, H., Wiedemann, H. G., *Z. anorg. u allgem. Chem.* 300, 142 (1959).
(16) Rynasiewicz, J., Flagg, J. F., ANAL. CHEM. 26, 1506 (1954).
(17) Sabatier, G., *J. chim. phys.* 52, 60 (1955).
(18) Simons, E. L., Newkirk, A. E., Aliferis, I., ANAL. CHEM. 29, 48 (1957).

RECEIVED for review May 26, 1960. Accepted September 1, 1960. Division of Analytical Chemistry, 137th Meeting, ACS, Cleveland, Ohio, April 1960.

33

Reprinted from *Vacuum Microbalance Techniques*, Vol. 3, K. H. Behrndt, ed.,
Plenum Press, New York, 1963, pp. 29–44

THE CAHN RECORDING GRAM ELECTROBALANCE

Lee Cahn and Harold R. Schultz
Cahn Instrument Company
Paramount, California

ABSTRACT

An automatic recording vacuum ultramicrobalance has been based on the authors' elastic-ribbon-suspension torque motor, reported at the last Symposium. The beam is rebalanced automatically by means of a phototube and servo amplifier; the balancing current required is used as the measure of sample weight change, and applied to a millivolt recorder through internal measuring circuits. The capacity has been increased to 1 g on the more sensitive loop and 2.5 g on the less sensitive one by means of a new beam design and improved assembly technique. The maximum sample weight variation remains 1 g. Sensitivity remains 0.1 μg on a 1-mv recorder, but precision as a fraction of total load is improved to 1 ppm. The range of measurable weight change can be changed at will while recording, permitting a precision of 1 part in 10,000 of the weight change, for all magnitudes of sample weight variation. Response time is less than 0.02 sec. A glass vacuum bottle has been made for the weighing mechanism, to cover the vacuum range attainable with either ground glass or Viton O-ring joints. The weighing mechanism may also be installed in other vacuum containers. The new balance offers the advantage of continuous automatic unattended recording of sample weight change in vacuum microbalance applications.

The Cahn Recording Gram Electrobalance was developed to measure weight changes in high vacuum, controlled atmospheres, and room air. It has a maximum capacity of 2.5 g, and an ultimate sensitivity of

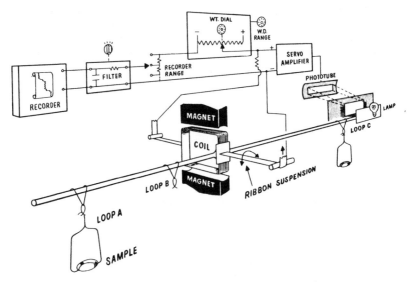

Fig. 1. Schematic of Recording Gram Electrobalance.

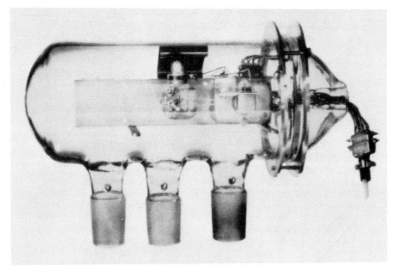

Fig. 2. Weighing mechanism installed in special glass vacuum bottle
(also used in bell jars, etc.).

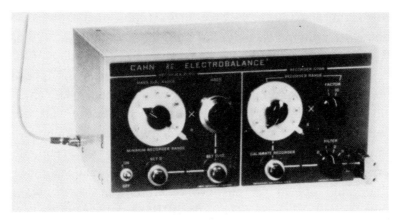

Fig. 3. Control unit, located outside the vacuum, up to 30 ft from the weighing mechanism, and usually next to the recorder.

0.1 μg, meeting most requirements for surface chemistry, thermogravimetry, and space materials research. It is based on our previous Gram Manual Electrobalance and RM Automatic Electrobalance.[1]

The system is shown in Fig. 1, and the instrument itself in Figs. 2 and 3. The beam is of hollow aluminum tubing and actually consists of a V-shaped structure for greater strength and rigidity, although only a single tube is shown. The beam is suspended by an elastic metal ribbon, as in the previous Gram design. Samples up to 1 g are suspended from the sample loop A, and are normally counterbalanced closely on the counter-weight loop C. A third sample suspension loop B is located at one-fifth the radius from the axis of rotation of the beam, and can accommodate samples up to 2.5 g. These values refer to the total load on the sample loop, including the container but not including the counterweight on loop C. Thus, the maximum total central bearing load is 2 g for loop A and 3 g for loop B. Above these values, the suspension deflects downward into cylindrical protective stops, so that the ribbon suspension is protected against substantial further overloads. It is quite obvious when the suspension engages the stops, because so much "noise" will appear on the recorder chart that data can no longer be taken.

If the sample weight decreases, the beam will tend to rotate so that the flag on the right-hand end uncovers more of the photocathode of the vacuum phototube, increasing the phototube current. This current is amplified and applied to the coil with polarity to rebalance the beam. The two-tube amplifier can provide both positive and negative currents. The loop gain of this servo system is in excess of 2000, so that the effects of nonlinearity and drift in the electronics are reduced by this factor. The high servo loop gain also results in the beam being returned to essentially the same position after a weight change. Beam deflection for a sample on loop A is less than 10μ per milligram of sample weight change. A sample on loop B moves less than 0.5μ per milligram of weight change. This is useful when the sample is in a nonhomogeneous environment, such as an arc furnace, etc. Also, since the coil remains in essentially the same position in the permanent magnetic field, the electromagnetic moment is exactly proportional to the current causing it. The field is also made symmetrical, so that its gradient with coil angle is as small as machining tolerances will permit, to further improve linearity.

Coil voltage is thus proportional to sample weight change, and is taken as a measure of it. It is linear with sample weight differences to better than 1 part in 10^4. Chart recorders are only accurate to 0.2-1%, so that it is necessary to provide a circuit for measuring to 0.01% or better with a 1% recorder. This circuit includes a very stable reference power supply and an accurate potentiometer, which can subtract a known fraction of the coil voltage. A dial on the potentiometer is calibrated in milligrams. The excess of coil voltage over potentiometer voltage is applied to a voltage divider, which establishes the calibration between voltage and weight, and then to a chart recorder.

The reference power supply for the potentiometer is stable to 1–2 parts in 10^5, and precision to this level could be obtained. However, it is seldom necessary to determine changes to such a fine fraction of the total change, so that precision is normally limited to 1 part in 10^4 by readability of the dial and resolution of the potentiometer itself. When required, higher precision could be obtained by substituting a better potentiometer at this point in the circuit.

The range of sample weight change desired can vary greatly between one experiment and another. Thus the reference voltage can be attenuated very accurately, before being applied to the potentiometer, so that full scale on the potentiometer and dial can be as great as 200 mg, and as fine as 1 mg, on loop A. Intermediate ranges of 10, 20, and 100 mg can also be selected. The attenuating networks are accurate to 0.025%, so that to within this accuracy one can change the range of weight variation which is selected and readable on the dial, without recalibration or breaking vacuum. Thus, it is not necessary to predict exactly how much sample weight variation will occur; one can always switch ranges during operation over the full available range. The corresponding ranges for loop B are five times as great, namely, 5, 50, 100, 500, and 1000 mg of variation.

The excess of coil voltage over reference voltage can be attenuated with a recorder range selector switch, whose resistors are accurate to 0.1%. This is more accurate than the best recorders. Recorder ranges can be selected from 0.02 to 200 mg full scale in a 1-2-4-10 sequence for loop A, and 0.1 mg to 1 g in 1-2-5-10 sequence for loop B. Thus all recorder ranges are accurately calibrated if one of them is, and they can be changed at will while recording. This permits presenting any weight change, from the finest to the coarsest possible, on a single traversal of a recorder chart, if desired.

The exact factor desired between weight and voltage will depend on the actual voltage span of the recorder. It is made adjustable, so that one can match the balance calibration to any recorder of nominally correct span. The balance is adjusted so that the recorder agrees with the calibrating weights and reads directly in milligrams, regardless of its actual voltage/deflection sensitivity.

Due to the high servo loop gain mentioned above, the balance responds very quickly with a damped period of about 0.040 sec. While useful for some fast-reaction studies, this speed also indicates rapid force changes due to vibration, table sway, etc., which would not appear in a slower system. It is thus desirable to be able to reduce the speed of response for severe combinations of vibration, sensitivity, and fast recorder speed. An adjustable electrical filter is provided before the recorder with single-capacity time constants of 0.07, 0.17, and 0.5 sec. In the worst possible cases, when operating at 0.1 μg sensitivity on shaky tables with 0.5 sec recorders, additional external filtering has been found desirable, to as much as a 2.5-sec time constant. This is obtained by connecting 40 μf in parallel with the internal 10 μf across the output terminals. It would not be required for a normal installation. Even so, a 2.5-sec time constant is still substantially faster than would be obtained with a conventional deflection-type recording balance.

As implied above, the balance has been designed for use with any make of potentiometric recorder. It has been tested satisfactorily with nearly all of them. The controls are labeled for a 1-mv recorder, and this recorder sensitivity is required for full sensitivity of the balance. One-tenth of a microgram on loop A corresponds to 5 μv, or 0.5% on a 1-mv recorder. This level can be handled conveniently on all commercial 1-mv recorders. Some of them do better. For work to less than full sensitivity, it is possible to use less sensitive recorders, in proportion.

Since the basic output is a voltage, other voltage-measuring readout devices can also be used and sometimes offer advantages. A sensitive galvanometer can be used as a null meter to bring the reference voltage to equal the coil voltage. In this case, one would read the total sample weight change from the dial. A digital voltmeter of suitable sensitivity can also be used; some of them provide four digits directly, so that the reference system would only be used in the initial setup.

OPERATION

Initially, the balance is set up with the empty sample container suspended from the loop selected. If the sample is not expected to change by a large fraction of its total weight, a substitution weight representing the constant portion of sample weight may also be added to the container. This load is then counterbalanced as closely as possible on loop C. Accurate calibration of all dials is ensured by adding and removing calibrating weights, and adjusting potentiometers in the circuit. The substitution weight is then removed, and the sample placed in its container. If the balance is to be evacuated, this is done, and the zero point in vacuum is observed. The absence of air buoyancy and aerodynamic currents will cause the vacuum zero point to differ from that in air, by as much as 50 μg, depending on the sample and counterweight.

As the sample weight changes, the pen will record it. The total sample weight is always the sum of the recorder reading, in milligrams, the reading of the dial on the reference potentiometer, and the substitution weight. When the pen goes off scale, it can be returned on scale with the reference potentiometer. One would note the new dial reading on the chart. One hundred traversals of the chart can be obtained in this way. If the sample should vary beyond the limits of the dial, the dial range can be changed, as described above, so that further changes can still be measured, up to the limit of the balance. The recorder range would normally be selected to exceed the sample weight change anticipated during a period of unattended operation. The recorder zero and span can be checked and readjusted during operation, without impairing the experiment, and this should be done as required on long runs.

ZERO STABILITY

The electrical system is only used for determination of weight changes. Where a heavy sample will change by a small fraction of its own weight, nearly all of its weight will be counterbalanced mechanically on loop C. The first performance parameter of interest is thus the zero stability of the system, which is primarily a matter of the mechanical stability of the beam and bearings, although there is some influence from the zero stability of the servo system. It can be measured by operating the balance near the condition of zero current in the coil, where nearly all electrical errors disappear.

At light loads on loop A, the balance will report that a standard weight is constant to within $\pm 0.1\,\mu g$ after warm-up. The zero shift during warm-up may be as large as 10-20 μg, and warm-up may require several hours. For the most critical work it seems advisable to let the balance warm up for 8 hr or more. There is no advantage to turning it off when it will be used frequently. When the room temperature is deliberately varied, a residual uncompensated temperature coefficient of about $+0.5\,\mu g/°C$ is observed. This is in air; it is possible that this coefficient would be further reduced in vacuum, due to the absence of aerodynamic forces. It is reproducible, and may also be corrected for.

With 1 g on loop A and 1 g on loop C, the temperature coefficient will be found to have changed sign, and will be about $-1\,\mu g/°C$ or 1 ppm of sample loop load per °C. For some intermediate load this temperature coefficient goes through zero. The variation in balance reading uncorrelated with temperature does not exceed 1 ppm of load per day.

Thus the zero stability is broadly consistent with measurements to $\pm 0.1\,\mu g$, or 1 ppm of load, whichever is greater. Experiments of shorter duration than a day might permit even finer precision, as a fraction of total load, while those lasting for many days will not do quite as well.

For loop B at light loads the performance is of course five times coarser than on loop A. At a 2.5-g load, the temperature coefficient is about 5 ppm/°C, and variations in indication uncorrelated with temperature may also be as great as 3-5 ppm/day. Thus loop B should only be used for work to about 5 ppm of load. For those experiments requiring the ultimate in precision as a fraction of total load, it is best to scale the sample down to 1 g and use loop A.

PRECISION AS A FRACTION OF WEIGHT CHANGE

Another important performance parameter for any vacuum balance is the precision as a fraction of the weight change that can be measured. In some applications it can be the limiting factor. In this instrument it can be affected by the constancy of the weight/voltage coefficient, the stability of the reference power supply, the resolution of the reference potentiometer, and one's ability to read the dial.

The first two factors combined have a temperature coefficient of about 1 part in 10^5 per °C. Stability in the presence of line voltage variations is also of the order of 1 part in 10^5, as is stability in the absence of any disturbance. The present potentiometer and dial limits this parameter to 1 part in 10^4; if better performance were desired, it could be obtained readily by substituting a better potentiometer and dial.

This is 10^{-4} of the dial range selected. Since the dial range can be varied at will, one can select a range just larger than the change which occurs, and thus measure to about 10^{-4} of the change, which is much more useful than 10^{-4} of some full scale which may not approximate the change. The multiple ranges of weight change also make it much easier to design an experiment.

ACCURACY

Besides being limited by the achievable precision, the accuracy is dependent on the linearity of the weight/voltage characteristic of the coil, the linearity of the servo system, the linearity of the potentiometer, and the weights used.

The coil has been found to be linear as far as it has been measured, which is to 1 part in 10^4. Similarly, the electronics have also been found to be perfectly linear to this level, for the reasons given above.

The potentiometer is again the limiting factor, with 0.05% linearity. Again, more linear ones could be used if desired.

The finest standard weights made are the U.S. National Bureau of Standards Class M,[2] and they have a tolerance in this range of ±5.4 μg. The Bureau will certify them only to ±1μg. In this range practical weights become so thin, with such a high surface/weight ratio, that surface moisture adsorption becomes a significant source of error. Alternatively, if one made cubical or spherical weights to minimize this ratio, they would be so small that they would be very difficult to handle, and would readily be damaged by handling. It seems that we have come to the "end of the line" with

standard weights. An attractive alternative is to calibrate the balance on a high-weight range, where 5.4 μg
is a negligible fraction of the total, and then switch
down to the lower mass dial ranges to work. In this way
one can preserve a calibration accuracy to 0.025% all
the way down. At a 1-mg range this corresponds to 0.25
μg, which would strain the credibility of a 1-mg weight.

VACUUM

The separate weighing mechanism is intended for
installation in a bell jar, or in a separate glass bottle
we have designed, shown in Fig. 2, or in any other
suitable vacuum vessel. The beam and magnet assembly
and preamplifier electronics are all mounted and aligned
on a frame, with eight tapped holes for mounting to a
wide variety of brackets. The only connections into the
vacuum are eight electrical leads.

Care was taken to use only materials consistent
with high vacuum service in the weighing mechanism
design. The main material is aluminum, with smaller
amounts of glass, and traces of beryllium copper,
nichrome, and ceramic. Wire insulation is teflon,[3]
rated at 200°C. Construction is similar to that of the
previous model RM, which has been reported[4] used to
$1 \cdot 10^{-7}$ torr, and also reported[5] to hold constant for
hours in the 10^{-6} torr range without pumping.

In neither of the above cases was the balance the
limit to attainable vacuum. Work is now planned to investigate the actual limits imposed by the balance on
system pressure.

TEMPERATURE

The balance imposes no limit to sample temperature.[4] The temperature of the weighing mechanism should not exceed 100°C. For the most critical work, the weighing mechanism should be well baffled against any furnace used, and if possible kept in the range of ambient temperature. The 100°C limit makes it desirable to use a vacuum system requiring minimum bakeout for ultrahigh vacuum applications.

CORROSIVE ATMOSPHERES

The materials of construction were also chosen with the objective of minimizing the effect of reactive atmospheres. This was also the case with previous types. Atmospheres reported for previous manual and recording types included O_2, H_2, and SO_3. So far no balance has ever been damaged by any atmosphere to which it was exposed.

ELECTRICAL

The circuit includes a magnetic regulator for all ac voltages. Critical circuits are also further regulated by silicon diodes. In addition, the feedback system minimizes the effect of supply voltage changes. No change in balance reading is observed for a 10% change in supply voltage.

Only two electron tubes are used, besides the phototube. Both are 5814A's, a premium long-life version of the familiar 12AU7. There are also six silicon diodes. Electrolytic capacitors are all "computer grade." Experience with the previous RM suggests that the probability of failure will be less than 1% per year, or a mean time to failure of over 100 years.

EXTERNAL MAGNETIC FIELDS

The effect of external magnetic fields is of interest for magnetic susceptibility studies. There is no effect when operating the balance at the condition of zero electromagnetic force, that is, when loops A and C have equal loads. The effect of external fields will be proportional to the electromagnetic force, and thus only to the change in sample weight from a balanced condition.

This was tested by placing a weighing mechanism 32 in. above a 25,000-gauss magnet. A change of 0.16% of the balance reading was observed when the field was turned on, and it reversed when the field was reversed. This might be further reduced, if desired, by

1. Calibrating the balance with the field on.
2. Shielding the weighing mechanism from the external magnet.
3. Correcting the observed data by computation.

CONSECUTIVE SAMPLE WEIGHING

The balance can also be used for the automated weighing of consecutive samples, in laboratories or factories,[6] where this is desired. Readout can be on a recorder, null meter, or digital voltmeter.

Performance is the same as with changing samples, with some exceptions. It is possible to weigh samples up to 20-30 mg to a standard deviation of 0.1 μg. One must wait about 30 sec after closing the window for the aerodynamic currents to stabilize, but they do, to within the necessary 0.1 μg. The operator can readily re-check the zero and adjust it as required to maintain 0.1 μg accuracy, even in the presence of substantial variations in ambient temperature.

Samples above about 30 mg can be weighed to a standard deviation of about 3 parts in 10^6. Above about 100 mg on either loop, the sample will swing too long for quick measurements, and the precision deteriorates further to 10-15 ppm. We have designed an arrestment mechanism for such operation, which stops the pan swing, and improves precision significantly by reducing the shock to the beam of adding and removing these heavier samples. With this mechanism, it is convenient to weigh samples up to 1 g on loop A with a standard deviation of about 3 parts in 10^6 of load. Loop B is not as precise and is not recommended for critical work with consecutive samples. It is convenient for less critical work, using the arrestment mechanism, down to about 25 ppm of load. The arrestment mechanism does not appear to offer any advantages for samples below about 100 mg. Precision as a fraction of the range of sample variation remains 1 part in 10^4, as with changing samples.

CONCLUSION

An automatic recording ultramicrobalance has been developed for use in high vacuum, controlled atmospheres, and room air. Applications include thermo-gravimetry, surface chemistry, corrosion research, space materials studies, magnetic susceptibility studies, and the automated weighing of consecutive samples.

REFERENCES*

1. Lee Cahn and H. R. Schultz, Vacuum Microbalance Techniques, Vol. 2, Plenum Press, New York, 1962, p. 7.
2. Circular 547, National Bureau of Standards, Washington 25, D. C.
3. W. A. Riehl, Considerations on the Evaporation of Materials in Vacuum, presented at 46th National Meeting, A.I.Ch.E. Scheduled for publication in their Aerospace Symposium Series.
4. D. H. Buckley, M. Swikert, and R. L. Johnson, Friction, Wear, and Evaporation Rates of Various Materials in Vacuum to 10^{-7} mm Hg, ASLE, preprint 61 LC-2.
5. A. L. Houde, Discussion at Third Symposium on Vacuum Microbalance Techniques, Los Angeles, 1962.
6. Techniques with the Cahn Electrobalance, no. 3, Cahn Instrument Company, Paramount, California.

*Reprints of references 1, 3, 4, and 6 are available from the authors.

34

Reprinted from *Talanta,* **11,** 549–571 (1964), by permission of Microforms International Marketing Corporation as exclusive copyright licensee of Pergamon Press Journal back files

NEW STUDIES ON CALCIUM OXALATE MONOHYDRATE

A GUIDE TO THE INTERPRETATION OF THERMOGRAVIMETRIC MEASUREMENTS

E. L. SIMONS and A. E. NEWKIRK

Research Laboratory, General Electric Company, Schenectady, New York, U.S.A.

(*Received* 21 *August* 1963. *Accepted* 20 *October* 1963)

Summary—The pyrolysis of calcium oxalate monohydrate in air occupies a unique place in the literature of thermogravimetry. Not only was a thermogram for this reaction the first pyrolysis curve published by Duval and his collaborators, but Duval and others have suggested that it can be used as a reference substance for judging the performance of a thermobalance. However, the pyrolysis of calcium oxalate monohydrate under a variety of conditions gives rise to considerable differences in its thermograms. The effects of sample size, heating rate, atmosphere and container geometry are presented in a series of paired thermograms, and the differences are accounted for with the aid of additional evidence from differential thermal analysis and from combustion train experiments. Variations in atmosphere are particularly important, and the atmospheres studied are dry nitrogen, humidified nitrogen, dry air, humidified air, dry oxygen, dry carbon dioxide and dry carbon monoxide. Even subtle variations in the shape of a thermogram obtained on a reliable balance may reflect the complexity of the reactions that produce the more noticeable over-all weight changes. Results presented in this report thus delineate conditions for the use of calcium oxalate monohydrate as a thermogravimetric reference substance, and show that its behaviour under controlled conditions in a thermobalance can provide an unusually versatile guide to the interpretation of thermogravimetric measurements.

INTRODUCTION

THE pyrolysis of calcium oxalate monohydrate in air occupies a unique place in the literature of dynamic thermogravimetry. A thermogram for this reaction (Fig. 1) was the first pyrolysis curve published by Duval and his collaborators[42] in their comprehensive study of the thermogravimetric behaviour of analytical precipitates, and was

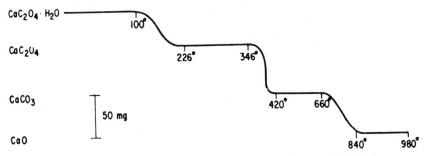

FIG. 1.—Pyrolysis curve of calcium oxalate monohydrate as published by Peltier and Duval.[42]

singled out by Duval for special mention in his book.[13] Because the thermogram's unusual sequence of consecutive reactions was characterised by four parallel, horizontal plateaux, corresponding successively to calcium oxalate monohydrate, calcium oxalate, calcium carbonate and calcium oxide, the material was used by Duval[12] "to adjust the thermobalances after they have been set up, cleaned, or repaired." He further claimed that the thermogravimetric measurements on this material with a photographic recording Chevenard thermobalance were accurate and reproducible enough to permit students to obtain reliable values for the atomic weight of carbon

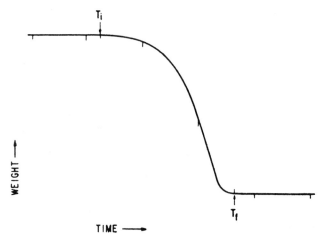

FIG. 2.—Schematic thermogram for general single-stage reaction:
$$A_{(solid)} = B_{(solid)} + C_{(gas)}.$$

from the dimensions of thermograms obtained as an instructional exercise.[13] Subsequently, Barcia Goyanes[2] and Wilson[57] suggested that the extremities of the plateaux in this thermogram be used as temperature calibration points for a thermobalance, as if they were characteristic transition temperatures.

Previous work in this Laboratory has demonstrated that the accuracy and precision of measurements with a Chevenard pen-recording thermobalance (which has the same suspension and sensitivity as the photographic recording balance) are below the level required for atomic weight work[48] and that the temperature limits of a plateau are not a property of the sample alone.[36] This report presents thermograms for calcium oxalate monohydrate pyrolysed under a variety of conditions and attempts to provide a consistent explanation for all of the observed differences, using additional data from differential thermal analysis measurements and from combustion train experiments. The results of these studies show that if calcium oxalate monohydrate is pyrolysed under carefully controlled conditions in a thermobalance, it can serve as a yardstick for judging the performance of the instrument and can provide an unusually versatile guide to the interpretation of thermogravimetric measurements.

CHARACTERISTICS OF SINGLE-STAGE REACTIONS
The thermogram of Fig. 2 has been drawn to represent the general reaction

$$A_{(solid)} = B_{(solid)} + C_{(gas)}.$$

Two temperatures may be selected as characteristic of any single-stage non-isothermal reaction: T_i (initial temperature) is the temperature at which the cumulative weight-change reaches a magnitude that the thermobalance can detect, and T_f (final temperature) is the temperature at which the cumulative weight-change first reaches its maximum value, corresponding to complete reaction. Although T_i may be the lowest temperature at which the onset of a weight-change can be observed in a given experiment, it is neither a transition temperature in the phase rule sense,[35] nor is it a true decomposition temperature below which the reaction rate suddenly becomes zero.[20,21,24,25] Because the value of T_i depends upon the interaction of many variables, it shall be called the *procedural decomposition temperature*,[36] a term introduced by Doyle[10] in reporting his thermogravimetric studies of the pyrolysis of polymers. At a linear heating rate, T_f must be greater than T_i, and the difference $(T_f - T_i)$ will be called the *reaction interval*. For an endothermic decomposition, T_i and T_f both increase with increasing heating rate, the effect being greater for T_f than for T_i, as has been clearly shown by Richer[45] and others.[8,34,36]

SIGNIFICANCE OF TEMPERATURE IN DYNAMIC THERMOGRAVIMETRY

At this point it is advisable to comment upon the significance of temperature as a variable in dynamic thermogravimetry. The fundamental datum produced by a recording thermobalance is a record that shows the variation with time of some quantity whose changes may be related in magnitude to changes in the weight of the sample. The foundations of modern dynamic thermogravimetry were laid in the early 1920's by Guichard,[22,23] who proposed that the method of continuous weighing be applied to "the study of all reactions that cause the weight of any solid to vary in one way or another . . . in a selected atmosphere, while the temperature is raised at a rate nearly proportional to the time, and slowly enough." Although Guichard did not explicitly specify the temperature to be measured, it is apparent from his work and that of his students that he referred to the temperature of the furnace atmosphere and not to that of the sample.[11,53,54]

Guichard had insisted, since the beginning of his studies, "on the necessity of realising a very regular rise of temperature." He pointed out that "there is a risk that every irregularity in this increase may appear on the curve of weight *vs.* time and complicate its interpretation." Such regularity in heating rate can be achieved in the furnace atmosphere but not in the sample, whose rate of temperature rise changes with the onset of each exothermic or endothermic phase transition or chemical change. These internally generated changes in the rate of sample temperature rise, of course, provide the basis for the widely used method of differential thermal analysis. Both differential thermal analysis and dynamic thermogravimetric analysis require a linear furnace heating rate. In differential thermal analysis one detects the onset of reactions or transitions in a sample by observing the perturbations in what would otherwise be a regular rate of increase in sample temperature. In thermogravimetry one detects and measures those transformations that produce a change in the weight of a sample. The only independent temperature variable in either technique is the heating rate of the furnace. The changes in sample weight and temperature are both consequences of the chemical and physical changes produced in a sample when it is brought into contact with a given atmosphere whose temperature is changing at a predetermined rate.

289

In this investigation an independent record was kept of the furnace temperature as a function of time, which record also served to verify the linear heating rate that had been set on the furnace controller. The temperature markings on the thermograms were established by coupling the temperature and weight recorders so that a pip was automatically introduced into the pyrolysis curve at each integral 100-degree reading of the furnace temperature. The pips could also have been introduced after the run by juxtaposition of the weight-time curve with the separate record of furnace temperature as a function of time. Values of T_i and T_f were interpolated between 100-degree markings.

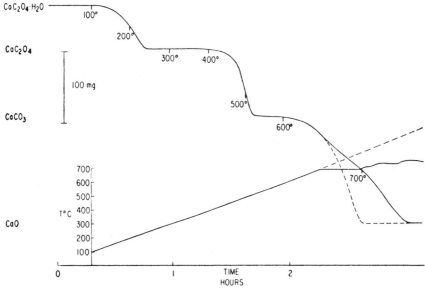

FIG. 3.—Effect of perturbation in rate of heating upon thermogram for calcium oxalate monohydrate (500 mg, platinum dish, flowing dry nitrogen):
– – – unperturbed rate of heating.

It is particularly risky to try to record a thermogram directly by feeding the weight and temperature signals to the two inputs of an X-Y recorder. Unexpected disturbances that sometimes occur in the heating rate or thermocouple response can produce artificial perturbations in the weight-temperature record. Only a separate temperature-time record can disclose these adventitious effects and permit effects caused solely by changes in sample weight to be distinguished from them.[27] Fig. 3 shows the thermobalance record of a pyrolysis of calcium oxalate monohydrate during which the temperature controller became stuck at 695°, for a period of time, after which it produced an irregular temperature rise up to about 750°. The dashed curve shows the course that the thermogram would have taken had the linear heating rate of 300 degree/hr been maintained above 695°. The perturbation in the recorded curve can be properly understood only in the light of the independent record of temperature as a function of time. Had an X-Y recorder been used for this run, the period of constant temperature would have been represented by a vertical line at 695°. Although such an abrupt change in the slope of the curve would probably have been suspect, more subtle, but nonetheless significant changes in heating rate could pass completely unnoticed in a direct recording of weight vs. temperature.

CHARACTERISTICS OF MULTI-STAGE REACTIONS

Successive, non-overlapping reactions

Under certain conditions, the thermogram for a sequence of successive decompositions, such as occurs during the pyrolysis of calcium oxalate monohydrate

$$CaC_2O_4 \cdot H_2O = CaC_2O_4 + H_2O$$
$$A_{(s)} = B_{(s)} + gas \tag{1}$$
$$CaC_2O_4 = CaCO_3 + CO$$
$$B_{(s)} = C_{(s)} + gas \tag{2}$$
$$CaCO_3 = CaO + CO_2$$
$$C_{(s)} = D_{(s)} + gas \tag{3}$$

can be considered as the composite of the thermograms for the individual stages.
This is shown in Fig. 4, in which are superimposed the thermograms that were
obtained on separate days by heating in air, at 300 degree/hr, 500 mg of calcium

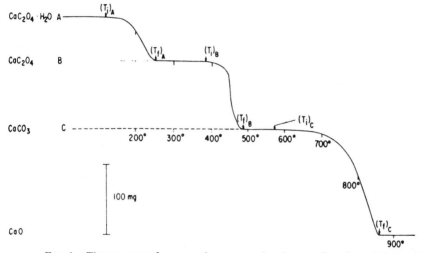

FIG. 4.—Thermograms for successive non-overlapping reactions (porcelain crucible,
ambient air, 300 degree/hr):
—————— 500 mg of calcium oxalate monohydrate,
. . . . . . 438 mg of calcium oxalate,
– – – – 343 mg of calcium carbonate.

It is not evident from this reproduction that the true value of $(T_i)_B$ is about 390°. The position of
$(T_i)_B$ shown in this figure was chosen for illustrative purposes only.

oxalate monohydrate and equivalent quantities of the anhydrous oxalate and of
calcium carbonate, each in a porcelain crucible. The initial stages of the latter two
curves are indicated by broken lines. Thermograms 4(B) and 4(C) are identical with
4(A) at temperatures above their intersections.

Although temperature $(T_f)_A$ marks the beginning of the plateau for calcium
oxalate, its value is determined by the interaction between the heating rate, the sample
size and the rate of dehydration of calcium oxalate monohydrate; it cannot be
interpreted as setting a lower limit to the thermal stability of the anhydrous salt.
Temperature $(T_f)_A$ lies on the calcium oxalate plateau only because, under the conditions of the experiment, the dehydration reaction was complete before the temperature
reached the value of $(T_i)_B$, the procedural decomposition temperature of the anhydrous

salt. When calcium oxalate, previously dehydrated at 175°, was the starting material, the thermogram, beginning at level B, showed no perturbation as it passed through $(T_f)_A$. The calcium carbonate plateau, of course, bears the same relationship to its precursor as the calcium oxalate plateau bears to the monohydrate.

The exact superposition of the independently determined thermograms of Fig. 4 could have been achieved only if the reactions being studied were indeed successive and non-overlapping, and only if the characteristics of the thermobalance remained constant over the period of days covered by the experiments.

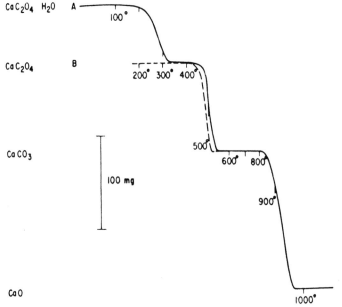

FIG. 5.—Thermograms for successive, partially overlapping reactions (porcelain crucible, ambient air, 600 degree/hr):
——— 500 mg of calcium oxalate monohydrate,
– – – – 438 mg of calcium oxalate.

Successive, partially overlapping reactions

The exact superposition of Fig. 4 is not characteristic of the pyrolysis of calcium oxalate monohydrate under all conditions. As noted earlier, both T_i and T_f for a given reaction rise as the heating rate is increased, the effect being more pronounced for T_f. The consequences of this are shown in Fig. 5, in which are superimposed the thermograms that were obtained by repeating two of the pyrolyses of Fig. 4, but at a heating rate of 600 degree/hr. The intermediate plateau for calcium carbonate is still clearly defined, but it begins at a higher temperature.[36] On the other hand, the formation of anhydrous calcium oxalate as an intermediate in the pyrolysis of the monohydrate is no longer marked by a true plateau, but simply by a change in the slope of the thermogram. At the higher heating rate the dehydration reaction on curve 5(A) was not complete by the time the furnace temperature had reached about 350°, which is the procedural decomposition temperature indicated for the anhydrous salt on curve 5(B). During the latter part of pyrolysis 5(A), therefore, water and carbon monoxide were being evolved simultaneously,[43] and at the point where the cumulative weight loss reached level B the sample crucible contained anhydrous calcium oxalate and a small amount of both the monohydrate and calcium carbonate.

A careful examination of the original thermograms discloses that curve 5(A) lies above curve 5(B) even in the region between 300° and 400°, and does not become coincident with it until the calcium carbonate plateau is reached. By increasing the heating rate from 300 to 600 degree/hr the initial sequence of two successive decompositions has been transformed to a sequence of two partially overlapping or simultaneous decompositions, the thermogram for which is not an exact composite of the individual stages.

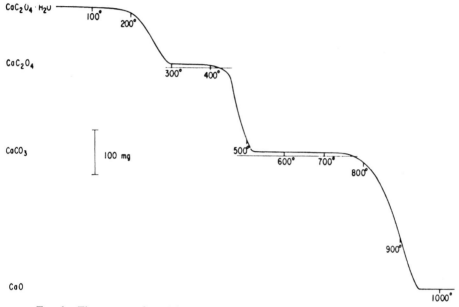

FIG. 6.—Thermogram for calcium oxalate monohydrate (1·019 g, porcelain crucible, ambient air, 300 degree/hr).

The transformation of a true plateau for anhydrous calcium oxalate into a curve like that of Fig. 5(A) was also produced by keeping the heating rate at 300 degree/hr, but increasing the sample weight from 0·5 g to approximately 1 g (Fig. 6). With more water in the initial sample, complete dehydration was not achieved before the procedural decomposition temperature for the anhydrous salt was reached, even at 300 degree/hr. This parallelism between the effects of heating rate and sample weight will be examined more closely in a subsequent section of this report.

At this point the following general conclusions may be drawn from the foregoing experiments on the multi-stage pyrolysis of calcium oxalate monohydrate:

1. *The appearance of a plateau for a compound on a dynamic thermogram does not necessarily imply that the compound is isothermally stable, either in a thermodynamic or practical sense, at all or any temperatures that lie on that plateau.* The temperature limits between which it extends apply only to the particular conditions of that pyrolysis.[5,20,44]

Certainly, none of the curves in Figs. 4, 5 or 6 can be used to set an upper limit to the isothermal stability of calcium carbonate in air. Similarly, anhydrous calcium oxalate undergoes slow isothermal decomposition in air at 300°, even though this temperature lies on the true plateau observed for that compound in Fig. 4.[43]

2. *If the thermogram obtained for a multi-stage reaction has no intermediate portion in which the sample weight remains constant with time over a range of temperature, one can make the reasonable inference that the reactions leading to the formation and to the subsequent decomposition of the intermediate are not independently sequential, but at least partly overlap.* The weight level at which a bend or change of slope occurs in the thermogram for such a pair of successive but overlapping reactions depends upon the relative amount of product B that has been formed by the time its procedural decomposition temperature has been reached and upon the relative rates of decomposition of A and B.

3. *In the absence of a true plateau, which can appear only if* $(T_f)_A \ll (T_i)_B$, *one cannot determine from a thermogram for successive reactions exact values for either* $(T_f)_A$, $(T_i)_B$, *or the stoichiometric weight level of B,[4] although a reasonable inference as to the latter can often be made.* There is little doubt, for example, that the gently sloping portion of Fig. 5 indicates the formation of anhydrous calcium oxalate.

CHARACTERISTICS OF INDIVIDUAL REACTIONS OF $CaC_2O_4 \cdot H_2O$

Among the factors that affect the thermogravimetric behaviour of calcium oxalate monohydrate are the reversibility of each stage of the pyrolysis and the enthalpy change that accompanies each of the individual reactions. The pertinent data are summarised in Table I.

TABLE I

Reaction	Heat of reaction $(\Delta H^\circ_{298^\circ K} kcal/mole)$[17]	Reversibility
(1) $CaC_2O_4 \cdot H_2O = CaC_2O_4 + H_2O$	13·3*	Reversible[43]
(2) $CaC_2O_4 = CaCO_3 + CO$	15·5*	Irreversible[13,58]
(3) $CaCO_3 = CaO + CO_2$	42·5	Reversible

* Heat of formation of CaC_2O_4 estimated from tabulated values for the di- and monohydrates.

Reactions (1) and (3)

Both of these reactions are endothermic and reversible. In their thermogravimetric work, Peters and Wiedemann[43] showed that calcium oxalate monohydrate, after dehydration at about 280°, could be regenerated by allowing the anhydrous salt to cool to room temperature in humidified air. The effect of carbon dioxide in markedly raising the procedural decomposition temperature and decreasing the reaction interval for calcium carbonate has been clearly demonstrated by these workers and by Richer and Vallet.[44,45] For any given atmosphere containing carbon dioxide, the procedural decomposition temperature must be equal to or greater than the temperature at which the dissociation pressure of calcium carbonate reaches the partial pressure of carbon dioxide in that atmosphere.[28]

Reaction (2)

Thermal measurements by several investigators[14,41,49,56] have shown that the decomposition of anhydrous calcium oxalate is endothermic in an inert atmosphere, but becomes strongly exothermic in an oxidising atmosphere. These results have been confirmed in a Chevenard thermobalance. With the balance mechanism locked and a Pt-Pt, 10% Rh thermocouple inserted into the powdered sample, the difference between the sample temperature and the furnace temperature was recorded as a function of the furnace temperature, and the tracings shown in Fig. 7 were obtained from experiments made in flowing dry nitrogen and oxygen.

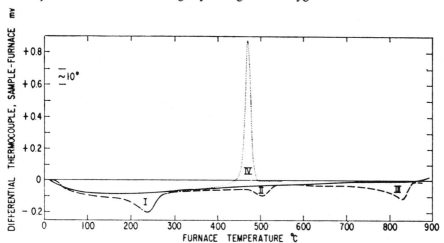

FIG. 7.—Differential thermal measurements on calcium oxalate monohydrate (500 mg):
——— inert sample (ignited calcium oxide),
– – – – calcium oxalate monohydrate in dry nitrogen,
. . . . . . calcium oxalate monohydrate in dry oxygen.

The exothermic peak observed in oxygen (and, with a more complex shape, in air) has been attributed by the investigators previously cited to the formation of carbon dioxide by oxidation of the carbon monoxide produced in the decomposition of the oxalate. The heat evolved/mole of carbon monoxide oxidised in the reaction

$$CO + 0.5O_2 = CO_2 \qquad (4)$$

is 67.7 kcal at 700°K,[7] which is about four times larger than the heat absorbed by reaction (2). Reaction (4) has also been postulated to explain the exothermic effects noted when thorium oxalate was decomposed under oxidising conditions.[6,38]

The evidence to support the hypothesis that reaction (4) can also occur during the pyrolysis of calcium oxalate is presented in the following section.

DECOMPOSITION OF ANHYDROUS CALCIUM OXALATE

A weighed sample of calcium oxalate monohydrate in a platinum boat was loaded into a combustion tube and heated overnight at 125° in a stream of dry nitrogen to effect complete dehydration. Then, as in the method used by D'Eye and Sellman[9] for studying the thermal decomposition of thorium oxalate, the modified combustion train shown in Fig. 8 was assembled, and isothermal decompositions of the anhydrous salt were carried out for 3 hr at 420° with a stream of suitable carrier gas passing over the sample at a rate of 20 ml/min.

The gas, after leaving the pyrolysis chamber, passed through a series of tubes filled, successively, with Ascarite, hot copper oxide at 700° and Ascarite. Any carbon dioxide in the exit gas stream was trapped in the first Ascarite tube; any carbon monoxide in the exit gas stream was oxidised by the hot copper oxide and subsequently absorbed as carbon dioxide in the last absorption tube.

The results obtained with dry air and dry nitrogen as the carrier gases are given in Table II, to which the following comments are pertinent:

1. The dehydration of the samples was complete at 125°.

CARRIER
GAS
INLET

A   B   C   D   E

FIG. 8.—Schematic illustration of modified combustion train for experiments with
dry nitrogen and dry oxygen:

A—furnace with combustion tube   C—furnace with tube of copper oxide,
    and platinum boat,   D—Ascarite absorption tube,
B—Ascarite absorption tube,   E—Ascarite tare tube.

2. The amount of calcium oxalate that decomposed in 3 hr at 420° was significantly greater in dry air than in dry nitrogen.

3. In dry air (run 1), the primary gaseous product of the decomposition of anhydrous calcium oxalate at 420° was carbon monoxide. Virtually all of this product had been oxidised to carbon dioxide by the time the gas stream reached the first Ascarite tube, and the total moles of carbon dioxide trapped by both Ascarite tubes is in excellent agreement with the weight loss of the sample, calculated as moles of carbon monoxide.

4. In dry nitrogen (run 2) most of the carbon monoxide produced during the decomposition of the anhydrous calcium oxalate left the reaction zone unchanged and appeared as carbon dioxide only after passing through the hot copper oxide tube. However, a significant amount of carbon dioxide was absorbed in the first Ascarite tube, between the pyrolysis furnace and the copper oxide furnace. In a nitrogen atmosphere, this carbon dioxide could not have been produced by oxidation of carbon monoxide, but must have arisen from the disproportionation reaction

$$CO = 0.5CO_2 + 0.5C \qquad\qquad \Delta H_{700°K} = -20.7 \text{ kcal}[7] \qquad (5)$$

On this basis, the following material balance can be made:

CO (mmole) undergoing disproportionation = 2(0.36) = 0.72
CO (mmole) undergoing oxidation by CuO   = 2.30
                                         Total   = 3.02
CO (mmole) calculated from weight loss   = 2.90

Thus, in dry nitrogen as in dry air, the primary gaseous product of the decomposition of anhydrous calcium oxalate at 420° was carbon monoxide.

The disproportionation reaction has been observed by Glasner and Steinberg[17-19] in their studies of the thermal decomposition of rare earth oxalates in vacuum. Qualitative evidence for its role in the vacuum decomposition of sodium oxalate and of alkaline earth oxalates has been reported by Günther and Rehaag[26] and by Wöhler and Schuff,[58] and it has been observed by D'Eye and Sellman[9] when thorium oxalate was decomposed in a stream of nitrogen.

Further evidence for disproportionation in these experiments was the grey discoloration observed in the sample that had been partially decomposed in a stream of nitrogen; the residue from the air run was pure white. The intensity of the discoloration could be reduced by passing air over the hot residue. A greyish-brown discoloration was also observed in the residues from thermogravimetric runs made in dry nitrogen.

To confirm the reported irreversibility of the decomposition of anhydrous calcium oxalate,[43,58] runs were made at 420° in a stream of dry carbon monoxide; this

TABLE II.—PYROLYSIS OF $CaC_2O_4 \cdot H_2O$ IN MODIFIED COMBUSTION TRAIN

Run no.	1	2	3(a)	3(b)
Atmosphere	Dry air	Dry N_2	Humid N_2	Dry O_2
Wt. $CaC_2O_4 \cdot H_2O$, mg	1331·5 (9·11 mmole)	1105·1 (7·56 mmole)	1109·0 (7·59 mmole)	Run 3(b) was made on residue remaining from 3(a)
Wt. loss in N_2 at 125° (stoich. = 12·33%), mg	165·2 (12·40%)	139·0 (12·58%)	140·0 (12·62%)	
Wt. loss in carrier gas at 420° for 3 hr, mg	93·2 (≡ 3·33 mmole CO)	81·2 (≡ 2·90 mmole CO)	99·6 (≡ 3·56 mmole CO)	48·6 (≡ 1·74 mmole CO)
Wt. CO_2 absorbed in 1st Ascarite tube, mg	140·0 (≡ 3·18 mmole CO_2)	15·7 (≡ 0·36 mmole CO_2)	89·9 (≡ 2·04 mmole CO_2)	77·8 (≡ 1·77 mmole CO_2)
Wt. CO_2 absorbed in 2nd Ascarite tube, mg	1·7 (≡ 0·04 mmole CO_2)	101·2 (≡ 2·30 mmole CO_2)	39·5 (≡ 0·90 mmole CO_2)	0·0
Total CO_2 absorbed, mg	141·7 (≡ 3·22 mmole CO_2)	116·9 (≡ 2·66 mmole CO_2)	129·4 (≡ 2·94 mmole CO_2)	77·8 (≡ 1·77 mmole CO_2)
Wt. H_2O absorbed in Anhydrone tube, mg			12·8 (≡ 0·71 mmole H_2O)	0·7 (≡ 0·04 mmole H_2O)

atmosphere neither prevented the decomposition of the pure salt nor reversed the decomposition of a partially decomposed sample. This observation means, of course, only that if the reaction is reversible, its equilibrium pressure of carbon monoxide at 420° is greater than 1 atmos.

The following conclusions may be drawn from the foregoing combustion train experiments and from the thermal measurements described in the preceding section:

1. In both oxidising and inert atmospheres the primary endothermic formation of carbon monoxide from anhydrous calcium oxalate is accompanied, to varying degrees, by one or more secondary exothermic reactions that produce carbon dioxide from carbon monoxide.

2. The direct oxidation of carbon monoxide by oxygen occurs more readily than the disproportionation reaction.

3. The heat evolved during direct oxidation of carbon monoxide by oxygen raises the temperature of the remaining sample and solid product at a rate greater than the heating rate of the furnace, and, as a result, the decomposition is more rapid in dry air than in dry nitrogen.

EFFECT OF VARIABLES UPON A DYNAMIC THERMOGRAM

The characteristics of the separate stages in a multi-stage pyrolysis have an important bearing on the interpretation of thermogravimetric data. In particular, the pyrolysis of calcium oxalate monohydrate, with its sequence of alternating reversible and irreversible reactions, coupled with secondary reactions between components of the atmosphere and a primary decomposition product, provides an ideal vehicle for demonstrating the effects of many variables upon the shape of a dynamic thermogram. In the remainder of this report, a series of comparison thermograms is presented to show these effects.

Geometry of sample container

Several investigators have recommended that for thermogravimetric measurements the sample should be loosely packed in a shallow dish in order to facilitate gaseous exchange between the sample and the atmosphere that surrounds it.[3,8,16,37,41] Indeed, Garn and Kessler[16] have declared, "That the traditional crucibles are useless in thermogravimetry is essentially true. In a few cases crucibles are acceptable ... In controlled atmosphere work, where the atmosphere is solely the gas involved in the reaction, the geometry of the container is immaterial."

As can now be shown with calcium oxalate, the geometry of the container is also immaterial if no interaction is possible between the solid phase and the gaseous atmosphere or products. The pyrolysis of 500 mg of calcium oxalate monohydrate was conducted in flowing dry carbon dioxide using both a porcelain crucible and a quartz dish as sample holders. The thermograms in Fig. 9 are identical above about 275°, which marks the completion of the dehydration reaction in the crucible. As expected, the loss of water occurred more readily from the shallow dish than from the crucible. On the other hand, the shape of the container had no effect upon the decomposition of anhydrous calcium oxalate because this reaction is not reversible, and in a carbon dioxide atmosphere no important diffusion-controlled secondary reactions can occur. The shape of the container also had no effect upon the dissociation of calcium carbonate because this reaction is reversible, and the atmosphere used was

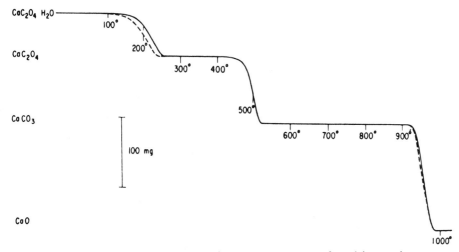

FIG. 9.—Effect of sample container upon thermogram for calcium oxalate monohydrate in flowing dry carbon dioxide (500 mg, 300 degree/hr):
– – – – quartz dish,
——— porcelain crucible.

solely the gas involved in the reaction. When the reversible pyrolysis of calcium carbonate was performed in a given container, the higher the pressure of carbon dioxide in the atmosphere, the higher the value of the procedural decomposition temperature and therefore, the smaller the reaction interval, $(T_f - T_i)$. (Compare Fig. 9 with Figs. 5, 4, 6, and 10.)

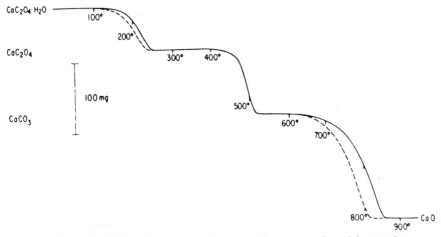

FIG. 10.—Effect of sample container upon thermogram for calcium oxalate monohydrate in flowing dry nitrogen (500 mg, 300 degree/hr):
– – – – quartz dish,
——— porcelain crucible.

When the pyrolysis was carried out in a flowing stream of dry nitrogen (Fig. 10), both the loss of water and the loss of carbon dioxide were affected by the shape of the container; the loss of carbon monoxide was unaffected. There is a thermodynamic possibility that the loss of carbon dioxide, illustrated by the dashed curve of Fig. 10, was caused by reaction between the calcium carbonate and the quartz dish. However,

Longuet[59] has shown that this reaction becomes significant below 900° only if both the silica and the calcium carbonate are finely divided and intimately mixed. We have repeated the pyrolysis of Fig. 10 using a platinum-lined quartz dish, and the results were the same as those obtained with the unlined dish. Both curves shown in Fig. 10, therefore, represent only the thermal dissociation of calcium carbonate.

The marked effect of container shape that is illustrated in Figs. 9 and 10 provides evidence that a significant pressure of water vapour and carbon dioxide must have existed in the interior of the crucible during dissociation, even when the atmosphere that flowed over the crucible entered the thermobalance free from either water or carbon dioxide or both.

In the study of reversible reactions or of reactions in which a component of the atmosphere can react either with the original sample or with a solid or gaseous decomposition product, one must recognise the possible existence of such partial pressure gradients throughout a mass of powdered sample. These gradients can effect both the shape of thermogravimetric curves[31,51,52] and the magnitude of thermal effects that accompany the reactions.[30] They can be reduced by packing the powder loosely in a shallow dish,[16] by using crucibles with micro-porous[15,29] or macro-porous[1] walls, or by passing a controlled atmosphere through the bed of powdered sample. The effectiveness of the latter technique for differential thermal analysis has been vividly demonstrated by Stone,[50] who has also announced the development of thermogravimetric equipment to provide dynamic gas flow through the specimen under study.[46] Papailhau[39,40] has designed a crucible, for use with a null-type balance, that permits an externally generated atmosphere to be passed through the entire bed of powdered sample.

Except for the experiments of Figs. 9 and 10 that were made with a quartz dish, all differential thermal and thermogravimetric measurements described in this report were made with Coors high form-000 porcelain crucibles, in which both thermal gradients and partial pressure gradients undoubtedly existed. Reproducible results were obtained by careful control of experimental conditions, but, as illustrated by Fig. 10, these results would have been different with more efficient interaction between solid sample and gaseous atmosphere. The interpretation of the experiments has been made with this restriction in mind.

Comparison of dry nitrogen and dry oxygen

Thermograms that were obtained by heating calcium oxalate monohydrate in flowing atmospheres of dry nitrogen and of dry oxygen are shown in Fig. 11. The dehydration step is unaffected by the change in atmosphere, and the two curves are nearly identical in this region because both gases are equally effective in sweeping evolved water vapour away from the sample surface. On the other hand, the thermograms diverge at the intermediate stage because in oxygen or air the secondary oxidation of carbon monoxide raises the temperature of the unreacted solid at a rate greater than the heating rate of the furnace, producing a marked acceleration in the decomposition rate. The decomposition of calcium oxalate occurs more rapidly and is completed at a lower furnace temperature in an atmosphere of dry oxygen (or air) than in an atmosphere of dry nitrogen.

Although the product of this decomposition in either atmosphere is calcium carbonate, the procedural decomposition temperature for the subsequent loss of carbon dioxide was higher in the run made in an oxygen atmosphere than for that made in a nitrogen atmosphere (Fig. 11). This small difference is not an artifact of the instrument. Indeed, thermograms that we obtained by heating separate portions of the same sample of *calcium carbonate* in nitrogen and in air were identical in the region of weight loss because nitrogen and air were equally effective in sweeping away the evolved carbon dioxide. However, the calcium carbonate produced by the decomposition of a given sample of calcium oxalate in an oxygen atmosphere is *not* identical with that produced by decomposition of the same material in a nitrogen atmosphere.

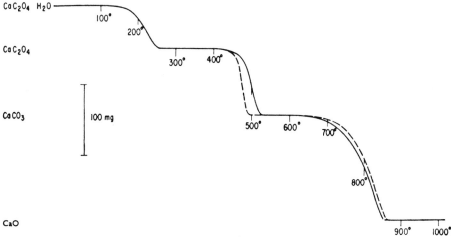

FIG. 11.—Effect of atmosphere upon thermogram for calcium oxalate monohydrate
(500 mg, porcelain crucible, 300 degree/hr):

– – – – dry oxygen,
———— dry nitrogen.

The dry oxygen run shown in Fig. 11 was repeated, but at a temperature of about 540° the furnace was turned off and the flowing oxygen was replaced by flowing nitrogen. After the temperature had dropped to about 150°; the heating rate of 300 degree/hr was resumed. The resulting thermogram, although obtained in a nitrogen atmosphere, was identical above 500° with the dashed oxygen curve of Fig. 11. The thermogravimetric behaviour of the calcium carbonate in these experiments was determined by the atmosphere in which it had been formed and not by the atmosphere in which it was decomposed. Thermogravimetry cannot disclose whether this difference in pyrolytic behaviour is a reflection of differences in particle size, surface area, lattice imperfections, or some other characteristics, but these experiments and those shown in Fig. 10 do illustrate the usefulness of thermogravimetric measurements, made on a reliable balance, for quickly surveying a variety of experimental conditions and thus disclosing those areas in which more detailed studies should be made, possibly with other supplementary techniques. As will be shown later, the shape of even a single thermogram can indicate complexities in the reaction under study.

Sample weight

In the earlier discussion of successive, partially overlapping reactions, certain similarities were noted between the effects of heating rate and sample weight upon the

301

shape of a dynamic thermogram. This parallelism cannot be carried too far. Unlike heating rate, sample weight is not a completely independent variable in thermogravimetry. For a given sample container a change in the weight of a powdered solid sample simultaneously changes the thickness of the sample bed, its total heat capacity, and its area of contact with the walls of the container, through which much of the heat transfer with the furnace atmosphere occurs.

Some of these factors can change during a pyrolysis, as illustrated by Fig. 12, in which are superimposed the thermograms obtained in ambient air with approximately 1 g of calcium oxalate monohydrate (Fig. 6) and those obtained with equivalent quantities of the anhydrous salt and calcium carbonate. The relationship between

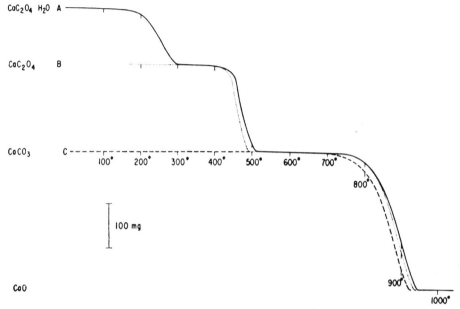

FIG. 12.—Thermogram for successive partially overlapping reactions (porcelain crucible, ambient air, 300 degree/hr; curve A as for Fig. 6):
———— $1 \cdot 019$ g of calcium oxalate monohydrate,
. . . . . . $0 \cdot 893$ g of calcium oxalate,
– – – – $0 \cdot 699$ g of calcium carbonate.

curve 12B and the second weight-loss section of curve 12A is the same as shown in Fig. 5 and arises from the overlap of the dehydration reaction and the loss of carbon monoxide. However, even though this second stage, the loss of carbon monoxide, was complete before the procedural decomposition temperature of calcium carbonate was reached (and therefore a true plateau was observed for this compound), the value of T_i for the loss of carbon dioxide is markedly lower for curve 12C than for curves 12A and B. Examination of the crucibles at the end of each pyrolysis showed that the residues of calcium oxide from runs 12A and B were more loosely packed than that from run 12C, and were separated by an air gap from the wall of the sample crucible. Although the final weight of calcium oxide was the same in each run, its volume represented a larger shrinkage from the initial sample volume in runs A and B than in run C. The gaps produced by this shrinkage undoubtedly reduced the flow of heat between the sample and furnace and thus contributed, at least in part, to the higher value of T_i in runs A and B.

Undoubtedly the effect of sample shrinkage observed with the 1-g (and equivalent) samples of Fig. 12 was also present during the runs made with 0·5-g (and equivalent) samples of Fig. 4, but because the sample was smaller the magnitude was much less.

Thus, although the thermograms of Fig. 4, demonstrate that under certain conditions the curve for a multi-stage pyrolysis can be considered as the exact summation of the curves for the individual stages, those of Figs. 11 and 12 emphasise that under other conditions the thermogravimetric behaviour of a given substance may depend upon whether it is the material initially loaded into the sample container or whether it is formed in the container by decomposition of a percursor.

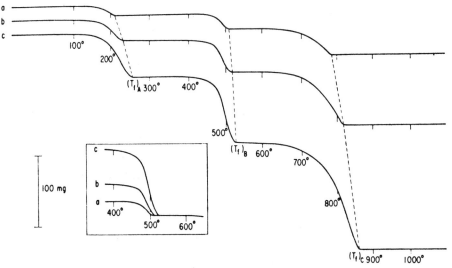

FIG. 13.—Effect of sample weight upon thermogram for calcium oxalate monohydrate in flowing dry nitrogen (porcelain crucible, 300 degree/hr):

a—126 mg of calcium oxalate monohydrate,
b—250 mg of calcium oxalate monohydrate,
c—500 mg of calcium oxalate monohydrate.

Insert shows three curves superimposed at calcium carbonate level to emphasise effect of sample weight on loss of carbon monoxide.

From the foregoing results it appears that no generalisations can be made about the effect of sample weight upon the procedural decomposition temperature, particularly for intermediate products. Even with powdered calcium carbonate as a starting material, Richer and Vallett[44,45] found that T_i was virtually independent of sample weight in the range 0·25 to 1 g, both in nitrogen and carbon dioxide. At 100 degree/hr, T_i was 517° in nitrogen and 914° in carbon dioxide. On the other hand, once the decomposition of a powdered solid has begun, it generally does not occur uniformly in every particle throughout the entire mass of sample.[31,51,52,55] Measurable temperature gradients exist even across relatively thin beds of powder.[32–34] Under such non-homogeneous conditions one would expect that the time required for complete decomposition of a powdered solid would increase with increase of sample weight. Because the furnace heating rate is linear there would be a resultant increase in the observed value of T_f. This expected increase was observed by Richer and

E. L. Simons and A. E. Newkirk

Vallet[44,45] for the pyrolysis of calcium carbonate and in this work for each stage in the pyrolysis of calcium oxalate monohydrate in nitrogen at 300 degree/hr (Fig. 13).

A perturbation in this generalisation can occur if the reaction is exothermic. The sample temperature then increases more rapidly than does the measured furnace temperature, and the resultant acceleration in specific reaction rate may compensate, at least in part, for the increase in sample weight. It has already been noted that in air the endothermic decomposition of calcium oxalate is accompanied by the highly exothermic oxidation of carbon monoxide, with a consequent acceleration of the reaction rate. As a result, as shown in Fig. 14, T_f for the decomposition of calcium

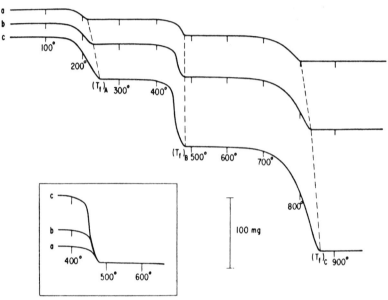

Fig. 14.—Effect of sample weight upon thermogram for calcium oxalate monohydrate in ambient air (porcelain crucible, 300 degree/hr):
a—126 mg of calcium oxalate monohydrate
b—250 mg of calcium oxalate monohydrate,
c—500 mg of calcium oxalate monohydrate.
Insert shows three curves superimposed at calcium carbonate level to emphasise that T_f for loss of carbon monoxide is independent of sample weight.

oxalate in air is less than in nitrogen and is virtually independent of the starting weight of monohydrate, at least in the range 125 to 500 mg. A comparison between Figs. 14 and 6 shows the normally expected increase in T_f between 500 mg and 1 g and the pronounced increase in T_i for calcium carbonate that was noted in the discussion of Fig. 12.

Water vapour

The value of a dynamic thermogram in the preliminary study of chemical reactions lies not just in its gross characteristics, such as plateaux or clearly marked changes of slope, from which one can draw inferences about the stoichiometry of the reaction, but also in the more subtle variations of shape that may reflect the complexity of the reactions that produce the more noticeable, over-all weight changes. This is illustrated by a re-examination of Fig. 4 for the pyrolysis of calcium oxalate monohydrate in ambient air. The over-all stoichiometry of the three distant stages is clearly defined by each of the plateaux. However, there is a significant qualitative difference between the shape of the curve that marks the loss of carbon monoxide and the shapes of the curves for the loss of water and of carbon dioxide. The latter curves are smooth and continuous between the two plateaux, but the loss of carbon monoxide is characterised by three rather abrupt changes in slope that have no apparent stoichiometric significance. This shape is reproducible and is not an artifact of the instrument, although it can be modified by using a platinum instead of a porcelain crucible. Clearly, the thermogram suggests that further investigation of this stage of the reaction is warranted.

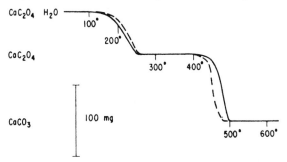

FIG. 15.—Effect of water vapour upon thermogram for calcium oxalate monohydrate in oxidising atmosphere (500 mg, porcelain crucible, 300 degree/hr):
———— dry air,
– – – – humid air.
In the region of carbon monoxide loss the curve obtained in ambient air (curve A of Fig. 4 and curve c of Fig. 14) is identical to the dashed curve in humid air.

Additional experiments were performed, which showed (Fig. 15) that the unusual shape, originally observed in ambient air, could be reproduced exactly in a flowing atmosphere of humidified air, but that in dry air the loss of carbon monoxide was characterised by the same smooth continuous shape noted in Fig. 11 for dry nitrogen and oxygen. Furthermore, the accelerating effect of water vapour upon the decomposition of CaC_2O_4 is clearly evident from the curves of Fig. 15. Finally, a comparison

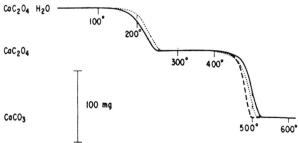

FIG. 16.—Effect of water vapour upon thermogram for calcium oxalate monohydrate in inert atmosphere (500 mg, porcelain crucible, 300 degree/hr):
———— dry nitrogen,
. . . . . . humid nitrogen,
– – – – dry air.

between thermograms obtained in dry air, dry nitrogen, and humid nitrogen (Fig. 16) shows that the loss of carbon monoxide begins more abruptly and at a lower furnace temperature in humid nitrogen than in dry air, although the total loss is completed at a lower furnace temperature in the latter gas.

That water vapour has a marked effect upon the decomposition of anhydrous calcium oxalate was confirmed by repeating the thermal experiments illustrated in Fig. 7 with humidified gases. The exotherm noted previously in dry oxygen was still present, but it had been shifted about 40° to a lower temperature. In humidified nitrogen, however, the magnitude of the endotherm for the oxalate decomposition was significantly reduced. The combustion train experiment was also repeated as a two-part experiment. During the first 3-hr period the carrier gas was humidified nitrogen; during the second 3-hr period it was dry oxygen. For the period in humidified nitrogen the combustion train was modified by inserting a large capacity water absorption tube at the exit end of the pyrolysis tube to ensure that no water vapour was present in the gas that entered the first Ascarite tube. In addition, an Anhydrone absorption tube was inserted in the train immediately after the copper oxide furnace. Any hydrogen in the gas stream leaving the reaction zone would be oxidised by the hot copper oxide and be absorbed as water in the Anhydrone tube.

The results are listed as runs 3(a) and 3(b) in Table II. The following comments are pertinent to the data:

1. The amount of calcium oxalate that decomposed during the first 3 hr at 420° was significantly greater in humid nitrogen than in dry nitrogen.

2. Unlike run 2 made in dry nitrogen, less than one-third of the total carbon dioxide that was absorbed by the Ascarite appeared in the tube that was mounted at the exit end of the copper oxide furnace, indicating that less than one-third of the evolved carbon monoxide left the reaction zone unchanged.

3. Part of the carbon monoxide evolved in the humidified nitrogen atmosphere was oxidised by the water vapour to carbon dioxide:

$$CO + H_2O = CO_2 + H_2 \qquad\qquad \Delta H_{700°K} = -9·1 \text{ kcal}^7 \qquad (6)$$

The hydrogen that was simultaneously produced appeared as water after it had passed through the copper oxide furnace.

4. Not all of the evolved carbon monoxide ended up as carbon dioxide. Unlike runs made in dry air and dry nitrogen, the total mole of carbon dioxide absorbed during run 3(a) was significantly less than the weight loss of the sample, calculated as mole of carbon monoxide. During this run in humid nitrogen a yellow deposit appeared at the cool exit end of the combustion tube. Preliminary pyrolysis experiments in humid nitrogen with both calcium and magnesium oxalates had produced similar deposits, which were soluble in benzene and whose infrared spectra showed them to contain aliphatic C-H bonds. In addition to hydrogen, various hydrocarbon species are thermodynamically possible products of the reaction between carbon monoxide and water vapour. Depending upon their volatility, such species can condense out of the gas stream at various points in the train before the copper oxide furnace and therefore would not appear ultimately as carbon dioxide and water. More volatile hydrocarbons might reach the hot copper oxide tube but be incompletely oxidised in passing through it. Experience in this Laboratory has shown that at the recommended temperature of 700° for operation of the copper oxide furnace, even the oxidation of methane may be slow enough to permit small amounts of the gas to pass

through the reaction zone unchanged. For the foregoing reasons a reliable material balance was not possible in this experiment.

5. To confirm the reliable operation of the train itself, run 3(b) was made by passing dry oxygen over the residue remaining from 3(a). As seen from the results in Table II, all of the evolved carbon monoxide was oxidised in the reaction zone and was absorbed as carbon dioxide in the first Ascarite tube.

No further work has been performed to explore and understand the effect of water vapour upon this reaction, and certainly thermogravimetry alone will not provide the answers. Thermogravimetry has provided, however, as few other techniques could, simple and graphic evidence for the existence of an effect that might warrant further investigation.

CONCLUSION

Calcium oxalate monohydrate is ideally suited for demonstrating many of the factors that affect the quality of thermogravimetric measurements. It is particularly useful for showing the great variety of effects that may be produced by three common atmospheric constituents, carbon dioxide, oxygen and water vapour, as well as three common experimental variables, heating rate, shape of sample container and size of sample. These variables and the interaction between them will determine whether the thermogravimetric curve for the pyrolysis of calcium oxalate monohydrate will appear as a succession of independent reactions or as a sequence of partially overlapping and non-independent reactions. Calcium oxalate monohydrate can be used as a standard for judging the performance of a thermobalance only if it is pyrolysed under carefully controlled conditions.

Acknowledgments—The authors acknowledge the skilful assistance of Adrian Breitenstein, Harley W. Middleton and Mrs. Renette Vincelette, in performing some of the measurements described in this report. They are also indebted to Dr. Paul D. Garn, University of Akron, for suggesting both the interpretation of the calcium carbonate portion of Fig. 11 and the revised experiments described in the text.

Zusammenfassung—Die Pyrolyse von Calciumoxalat-Monohydrat an Luft nimmt in der Literatur über Thermogravimetrie einen hervorragenden Platz ein. Ein Thermogramm dieser Reaktion war die erste von Duval und Mitarbeitern publizierte Pyrolysenkurve und Duval und andere schlugen diese Substanz als Bezugssubstanz zur Beurteilung der Anzeige einer Thermowaage vor. Die Pyrolyse von Calciumoxalat-Monohydrat gibt aber unter verschiedenen Bedingungen beträchtliche Unterschiede in den Thermogrammen. Die Einflüsse von Probengröße, Aufheizgeschwindigkeit, Atmosphäre und Gefäßgeometrie werden in einer Reihe von Thermogrammpaaren vorgelegt und die Unterschiede mit zusätzlichen Informationen aus Differentialthermoanalyse und Versuchen im Verbrennungs-schiffchen erklärt. Besonders wichtig sind Änderungen der Atmosphäre: untersucht wurden trockener und feuchter Stickstoff, trockene und feuchte Luft, trockener Sauerstoff, Kohlendioxyd und Kohlenmonoxyd. Selbst geringfügige Änderungen der Form eines Thermogramms von einer verläßlichen Waage können die Kompliziertheit der Reaktionen widerspiegeln, die die leichter bemerkbaren Gewichtsänderungen insgesamt hervorrufen. Die hier vorgelegten Ergebnisse grenzen auf diese Weise die Arbeitsbedingungen ab, unter denen Calciumoxzlat-Monohydrat als thermogravimetrische Bezugssubstanz verwendet werden kann, und sie zeigen, daß sein Verhalten in einer Thermowaage unter kontrollierten Bedingungen eine ungewöhnlich vielseitige Anleitun zur Interpretation thermogravimetrischer Messungen bieten kann.

Résumé—La pyrolyse dans l'air de l'oxalate de calcium monohydraté occupe une place unique dans la littérature de la thermogravimétrie. Non seulement la première courbe de pyrolyse publiée par Duval et ses collaborateurs a été un thermogramme de cette réaction, mais Duval et d'autres auteurs ont proposé ce composé comme substance de référence pour juger des possibilités d'une thermobalance. Toutefois, la pyrolyse de l'oxalate de calcium monohydraté dans des conditions variées conduit à des différences considérables dans les thermogrammes. Les influences de la taille de la prise d'essai, de la vitesse de chauffage, de l'atmosphère, et de la forme du récipient, sont présentées dans une série de thermogrammes comparatifs, et les différences sont confirmées au moyen de preuves supplementaires apportées par l'analyse thermique différentielle et des expériences de combustion en série. Des variations dans l'atmosphère sont particulièrement importantes, et les atmosphères étudiées sont l'azote sec, l'azote humide, l'air sec, l'air humide, l'oxygène sec, le gaz carbonique sec et l'oxyde de carbone sec. Même de très faibles variations dans la forme d'un thermogramme obtenu avec une balance fidèle peuvent refléter la complexité des réactions qui produisent les changements de poids globaux les plus remarquables. Les résultats présentés dans ce mémoire délimitent ainsi les conditions d'emploi de l'oxalate de calcium monohydraté comme substance de référence thermogravimétrique, et montrent que son comportement dans une thermobalance, dans des conditions contrôlées, peut constituer un exemple particulièrement souple pour l'interprétation des mesures thermogravimétriques.

REFERENCES

[1] M. Bachelet and M. Christen, *Compt. rend.*, 1960, **251**, 2961.
[2] C. Barcia Goyanes, *Inform. Quim. Anal. (Madrid)*, 1955, **9**, 159.
[3] G. G. Briggs, U.S. At. Energy Comm., NLCO-720, 1958.
[4] M. Capestan, *Ann. Chim. (France)*, 1960, **5**, 207.
[5] P. Caro and J. Loriers, *J. Rech. Centre Nat. Rech. Sci., Lab. Bellevue (Paris)*, 1957, **39**, 107.
[6] B. Claudel, M. Perrin and Y. Trambouze, *Compt. rend.*, 1961, **252**, 107.
[7] J. P. Coughlin, *U.S. Bur. Mines Bull.*, 1954, 542.
[8] J. Cueilleron and O. Hartmanshenn, *Bull. Soc. Chim. France*, 1959, 172.
[9] R. W. M. D'Eye and P. G. Sellman, *J. Inorg. Nuclear Chem.*, 1955, **1**, 143.
[10] C. D. Doyle, *Analyt. Chem.*, 1961, **33**, 77.
[11] P. Dubois, *Bull. Soc. Chim. France*, 1936, **3**, 1178.
[12] C. Duval, *Analyt. Chem.*, 1951, **23**, 1271.
[13] *Idem, Inorganic Thermogravimetric Analysis.* Elsevier Publishing Co., Amsterdam, 1953.
[14] L. Erdey and F. Paulik, *Acta Chim. Acad. Sci. Hung.*, 1955, **7**, 27.
[15] J. L. Evans and J. White, *The Thermal Decomposition (Dehydroxylation) of Clays* in *Kinetics of High Temperature Processes*, W. D. Kingery, ed. Technology Press of MIT, Cambridge, 1959, pp. 301–8.
[16] P. D. Garn and J. E. Kessler, *Analyt. Chem.*, 1960, **32**, 1900.
[17] A. Glasner and M. Steinberg, *J. Inorg. Nuclear Chem.*, 1961, **16**, 279.
[18] *Idem, ibid.*, 1961, **22**, 39.
[19] *Idem, ibid.*, 1961, **22**, 156.
[20] H. Guérin, J. Masson, P. Mattrat, R. Boulitrop, C. Duc-Maugé, R. Martin and R. Mas, *Bull. Soc. Chim. France*, 1953, 440.
[21] H. Guérin, *ibid.*, 1955, 1536.
[22] M. Guichard, *ibid.*, 1935, **2**, 539.
[23] *Idem, Ann. Chim. (France)*, 1938, **9**, 323.
[24] G. Guiochon, *Chimie et Industrie*, 1960, **84**, 734.
[25] *Idem, Analyt. Chem.*, 1961, **33**, 1124.
[26] P. L. Günther and H. Rehaag, *Ber.*, 1938, **71**, 1771.
[27] J. Haladjian and G. Carpenti, *Bull. Soc. Chim. France*, 1956, 1679.
[28] K. J. Hill and E. R. S. Winter, *J. Phys. Chem.*, 1956, **60**, 1361.
[29] H. E. Kissinger, H. F. McMurdie and B. S. Simpson, *J. Amer. Ceram. Soc.*, 1956, **39**, 168.
[30] D. Kolar, E. D. Lynch and J. H. Handwerk, *ibid.*, 1962, **45**, 141.
[31] R. Marcellini, thesis, Faculté des Sciences de Lyon, 1959.
[32] H. Mauras, *Bull. Soc. Chim. France*, 1954, 762.

[33] *Idem, ibid.*, 1959, 16.
[34] *Idem, ibid.*, 1960, 260.
[35] P. Murray and J. White, *Trans. Brit. Ceram. Soc.*, 1949, **48**, 187.
[36] A. E. Newkirk, *Analyt. Chem.*, 1960, **32**, 1558.
[37] K. J. Notz, U.S. At. Energy Comm., NLCO-814, 1960.
[38] V. M. Padmanabhan, S. C. Saraiya and A. K. Sundaram, *J. Inorg. Nuclear Chem.*, 1960, **12**, 356.
[39] J. Papailhau, Brevet D'Invention, N. Provisoire PV763, 520 (4/18/58).
[40] *Idem, Bull. Soc. Franc. Mineral. Crist.*, 1959, **82**, 367.
[41] F. Paulik, J. Paulik and L. Erdey, *Acta Chim. Acad. Sci. Hung.*, 1961, **26**, 143.
[42] S. Peltier and C. Duval, *Analyt. Chim. Acta*, 1947, **1**, 345.
[43] H. Peters and H. Wiedemann, *Z. Anorg. Chem.*, 1959, **300**, 142.
[44] A. Richer and P. Vallet, *Bull. Soc. Chim. France*, 1953, 148.
[45] A. Richer, Inst. Recherches de la Sidérugie, Ser, A, No. 187, 1960.
[46] Robert L. Stone Co., Austin, Texas, Model TGA-2, 1962.
[47] F. D. Rossini, D. D. Wagman, W. H. Evans, S. Levine and I. Jaffee, *Nat. Bur. Std. (U.S.) Circ.*, **500**, 1952.
[48] E. L. Simons, A. E. Newkirk and I. Aliferis, *Analyt. Chem.*, 1957, **29**, 48.
[49] J. R. Soulen and I. Mockrin, *ibid.*, 1961, **33**, 1909.
[50] R. L. Stone, *ibid.*, 1960, **32**, 1582.
[51] S. J. Teichner, R. P. Marcellini and P. Rue, *Adv. Catalysis*, 1957, **9**, 458.
[52] S. J. Teichner, J. Aigueperse, B. Arghlropoulos and R. P. Marcellini, 1959 *XVII Intern. Congr. Pure Appl. Chem., Munich, Abstracts*, **1**, 207.
[53] P. Vallet, *Bull. Soc. Chim. France*, 1936, **3**, 103.
[54] P. Vallet and M. Bassière, *ibid.*, 1938, **5**, 546.
[55] P. Vallet and A. Richer, *Compt. rend.*, 1954, **238**, 1020.
[56] W. W. Wendlandt, *J. Chem. Educ.*, 1961, **38**, 571.
[57] C. L. Wilson, *J. Roy. Inst. Chem.*, 1959, **83**, 550.
[58] L. Wöhler and W. Schuff, *Z. anorg. Chem.*, 1932, **209**, 57.
[59] P. Longuet, *Rev. materiaux construit, et trav. publ. C*, No. **537**, 139; No. **538–539**, 183; No. **540**, 233 (1960); *Bull. soc. franc. ceram.*, No. 48, 69 (1960).

35

This article was originally published in German in *Chem.-Ing.-Tech.*, **36**, 1105–1114 (1964), and this English translation is a special reprint prepared by the staff at Mettler Instruments AG, Switzerland

Universal Measuring Instrument for Gravimetric Investigations Under Variable Conditions

Thermogravimetric Investigations VI

by H. G. WIEDEMANN *

Mettler Instrumente AG, Stäfa/ZH, Switzerland

In recent decades thermogravimetric instruments — thermo-balances — have been used in many fields of research and industry[1-32]. Continued development of these instruments has now made it possible to simultaneously record the changes in weight and heat of reaction of a substance as functions of a variety of conditions, such as pressure, temperature and gas atmosphere. The use of recording balances, in particular, permits processes to be followed and the mechanism of reactions to be detected. This gives the possibility of comparison and combination with other continuous analytical techniques.

So far it has been the practice to build special balances for most specialized fields; e. g., density balances for the determination of the density of gases and liquids, adsorption balances for determination of spezific surface, sedimentation balances for measuring grain size, thermo-balances for the various kinds of thermal analyses, etc.

The balance described here has been designed for application in various fields. The overall conception of the instrument, by reason of its modular construction principle, enables measurement techniques to be employed which are adapted to special situations. When making specific measurements, a simple system can be used which, nonetheless, remains capable of being expanded to cover the most diverse variants of a multi-purpose instrument.

When the fields of application of thermo-balances are considered, it can be seen that a compromise between an analytical balance and a micro-balance in regard to sample capacity and accuracy had to be found for

*) Paper presented at the Achema Congress 1964 in the section for "Modern physical methods of chemical analysis, especially trace analysis", in Frankfurt, 26. 6. 1964.

various experimental methods. The ratio of loading capacity (40 g) to sensitivity (± 0.01 mg) has been so selected that thermo-gravimetric dissociations, for example, can be measured with adequate sensitivity even where small quantities are weighed-in to insure optimum heat distribution throughout the sample. On the other hand, it is also possible to perform experiments in which a heavy experimental instrument is loaded onto the pan in addition to the sample; e. g., a measuring head in differential thermal analysis, a plummet in the case of density measurements, or a separating column or adsorber for gaschromatographic experiments.

Just as the ratio between sensitivity and loading capacity limits the weighing range in the case of most thermo-balances, so does the unilateral *arrangement of the weighing chamber* also constitute a bottleneck. *Top-loading* balances are very suitable for high temperature tests because of better temperature distribution in the reaction chamber and the small influence of temperature on the balance, but are difficult or impossible to use for measurements at very low temperatures or in conjunction with measuring apparatus requiring a downward layout. On the other hand, in the case of *underslung* designs, a great effort must be expended on protection of the balance against heating, and this renders operation, e. g., charging the sample, and combination with other measuring processes difficult. The present Thermoanalyzer is so designed that weighings can be made above or below the balance, in accordance with the temperature range and auxiliary apparatus required. It has been possible to dispense with the other preconditions for setting up balances of this type, as the present balance is provided with measuring instrumentation largely shielded from external influences such as shocks, corrosion, or temperature variation. It is, therefore, pos-

[*Editors' Note:* Figures 5 and 12 have been omitted because they are in color and cannot be rendered effectively in black and white.]

sible to carry out precise physical and chemical experiments on the balance pan within the following limits:

Temperatures from −200°C to +1600°C
Pressures from approx. 10⁻⁶ to 760 mm Hg
Stationary or flowing gaseous atmospheres of various compositions, including corrosive gases.

The Balance System

In overall construction the Thermoanalyzer consists of two units: the balance proper, and the electronic cabinet (cf. figs. 1 and 2). The Thermoanalyzer is designed as a modular unit and is housed in a metal cabinet together with the other elements, such as the vacuum system, furnace and furnace elevator and

control [33)] also performs balance damping which, in contrast to air damping, is effective both under normal conditions and in high vacuum.

Built-in weights compensate for weight variations which exceed the electrical range. The *weighing range* for topload measurements is 16 grams and for underslung measurements, 42 grams.

Two recording and compensation ranges of 1000 mg and 100 mg, respectively, are available. Mechanical and electrical taring devices permit the experiment to commence at any point desired within the range selected. The weight is simultaneously recorded in

Fig. 1. View of the Thermoanalyzer.
Left: balance stand with high temperature furnace.
Right: Electronic cabinet with recorder.

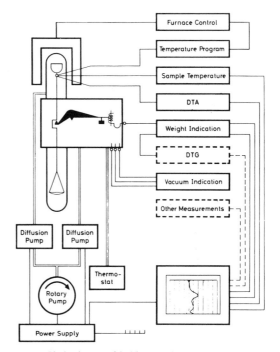

Fig. 2. Block schematic of the Thermoanalyzer.

water-jacket thermostat. A separate cabinet houses the electronic measurement panels and the recorder. All panels are standard 19″ elements.

The balance is a beam type operating on the *substitution* principle. The design is similar to that of a normal analytical balance but adapted, in respect to material and design, to the operating conditions of thermo-gravimetry. The balance has been equipped with electro-magnetic compensation (fig. 3) for continuous readout of results; i.e., the changes occurring in weight are converted into an electrical signal which takes the place of an optical scale readout. The beam position is sensed by photo-transistors and set by a control circuit. This proportional differential

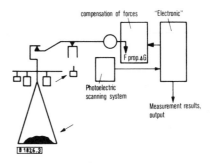

Fig. 3. Principle of the substitution balance with electro-magnetic compensation.
F = electrical force proportional to the variation in weight, ΔG, of the subject weighed.

two sensitivities in the ratio of 1:10. In the 1 g range the 10 inch chart width corresponds to 1000 mg, the additional expanded curve corresponding to 100 mg. In the 100 mg range the chart width corresponds to 100 mg, or 10 mg expanded. In the most sensitive range, a chart subdivision of 1/10th inch thus represents 0.1 mg.

Crucible holders are provided instead of a balance pan and these extend from the hangers into the upper or lower sample chamber. The crucible holders – four-hole capillary tubes of sintered alumina – accommodate the thermocouples. The measuring points of the thermocouples are in direct contact with the sample or its container. Special metal bands, having no influence on weighing accuracy, are used to connect the thermocouple leads from the balance hanger to the fixed housing.

A built-in set of calibration weights permits the sensitivity of the balance to be checked from the outside. Adjustment, if required, is made electrically.

Fig. 4 shows the *course of decomposition of strontium oxalate* during a linear temperature rise of 10°C/min. Four different curves are shown in the diagram. Curve 1 shows *weight* in relation to rise in *temperature* (1 chart division ≙ 10 mg). The decomposition is divided into three distinct phases: separation of water, release of CO and CO_2. *Weight curve* 2 is expanded 10 times (1 chart division ≙ 1 mg). If, as in this case, the curve takes up more than one chart width, an electrical weight is automatically added and the curve starts again at the left-hand edge of the diagram.

Curve 3 shows the various heats of the individual reactions: First, the transition to anhydrous oxalate can be seen (A), then the decomposition of the oxalate follows with simultaneous combustion of the liberated carbon monoxide (B), and finally the conversion (rhombic to hexagonal) of the strontium carbonate, which is transformed into the oxide form by liberation of carbon dioxide. The combustion of the

liberated carbon monoxide shows that the carbon dioxide used was not quite free from oxygen.

$$CO \xrightarrow[-Q]{+\frac{1}{2}O_2} CO_2$$

$$Sr(COO)_2 \cdot H_2O \xrightarrow[+Q]{-H_2O} Sr(COO)_2 \xrightarrow[+Q]{-CO} SrCO_3 \text{ (rhomb.)} \xrightarrow{+Q}$$

$$SrCO_3 \text{ (hex.)} \xrightarrow[+Q]{-CO_2} SrO$$

The *temperature program* during the experiment is shown by curve 4; the various reaction temperatures can be read off the recording chart directly with a temperature scale or determined by means of the thermo-electric voltage.

Where decomposition stages follow one another in rapid sequence, the *differentiated weight curve* provides results which are more easily evaluated. It is, therefore, possible to record a differentiated temperature/weight curve simultaneously with the other data using a special measuring panel with electro-mechanical differentiation. The differentiated weight is particularly helpful, for example, in sedimentation analysis in obtaining an indication of the distribution of the various particle sizes.

Balance housing, sample chamber, crucible holder, crucible

A stainless steel housing built into the balance stand holds the balance. It is a vacuum-tight compartment which, nonetheless, permits operation of the balance weights and arrestment system by means of various vacuum sealed shafts. To protect the balance against temperature variations, the housing is thermostatically kept at 25°C. Temperature variations within the balance housing do not exceed ± 0.1°C even during high temperature experiments. An independently cooled plate, mounted on the furnace and the

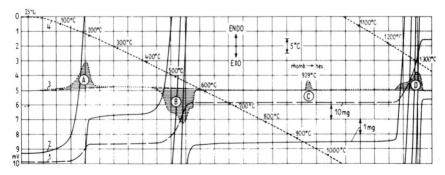

Fig. 4. Decomposition of strontium oxalate in carbon dioxide.
Curve 1: summary curve, weight as function of temperature (1 chart subdivision ≙ 10 mg). Curve 2: expanded weight range (1 chart division ≙ 1 mg). Curve 3: curve for differential thermal analysis. Curve 4: temperature program (10°C/min).

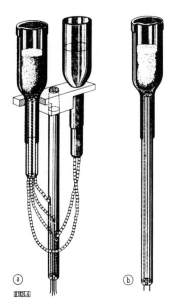

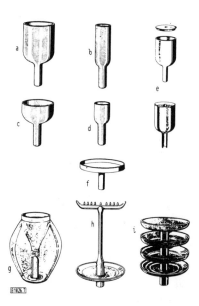

Fig. 6. Various crucible holders for thermoanalytical experiments.
Left: DTA crucible holder.
Right: crucible holder for thermogravimetric decompositions.

Fig. 7. Various crucibles for thermogravimetric investigations. a)–d): crucibles of various capacities; e) crucible for vapor pressure measurement; f) plate-type crucible for decompositions in high vacuum; g) crucible for effervescent substances; h) holder for plate and foil samples under corrosion test; i) extended plate-type crucible.

housing, shields the balance from direct radiation at high temperatures. Fig. 5 (see color plate page 9) shows the water circulation system of the entire apparatus.

Openings to the two *reaction chambers* are arranged above and below the balance. These openings can be closed with quartz, glass or alumina sleeves, which surround the sample and its container and serve as reaction or furnace chambers. The sleeves are held in place with quick lock "O" ring seals. Space above and below the balance is available for the addition of auxiliary experimental apparatus. The vacuum system connections are so arranged that the weighing operation is not affected by evacuation.

The passage between the upper reaction chamber and the housing is shielded by a baffle, so that no substance can fall into the balance housing either during the weighing-in or measuring operation. The baffle is so constructed that it acts as a radiation shield for high temperature experiments and also acts as a light tight diffusion baffle during high vacuum experiments.

Fig. 6 shows two crucible holders for thermogravimetric experiments. The arrangement a) is for measurements where thermogravimetry and differential thermoanalysis are combined, while arrangement b) is for thermogravimetric measurements alone.

The crucible holders are provided with a plug at the end of the capillary tubes which plugs into the balance stirrup hanger. The holders are thus readily interchangeable. The hot junction of the thermocouple is in contact with the crucible base, thus insuring accurate temperature readings even with small, weighed-in quantities or with crucible materials of poor conductivity.

Crucibles must be suitable for the required experimental conditions in respect to their material, capacity and shape. As in the techniques of the chemical laboratory, crucible materials are selected to obviate the possibility of reaction between crucible and sample material. The main materials used are the precious metals, oxide ceramics, quartz and graphite. The size of the crucible is determined by the volume to be weighed-in . . . it is usually between 0.1 and 5 cc.

Two procedures may be followed for *reactions in defined atmospheres.* The crucible can be closed with a lid and decomposition effected in the atmosphere of the liberated products [34], or an open crucible may be used in an appropriate gas atmosphere. In the case of decompositions in air, it is preferable for the substance to fill the crucible completely as otherwise the gases remaining temporarily within the reacting substances will modify the course of decomposition (arrangement similar to that with crucible and lid) [35]. Fig. 7 shows a variety of *crucible shapes.* The four crucibles, (a) to (d), may be made of platinum-rhodium, aluminum oxide or the like. Crucible (e) with lid is suitable for vapor pressure measurements by the Knudsen effusion method. The flat crucible

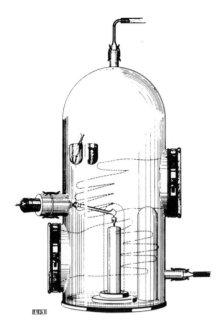

Fig. 8. Special device for weighing-in in controlled atmospheres on the Thermoanalyzer.

(f) is intended for high vacuum experiments. Thin layers of sample are weighed-in to prevent the substance from erupting during decomposition. Crucible (g) is designed for substances which spray heavily at normal pressures, and crucible (h) for suspending metal foils, while the arrangement (i) is an extended, flat, plate-type crucible.

It is usually possible to weigh-in the substance directly on the Thermoanalyzer or on a separate analytical balance. In the case of substances sensitive to air or humidity, a special device for *direct weighing-in* on the Thermoanalyzer can be used (cf. fig. 8). The substance is introduced in ampule form into the pigeonhole in advance and, after change of the atmosphere, can be weighed-in with a vibro-spatula. After the balance has been closed the device can be removed.

Furnace Equipment

In every thermoanalytical experiment, reproducibility of temperature measurements and temperature program rates is critical if experimental results are to be compared. In combined measuring procedures in particular, where thermogravimetric measurements are carried out simultaneously with differential thermal analysis and absolute readings and temperature program rates directly influence the measured result, high demands must be made on the capability of the furnace equipment.

At the present time, two *types of furnaces* have been provided for various fields of application: a low

temperature furnace for room temperature up to 1000°C, and a high temperature furnace for 25 to 1600°C. In designing these furnaces, special attention has been paid to homogeneous temperature distribution in the reaction chamber and to ease of regulation. Fig. 9 shows the vertical temperature distribution in a furnace at various temperature levels; the horizontal picture is similar.

The *low temperature furnace* (fig. 10) consists of a quartz sample chamber tube with a quartz, bifilar spiral tube is sealed in at the upper end. The heating filament set in this tube system is open to the outside atmosphere and is not exposed to the more or less corrosive decomposition products during decomposition processes. Reflectors insure insulation and good heat distribution. This furnace is capable of particularly good regulation. Cooling time from 1050°C to room temperature is 15 minutes.

The *high temperature furnace* is equipped with Kanthal-Super elements. The assembly is arranged on a pivotable elevator assembly and can thus be moved easily into place.

Platinum/platinum rhodium thermocouples are used as temperature sensors on the various sample holder assemblies.

An exactly linear temperature rise is insured by an electromechanical program control. The thermoelectric voltage of the furnace thermocouple is compared to the program voltage, which is generated by a cam system corresponding to the standard thermocouple voltage function. The temperature regulator is controlled by the differential voltage which results [36]. At the same time, the thermoelectric voltage is fed to the recorder where it is used as the temperature indication. The accuracy of the program control is ± 0.5°C, and furnace regulation accuracy is ± 1.5°C.

In addition to the temperature curve recorded, the appropriate temperature can be read off digitally on the programmer. An automatic preselector permits any desired figure between 25 and 1600°C to be set digitally. When the furnace reaches this temperature

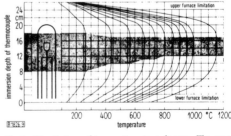

Fig. 9. Distribution of temperature in a furnace. The gray area shows the zone of homogeneous temperature in relation to the position of the crucible.

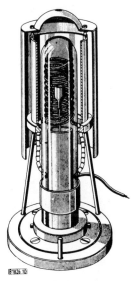

Fig. 10. Low temperature furnace (operating range: 25 to 1050°C).

it is either switched off or the temperature held constant, as desired. The same technique is used for decreasing temperatures.

In regard to temperature *heating* or *cooling rates*, a choice of ten rates is available between 0.5 and 25°C/min. Below 300°C the maximum controlled cooling rate depends on the natural cooling rate of the furnace used. A start and stop button enables the operator to stop, hold or restart the temperature rise at any desired temperature between 25 and 1600°C. This provides the facility for carrying out "*progressive isothermic heat additions*" [37]. The advantage of this is that the curves thus arrived at, by contrast to the normal, dynamic thermogravimetric curves (with

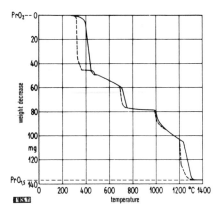

Fig. 11. Course of decomposition of praseodymium oxide under various heat treatments.
——— linear temperature rise (8°C/min) in air
------ isothermic heat addition in air.

linear temperature rise), show the end of the chemical reaction at a temperature at which the system has almost achieved thermodynamic equilibrium (see fig. 11).

Vacuum System

The fields in which the Thermoanalyzer can be applied depend on how many ways one can vary the conditions affecting the sample. The course of the reaction is particularly dependent on the ambient atmosphere and pressure. In the case of this balance design, it is possible to operate not only in a flow of air, but also in other defined atmospheres such as hydrogen, nitrogen, oxygen or gas mixtures.

Fig. 12 (see colored plate page 10) shows a section of the entire vacuum system and its arrangement in relation to the Thermoanalyzer. When the atmosphere is changed, the balance must be evacuated

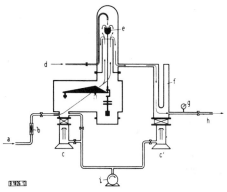

Fig. 13. Path of gas stream through the balance. The labyrinth arrangement below the sample prevents the penetration of gaseous decomposition products into the balance when working under gas flow.
a) gas inlet, b) flow meter, c) and c') diffusion pumps, d) inlet for corrosive gases, e) sample, f) cold trap, g) manometer, h) gas exit, i) rotary pump.

and then refilled with the appropriate washed gas by means of the needle valve H. When atmospheric pressure is again reached, outlet valves C and E are opened. From that point on it is possible to work in a gas at flow rates of up to 30 liters/h without affecting the weighing result. The path of the gas stream is indicated in fig. 13. The labyrinth arrangement shields the balance from corrosive decomposition products and condensable substances. For measurements in corrosive gases, an inert gas must be pumped in through the gas inlet and a separate inlet used for the corrosive gas.

Because the balance housing is separated from the reaction chamber (cf. fig. 12), one diffusion pump evacuates the balance housing only while the other evacuates the reaction chamber. The reaction cham-

315

ber is connected with the diffusion pump by a cooling trap. The decomposition products can be taken from the gas outlet for analysis or condensed on the cold trap by means of a coolant.

In the case of operations in the range between 760 and 5×10^{-2} mm Hg, the two-stage rotary pump is adequate. Ultimate vacuum, using the entire pump installation, is better than 8×10^{-6} mm Hg.

A precision aneroid pressure meter L is used for measurements in the $760 - 1$ mm Hg vacuum range. Thermocouple gauges (M 1 + M 2) are used in the $1 - 1 \times 10^{-2}$ range. A cold cathode ionization gauge (K) is used in the high vacuum range down to 8×10^{-6} mm Hg.

Important *applications* in the high vacuum range are surface area determinations, vapor pressure measurements and degassing operations. In the case of thermogravimetric dissociations, it is important to govern the sample mass so that the amount of gas given off during heating is adapted to the pumping speed of the vacuum system. Continuous recording of gas pressure can be made along with the weight curve during the experiment. This, in particular, leads to the possibility of better evaluations of many types of measurement (cf. fig. 17).

The *measuring point* K for the high vacuum range is located on the housing just below the cooling plate adjacent to the reaction chamber (cf. fig. 12). The measuring cell is calibrated so that it shows the actual vacuum in the sample chamber [1].

Differential Thermal Analysis

In general, the differential thermo-analytical and thermogravimetric results obtained by separate tests on a substance are not easily compared, since it is difficult to achieve absolutely equal experimental conditions where different instruments are used. For this reason a measuring device for the simultaneous performance of differential thermal analysis has been combined with the thermo-balance. The sample carrier for thermogravimetric analysis alone is interchangeable with the sample carrier for simultaneous analysis, so that it is possible to always locate the samples centrally in the furnace.

Fig. 6a shows a crucible holder with two crucibles inserted into a T-piece of sintered alumina. The crucible with lid holds the reference substance and the other crucible holds the sample. For quantities of substance between 3 and 25 mg a measuring head is used (cf. fig. 14) which was developed by *C. Mazières* [38]. In this case the hot junction of the thermocouple is welded to a crucible. This measuring head is designed to take carefully adapted small crucible liners which hold the sample substance. The full apparatus is made up of three crucibles: one for the

reference substance, one for the reacting sample, and a third, also filled with an inert substance, for temperature programming and measurement. The two measuring heads shown in fig. 14 are distinguished from one another by a platinum cylinder borne on a ceramic plate. The hooded device is for measurements in air or in other gases and serves to equalize the temperature in the crucible chamber. The hood is not used for measurements in vacuum or high vacuum, as the heat equalizes out too slowly after the reactions, which would result in drifting.

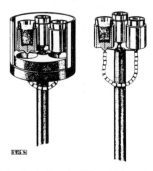

Fig. 14. Crucible holders for differential thermal analysis. Left: crucible holder for small weighed-in quantities (up to approx. 25 mg) for measurements under normal pressure. Right: crucible holder for measurements under vacuum.

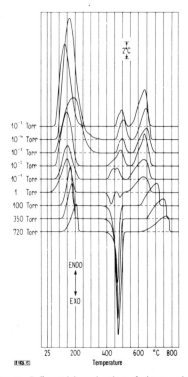

Fig. 15. Differential thermal analysis of calcium oxalate under various pressures.

Fig. 15 shows curves from the differential thermal analysis of calcium oxalate recorded in relation to temperature and various gas pressures. It can be seen that the decomposition temperatures vary with changes in pressure. The peaks lying between 150 and 200°C are caused by the liberation of water of hydration, while the second group of peaks shows the decomposition of the anhydrous oxalate to carbonate (450 to 500°C). The carbon monoxide given off is completely oxidized between pressures of 720 to 100 mm Hg, while at pressures from 1 to 10^{-2} mm Hg oxidation takes place only partially, and below 10^{-2} mm Hg not at all. In the last group of peaks showing the decomposition of calcium carbonate to calcium oxide, a shift of the decomposition temperatures (675 to 875°C) also takes place as a function of pressure.

The amplifier for differential thermal analysis has ranges of 20μV, 100μV and 500μV recorded over a 10 inch chart width. In each range it is possible to shift the signal a total of four calibrated chart widths. Continuous zero displacement permits beginning the experiment at any desired point within the range selected.

For quantitative evaluations of the thermal effects obtained from differential thermal analysis, *H. E. Schwiete* [39] and *K. Torkar* [40] have tried a method which they call *dynamic differential calorimetry*. Using a sample holder arrangement similar to that described above, a suitable calibration substance with known heat of reaction is used instead of the reference material. The calibration results obtained by this method (cal/sqc.) are used to evaluate the results from the substance under experiment. The temperature of the heat peak of the calibration substance should be close to the temperature of the thermal effect under study in the sample substance.

Sources of Error and Corrections

In general, the variations in weight of a substance can be considered as being functions of temperature, pressure and time. When kinetic reactions are being recorded, the relationship of weight to time must be measured at constant temperature and pressure so that only one independent variable is considered.

A more commonly used method is the method of measuring the loss of weight of a sample against linearly rising temperature. The weight losses thus continuously recorded include the time-conditioned mutations of the sample during heat addition. In measurements of this kind only approximate equilibrium is attained. This must be taken into account when interpreting the curves (cf. fig. 11).

With many substances preparatory treatment of the sample effects the reproducibility of the measured results. The purity of the substance is the most important factor, except where mixtures of substances are to be analyzed. Precipitation reagents, particle size, pre-drying, the material of the crucible and the shape of the crucible may also have an effect on the mechanism of decomposition [41]. In the case of the decomposition of substances with a high crystal water content, for example, the speed of H_2O liberation is dependent on the size of the crystals. In the case of substances of this kind it is, therefore, advisable to use samples of uniform particle size. For similar reasons the course of decomposition is, to a certain extent, dependent on the mass of the sample. To facilitate subsequent evaluation of the curves it is convenient to keep the quantity weighed-in to a definite fraction of the molecular weight of the substance.

When making an experiment, the crucible containing the substance is placed on the crucible holder and the sample chamber closed by locking the sample chamber tube in place. The balance is then tared and the recorder brought to a starting position corresponding to the expected gain or loss in weight, so as to take full advantage of the maximum recording range. In order to remove reaction products from the sample chamber during the course of the experiment, a gas stream can be passed through the balance housing and sample chamber at rates varying from 0 to 30 liters/h. Without this gas flushing, errors may arise, as condensable reaction products might precipitate onto the cooler parts of the balance, such as the crucible holder, and be included in the measured weight.

In the case of experiments in a specific gas atmosphere, for example, hydrogen or oxygen, the apparatus is first evacuated to 0.1 mm Hg and then filled with gas which has been purified and dried. To avoid excess pressure during filling, a pressure relief valve must be located in the filling circuit. During the period of the experiment a continuous gas stream is again passed through the reaction chamber. If an interferometer or other suitable apparatus is mounted in the gas flow beyond the balance, gas analysis can be carried out at the same time.

The heating rate or the temperature program which is selected, should be suitable to the course of the reaction as experience may dictate.

When weighing with normal analytical balances, one or two corrections are necessary (e. g., reduction to vacuum). In the case of work done on the thermobalance, however, there are, in addition to these corrections, certain other effects created which must be taken into account; for example, the buoyancy [35] of the sample substance and those parts of the balance which are at a high temperature must be borne in mind. In addition, as the temperature rises the *density of the gas phase* decreases considerably. At about 300°C the density and, therefore, the buoyancy exerted on the substance is only about half as great as at 25°C. In air this results in an apparent weight variation of 0.6 mg/cc. If ideal gas behavior is assumed, the change in buoyancy ΔA by a body of mass M and density d on addition of heat from temperature τ (room or balance temperature) to furnace temperature T

$$\Delta A = M\varrho_\tau \, \frac{1}{d} \left(1 - \frac{\tau}{T}\right) ,$$

where ϱ_τ is the density of the gas.

The ΔA can be calculated for the crucible and the substance, as both are at a constant temperature (cf. fig. 9). With the crucible holder the situation is different, as it is located in the temperature gradient area between the sample and the balance. Its buoyancy variation must, therefore, be determined empirically.

Fig. 16 shows the variation in gas density and buoyancy (mg/cc.) in relation to temperature at various pressures. The area lying between the curves roughly corresponds to the normal pressure fluctuations expected while working at atmospheric pressure. Buoyancy is different for every instrument. Departures from the buoyancy curves may be caused by the heat conductivity of the gas [42].

In the case of a thermo-balance whose housing can be evacuated or filled with gases at varying pressures or of different compositions, the zero point of the balance system is dependent on the density of the gas. As construction is always asymmetrical, components of the balance on the two sides of the central bearing have different volumes at equal weight. A change in gas density, resulting from alterations of the gas composition or from changes in temperature, thus has different effects on the buoyancy factor of these balance components and thus brings about a displacement of the zero point. It is possible to

calculate and compensate for the difference in volume between the two sides by measuring the dependence of the zero position on the gas density. In balances which have been compensated, it is only additionally necessary to take account of the effect caused by the variations in density of the sample substance and of the substitution weights.

During thermogravimetric decompositions in a *vacuum*, it has been observed that the mass of the sample substance and the thickness of the substance layer should not generally exceed certain limits. If the normal forms of crucible are used, the great thickness of the sample layer causes particles of substance to be ejected by the gases escaping during decomposition, even where the sample mass is small. If crucibles shown in fig. 6f are used, good results can be achieved with sample masses of the order of 0.1 g. It is important that the sample is distributed very evenly over the crucible.

By way of example, fig. 17 shows the dehydration of calcium oxalate monohydrate. The broken curve shows dehydration under normal pressure, and the solid line dehydration under high vacuum conditions. The change in pressure is recorded below, measured simultaneously with the decomposition in the reaction chamber. The lowest curve shows the temperature. The readings show that the amounts of gas given off by the substance must coincide with the

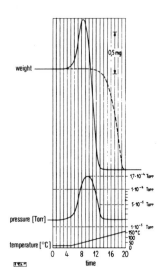

Fig. 17. The dehydration of calcium oxalate. (Ca(COO)₂.H₂O) under various conditions.
——— course of decomposition under high vacuum.
------ course of decomposition in flowing air (10 liters/h, 720 mm Hg).
In addition to the pressure changes in the reaction chamber, the temperature rise and the time-progression of the reaction are also shown.

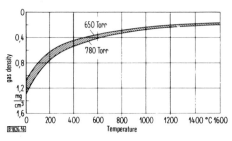

Fig. 16. Changes in gas density or buoyancy in relation to temperature at various pressures (for air). The zone lying between the curves corresponds roughly to pressure variations to be expected when operating at atmospheric pressure.

pumping speed of the vacuum system and that, for example, a gas effusion of approx. 1 mg H₂O under these conditions is enough to bring about a pressure variation of the order of one decade. The "apparent" gain in weight measured under high vacuum at the beginning of the decomposition stage is due to the fact that a percentage of the molecules emerging from the substance on decomposition collide with the crucible during evacuation*).

Fields of Application of the Thermoanalyzer

The term "thermo-balance" was introduced in 1915 by *K. Honda*[2]), describing a balance suitable for measuring the variations in weight of a sample as a function of temperature. At first, such balances were used for the solution of analytical problems. The work of *C. Duval*[43], in particular, led to greater understanding of the potential thermogravimetric studies. In the meantime, many new methods have been developed, of which it is possible to mention only a few here.

Simultaneous measurements with a thermo-balance can mean considerable simplification in many experiments. For example, calcium together with a very small quantity of magnesium (or vice versa) can be determined quantitatively if both are precipitated as oxalates and decomposed in a thermo-balance. From a comparison of resulting curves with those of the pure components, it is possible to calculate the quantities of the two elements contained in the sample. It is also possible to check the purity of reagent chemicals by comparing their decomposition curves. With the aid of the Thermoanalyzer, *heterogeneous equilibria* and the speed of *homogeneous reactions* can also be

investigated. The most frequent examples of this type study are oxidation and reduction processes using gases or mixtures of gases. Experiments on the oxides of heavy metals which reveal varying compositions according to the oxygen pressure used in the experimental chamber may be mentioned by way of example. It is possible to note every variation of the stoichiometric relationships by means of the weight changes in a single sample.

Fig. 18 shows, as example, measurements made on praseodymium oxide under varying partial pressures of oxygen, a temperature increase of 8° C/min. and a gas flow of approx. 10 liters/h. Fig. 11 shows the course of decomposition of praseodymium oxide under various thermal treatments. In this case a curve showing a linear temperature rise of 8°C/min. is compared with one showing "staged isothermic heat additions" at 10°C intervals *).

Adsorption equilibria for determining specific surfaces have also frequently been investigated. The gas adsorbed brings about an equivalent increase in the weight of the sample. Fig. 19 shows the apparatus for a measurement of this type.

The Thermoanalyzer is particularly suitable for research into the kinetic characteristics of heterogeneous *reactions,* e. g. for the investigation of dry corrosion (oxidation) of metals and alloys.

The *vapor pressure* of substances can also be determined at low and high temperatures by the *Knudsen* effusion method, the vaporization being followed thermogravimetrically [44].

By combining a thermo-balance with a gas *chromatograph,* the individual components can be determined

---

*) A summarized paper on these experiments will be published.

*). A summarized paper on these experiments will be published.

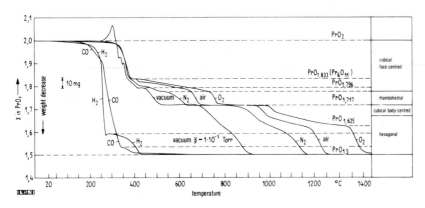

Fig. 18. Oxide system of praseodymium.
Sample weight: 2.95586 g PrO₂: speed of gas flow 10 liters/h. Temperature increase 8°C/min.

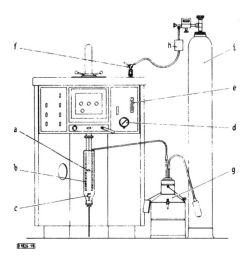

Fig. 19. Apparatus for surface area measurements with the balance.
a) sample container, b) liquid nitrogen, c) heat exchanger, d) manometer, e) flow meter, f) gas inlet, g) container for liquid nitrogen, h) gas cleaner, i) gas bottle.

quantitatively if they are trapped in special crucibles in the thermo-balance together with suitable absorbents [45].

The design and construction of the balance also permit combination with a device for magnetic measurement. This enables *paramagnetic and diamagnetic measurements*, as well as *measurements of susceptibility* to be carried out under vacuum. Thermomagnetic measurements permit the determination of the temperature dependence of the paramagnetic susceptibility of a substance. As a result, it is possible to determine the *Curie* temperature and thus the magnetic moment.

Of great interest are those procedures which permit the use of combined measuring techniques. Since it is possible to follow not only weight variations but also, for example, pressure variations during an experiment under vacuum, one can observe the degassing of metals or other materials used in high vacuum techniques.

The measuring arrangements for the simultaneous determination of *weight variations* (thermogravimetric analysis) and reaction heats (differential thermal analysis) on a single sample are especially worthy of mention. It is possible to record not only reactions which are associated with weight variations but also to obtain quantitative information as to transformation, fusion, vaporization and reaction enthalpy.

Using a gravimetric micro-distillation method by *F. Paulik, J. Paulik* and *L. Erdey* [46, 46] the equilibrium temperatures of liquid-vapor systems may be determined. The composition of mixtures of liquids can

be ascertained by means of the differentiated weight indication of the distillation curve.

Density measurements in fluid solutions under buffer gas, determination of surface tension, the observation of photochemical reactions, gravimetric measurements of vaporizations, thermogravimetric analyses of radio-active substances and materials tested in radiation fields are other regions in which this apparatus can be utilized.

A complete summary of every possibility of application is not possible in the space here available, but will form the subject of a special paper to be published.

(Received September 10, 1964 (B 1826))

Bibliography

[1] V. Mitteilung: *H. P. Vaughan* and *H. G. Wiedemann:* Vacuum Microbalance Techniques, Vol. 4, Plenum Press. New York, 1964.
[2] *K. Honda.* Sci. Repts. Tôhoku Imp. Univ., Ser. IV, *1915,* 97.
[3] *M. Guichard:* Bull. Soc. Chim. France, Ser. *37,* 62, 251, 381 [1925].
[4] *A. Stock* and *G. Ritter.* Z. anorg. allg. Chem. *119,* 321 [1931].
[5] *C. Wagner* and *K. Grünwald,* Z. physik. Chem., Abt. B *40,* 455 [1938].
[6] *E. A. Gulbransen.* Trans. elektrochem. Soc. *81,* 327 [1942].
[7] *P. Chevenard. X. Waché* and *R. de la Tullaye,* Bull. Soc. chim. France, Mém. Ser. (5) *11,* 41 [1944].
[8] *F. Paulik, J. Paulik* and *L. Erdey,* J. analyt. Chem. (russ.) *160,* 241 [1958].
[9] *A. Rahmel* and *K. Hauffe,* Z. physik. Chem. *199,* 152 [1952].
[10] *W. W. Wendlandt.* Analytic. Chem. *27,* 1277 [1955].
[11] *S. I. Gregg,* J. chem. Soc. *1946,* 561.
[12] *Ch. Eynaud,* J. Physique Radium *14,* 638 [1953].
[13] *P. Dubois.* Thesis. Paris (26. Juni 1935) Nr. 2428.
[14] *J. W. Mc. Bain* and *A. Bakr,* J. Amer. chem. Soc. *48,* 690 [1926].
[15] *O. Neunhoeffer* and *K. Hauffe,* Z. anorg allg. Chem. *262,* 300 [1950].
[16] *J. F. Corles,* this journal *30,* 342 [1958].
[17] *W. Forkel,* Z. Instrumentenkunde *69,* 215 [1961].
[18] *P. L. Waters,* Mesures Contrôle ind. *305,* 1233 [1962].
[19] *A. Blažek,* Hutnické Listy *12,* 1096 [1957].
[20] *I. G. Rabatin* and *C. S. Card,* Analytic. Chem. *31,* 1689 [1959].
[21] *M. I. Pope,* J. sci. Instruments *34,* 229 [1957].
[22] *C. Groot* and *V. H. Troutner,* Analytic. Chem. *29,* 835 [1957].
[23] *M. Linseis,* Keram. Z. *11,* 54 [1959].
[24] *S. Gordon* and *C. Campbell,* Analytic. Chem. *32,* 271 R [1960].
[25] *J. G. Hooley,* Canadian J. Chem. *35,* 374 [1957].
[26] *H. Peters* and *H. G. Wiedemann,* Z. anorg. allg. Chem. *298,* 202 [1959].
[27] *R. Splítek,* Hutnické Listy *8,* 697 [1958].
[28] *Z. Szabo* and *D. Kiraly,* Magyar Kémiai Folyóirat *63,* 158 [1957].
[29] *V. Šatava,* Silikáty [Praha] *1,* 188 [1957].
[30] *L. Simons, A. E. Newkirk* and *I. Aliferis,* Analytic. Chem. *29,* 48 [1957].
[31] *G. M. Schwab* and *J. Philinis,* Z. anorg. Chem. *253,* 71 [1945].

32) *P. D. Kalinin* and *A. K. Kuznetsov,* J. physic. Chem. (russ.) *32,* 1658 [1958].

33) *H. Steiner,* Mettler News *32,* 9 [1964].

34) *P. D. Gam* and *J. E. Kessler,* Analytic. Chem. *32,* 1563 [1960].

35) *H. Peters* and *H. G. Wiedemann,* Z. anorg. allg. Chem. *300,* 142 [1959].

36) *W. Schürmann,* Mettler News *32,* 15 [1964].

37) *M. Pettre* and others, Angew. Chem. *65,* 549 [1953].

38) *C. Mazières,* Analytic. Chem. *36,* 602 [1964].

39) *H. E. Schwiete* and *G. Ziegler,* Ber. Dtsch. Keram. Ges. *35,* 193 [1958].

40) *K. Torkar, K. Lasser* and *H. P. Fritzer,* Sprechsaal Keram., Glas, Email *10,* 95 [1962].

41) *H. G. Wiedemann* and *D. Nehring,* Z. anorg. allg. Chem. *304,* 137 [1960].

42) *H. G. Wiedemann,* Z. anorg. allg. Chem. *306,* 84 [1960].

43) *C. Duval:* Inorganic Thermogravimetric Analysis, Elsevier Publishing Company, Amsterdam, 1963.

44) *J. F. Cordes* and *S. Schreiner,* Z. anorg. allg. Chem. *299,* 87 [1959].

45) *S. C. Bevan* and *S. Thorbum,* J. Chromatog. Amsterdam *11,* 301 [1963].

46) *F. Paulik, J. Paulik* and *L. Erdey,* Z. analyt. Chem. *160,* 321 [1958].

47) *F. Paulik, L. Erdey* and *S. Gál,* Z. analyt. Chem. *163,* 321 [1958].

36

Copyright © 1971 by the American Chemical Society

Reprinted from *Anal. Chem.*, **43**(2), 223–227 (1971)

Automated Thermoanalytical Techniques:
An Automated Thermobalance[1]

W. S. Bradley and W. W. Wendlandt

Thermochemistry Laboratory, Department of Chemistry, University of Houston, Houston, Texas 77004

An automated thermobalance is described which is capable of recording the TG curves of eight samples in a sequential manner. The instrument consists of a recording top-loading balance, a furnace and temperature programmer, and an automatic sample changer. Each sample in the sample holder disk is positioned into the furnace automatically, heated to a preselected temperature, then removed. After the furnace is cooled back to room temperature, the cycle is repeated with a new sample. Operation of the thermobalance is completely automatic and it requires no operator attention, once the cycle is begun.

HONDA (*1*) WAS PERHAPS the first investigator to use the term "thermobalance" to describe an apparatus which was used to determine the continuous weight-change of a sample as the sample was heated to elevated temperatures in a furnace. Although the instrument was rather crude, it enabled him to obtain weight-change curves of a number of inorganic compounds and also to establish a Japanese school of thermogravimetry, the results of which have been summarized by Saito (*2*). In 1923, a similarly crude thermobalance was described by Guichard (*3*) which was to be the first of a large number of instruments used by French workers in this field.

The historical development of the modern thermobalance has been adequately described by Gordon and Campbell (*4*),

Duval (*5*), Wendlandt (*6*), Keattch (*7*), Saito (*8*), and others (*9, 10*). The modern instruments have been described in well-known textbooks in the field (*5, 6, 9*) and other sources (*7, 11*). By far the most sophisticated multifunction instrument is the Mettler thermobalance, as described by Weidemann (*12*). Besides recording the weight-change curves of a sample at two different sensitivities, it also records the derivative of the weight-change and the DTA curve. Another multifunction instrument, the Derivatograph, has previously been described by Paulik *et al.* (*13*).

The modern thermobalance is an automatic instrument in that the weight-change of a sample can be recorded over a wide temperature range. None of the instruments are capable of introducing a new sample automatically into the furnace chamber or studying multiple samples in a sequential manner. We wish to report here an automated instrument which is

---

[1] See reference (*14*) for Part I.

---

(1) K. Honda, *Sci. Rep. Tohoku Univ.*, **4**, 97 (1915).
(2) H. Saito, "Thermobalance Analysis," Gijitsu Shoin, Tokyo, 1962.
(3) M. Guichard, *Bull. Soc. Chim. Fr.*, **33**, 258 (1923).
(4) S. Gordon and C. Campbell, ANAL. CHEM., **32**, 271R (1960).

(5) C. Duval, "Inorganic Thermogravimetric Analysis," Second Ed., Elsevier, Amsterdam, 1963.
(6) W. W. Wendlandt, "Thermal Methods of Analysis," Interscience, New York, N. Y., 1964.
(7) C. Keattch, "An Introduction to Thermogravimetry," Heyden, London, 1969.
(8) H. Saito, "Thermal Analysis," R. F. Schwenker and P. D. Garn, Ed., Academic Press, New York, N. Y., 1969, pp 11–24.
(9) P. D. Garn, "Thermoanalytical Methods of Investigation," Academic Press, New York, N. Y., 1965, Chap. 10.
(10) H. C. Anderson, "Techniques and Methods of Polymer Evaluation," P. E. Slade and L. T. Jenkins, Ed., Dekker, New York, N. Y., 1966, Chap. 3.
(11) W. W. Wendlandt, *Lab. Management*, October, p 26 (1965).
(12) H. G. Wiedemann, Achema Congress paper, Frankfurt, Germany (June 26, 1964).
(13) F. Paulik, J. Paulik, and L. Erdey, *Z. Anal. Chem.*, **160**, 241 (1958).

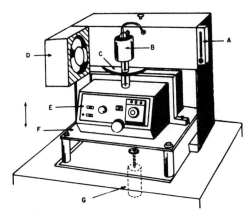

Figure 1. Schematic diagram of balance, furnace, and sample changer mechanism

    A.   Gas flowmeter
    B.   Furnace
    C.   Sample holder disk
    D.   Cooling fan
    E.   Cahn Model RTL recording balance
    F.   Balance platform
    G.   Platform motor

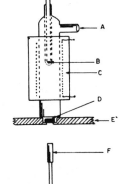

Figure 2. Schematic illustration of furnace and sample holder

    A.   Gas inlet tube
    B.   Thermocouples
    C.   Furnace heater windings and insulation
    D.   Sample container
    E.   Sample holder disk
    F.   Ceramic sample probe

capable of automatic changing of the sample, furnace atmosphere, and automatic temperature programmed heating rates. Eight samples, contained in the rotatable sample holder disk, can be studied in an individual manner.

EXPERIMENTAL

A line drawing of the balance, furnace, and sample changer mechanism is shown in Figure 1, while a detailed diagram of the furnace and sample holder configuration is given in Figure 2.

The thermobalance is conventional in design in that it consists of a top-loading recording balance (Cahn Model RTL balance), a Leeds and Northrup four-channel multipoint potentiometric recorder (0-5 mV full-scale), a small tube furnace, a sample changer mechanism, and an automatic furnace temperature programmer. Most of the thermobalance components were available in the laboratory. Per-

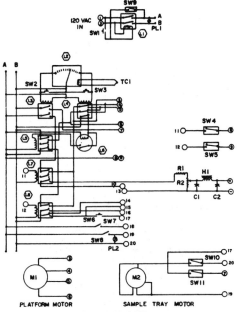

Figure 3. Relay circuits for the thermobalance

  L1.   Latching relay, DPDT, 120VAC coil
  L2.   Meter relay, Simpson Model No. 3324, 0-50 mV
  L3.   Relay, DPDT, 120VAC coil
  L4, L5, L7, L8.   Relay, 3PDT, 120VAC coil
  L6.   Relay, time delay, Amperite, 115 NO5T, 5-sec delay, N.O.
  C1, C2.   Capacitor, 200 mfd, 15 V
  SW1, SW2, SW3, SW6.   Push-button switch
  SW4, SW5, SW9, SW10, SW11.   Microswitch
  SW7, SW8.   Toggle switch, SPST
  PL1, PL2.   Pilot lamp, 120VAC, Neon
  M1.   Motor, Bodine, Type NSY34R, 90RPM
  M2.   Motor, Hurst, Type PCSM, 1/2RPM
  TC1.   Thermocouple, Chromel-Alumel
  R1.   Resistor, 15 KΩ, 1/2 W
  R2.   Potentiometer, 200Ω, 3 W
  L1.   Choke, 10 Henry

haps the most novel feature of the instrument is the automatic sample changer mechanism which operates in the following manner: The samples to be investigated are placed into small cylindrical platinum containers, Figure 2(D), (5.0 mm in diameter by 2.0 mm in height). Eight such containers are placed in the circular indentations cut in the periphery of the 0.25-in. thick by 8.0-in. diameter aluminum sample holder disk, Figure 2(E). The sample containers are positioned directly below the opening of the Vycor tube furnace (5/8-in. i.d. × 4 inches long, 16 ohms of Chromel A wire winding), Figure 2(C), by the rotation of a small electric motor connected to a microswitch which is tripped by an indentation in the circumference of the disk. The positioned sample is picked up by the ceramic sample probe, Figure 2(F), which is attached to the beam of the balance. Movement (approx. 2.5 inches) of the entire balance and balance platform, Figure 1(E and F), is controlled by a motor-driven (G) screw in the base of the platform. The motor is reversible so that the platform can be raised or lowered with limits of movement in both directions controlled by microswitches. After the sample is positioned in the central part of the furnace, the furnace is flooded with nitrogen or some other gas, and the furnace temperature programmer activated. On attaining a

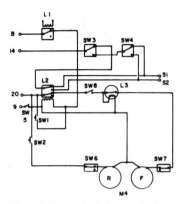

Figure 4. Relay circuit for changing heating rate and furnace atmosphere.

L1. Meter relay (see Figure 3)
L2. Latching relay, DPDT, 120 120VAC coil
L3. Relay, variable time delay, Magnecraft Electric Co., Model No. W211ACPSOX-7, 1.0 sec to 180 sec, 100-130 VAC—60 Hz
SW1, SW2. Push button switch
SW3. Toggle switch, SPDT, center off
SW4, SW5, SW8. Toggle switch, SPDT
SW6, SW7. Microswitch, N.C.
M4. Bristol Dual, Model 420C, Mod. 425C
S1, S2. Solenoid valve, 120VAC coil

preselected furnace maximum temperature limit, the balance is lowered and the sample container retained by the sample holder disk. The disk then rotates to position a new sample at the base of the furnace. A cooling fan, Figure 1(D), is activated which cools the furnace to a preselected lower temperature limit at which point the entire cycle is repeated, using a new sample.

The furnace atmosphere can be automatically changed (e.g., N_2 to O_2) each time the lower temperature limit is reached, and alternation of gases occurs throughout the eight successive samples. Increase of the furnace heating rate in increments of 5 °C/min to 25 °C/min can also be accomplished automatically on every other sample when the lower temperature limit is reached.

Each sample is preweighed into the sample containers using a Mettler semimicro printing balance. The individual sample containers are tared to within ±1 mg (empty weight is about 130 mg); each sample is kept under 10.0 mg so that the recorder pen deflection remains on the recorder scale. The recorder mass range is 0–10 mg at 1.00 mg per in. on a 10-in. wide chart; a chart speed of $^1/_{15}$ or $^1/_6$ inch per min was normally used.

Schematic diagrams of the relay circuits employed and the balance system are shown in Figures 3, 4, and 5, respectively.

In Figure 3, all power to the various circuits is through the latching relay, L1. The microswitch, SW9, is activated by the sample holder disk so that after the eighth sample is run, it will activate the latching relay and turn off all power to the various circuits.

The meter relay, L2, is used to activate relays L3, L4, and L6, which operate the balance platform motor, M1. Push-button switches, SW2 and SW3, permit manual override of the meter relay. Lower and upper temperature limits are set on the meter relay in order to control the furnace temperature programmer, raise and lower the balance platform, start and stop the chart drive and printing mechanism of the multipoint recorder, and to activate the sample holder disk motor. Manual override switches, SW7 and SW8, are used in the fan and sample disk motor circuits, respectively.

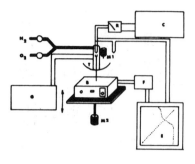

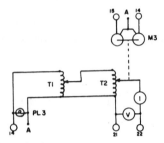

Figure 5. Schematic diagram of thermobalance system

B. Cahn Model RTL balance
M1. Tray motor
T. Sample tray
R. Meter relay
C. Control panel
M2. Platform motor
F. Filter
E. Multipoint recorder
G. Furnace temperature programmer

(Bottom) Furnace temperature programmer circuit

TC1, TC2. Thermocouples, Chromel-Alumel
M3, M4. Motor, dual, Bristol
T1, T2. Transformer, variable voltage, Ohmite, Model No. VT3N
I. Ammeter, 0-5 A (AC)
V. Voltmeter, 0-50 VAC
PL3. Pilot lamp

Input terminal numbers and letters in Figure 4 refer to the corresponding output terminal positions in Figure 3. Power to the latching relay, L2, is applied through the lower set point of the meter relay, L1 (L2 in Figure 3). When L2 steps (SW5, SW8, and L3 normally closed), one gas solenoid (S1 or S2) is opened while the other remains closed, and power is applied to time-delay relay L3. The forward motor (F) of temperature programmer motor M4 (see also Figure 5) advances transformer 1, T1 (Figure 5), until L3 opens the circuit. Since the time delay of L3 is adjustable, numerous incremental heating rate increases are possible. Normally, heating rate was set for 5 °C/min so that eight identical samples could automatically be studied at four different heating rates (e.g., 10, 15, 20, and 25 °C), and the furnace atmosphere changed automatically after every sample. Switches SW3 and SW4 permit manual selection of static or dynamic gas atmosphere while SW1 allows manual advance of heating rate. Microswitches SW6 and SW7 limit travel of M4, and push switch SW2 is used to manually return M4 to its lower set point.

324

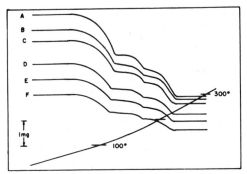

Figure 6. TG curves of Co(py)₂Cl₂ in N₂ atmosphere. Furnace temperature curve illustrated

Curve	Sample weight, mg
A	6.21
B	5.13
C	4.89
D	4.00
E	3.17
F	2.72

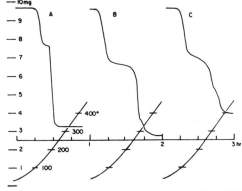

Figure 7. Successive TG curves of some hydrated nickel(II) salts

A. Ni(CHO₂)₂·4H₂O, 9.87 mg
B. Ni(C₂H₃O₂)₂·4H₂O, 9.24 mg
C. NiSO₄·(NH₄)₂SO₄·6H₂O, 9.60 mg
Furnace temperature indicated; nitrogen atmosphere

Connection of the thermocouple, *TC1*, to meter relay, *R*, and control circuit, *C*, is shown in block form in Figure 5. Thermocouple *TC2* is maintained at 0 °C by means of an ice-water bath. Filter, *F*, is used to condition the balance, *B*, output signal. Motor *M1* rotates the sample turntable, *T*, while *M2* raises and lowers the balance and balance platform. Heating of the furnace is provided by the temperature programmer, *G*, shown in greater detail in the lower portion of Figure 5. Details on the furnace temperature programmer have been described previously in Figure 4 and also (*14*). An output voltage of 40 V from *T1* gave a linear furnace heating rate of 10 °C per min. above 200 °C.

Procedure. The eight samples were weighed out into the previously tared platinum containers, using a semimicro balance. Sample weights usually ranged from 5 to 9 mg,

(14) W. W. Wendlandt and W. S. Bradley, *Anal. Chim. Acta*, in press.

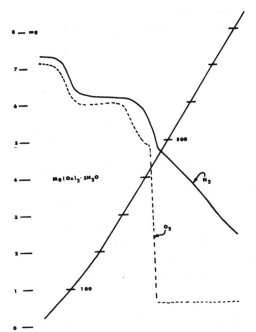

Figure 8. TG curves of Mg(Ox)₂·3H₂O in oxygen and nitrogen atmospheres

Furnace temperature illustrated

although larger samples could be employed if the tare control on the balance was used to keep the recorder pen on scale. The sample containers were then loaded into the sample holder disk, the positioning of the latter being controlled by the manual override switch, *SW8* (Figure 3). The number 1 sample was then positioned under the furnace and the lower-limit set point of the meter-relay activated. This caused the balance platform to rise, activated the recorder chartdrive and recording mechanism, and opened one of the gas solenoids to purge the furnace. Elevation of the balance platform was limited by a microswitch which also deactivated the platform motor circuits, and turned on the furnace temperature programmer. After the furnace reached a preselected upper furnace temperature limit, as set by the meter-relay, the balance platform was automatically lowered, the gas flow stopped, and the temperature programmer reset. The spent sample and container were retained by the sample holder disk, and the cooling fan was activated. A new sample was then positioned beneath the furnace. This entire cycle was then repeated for all of the samples retained in the sample holder disk. Operation of the instrument was completely automatic and it required no operator attention after it was started. All samples were studied under a dynamic nitrogen or oxygen atmosphere whose gas flow rate was 50 ml per min.

The precision and accuracy of the weight recording were estimated at about ±2%, while the temperature recording was estimated at about ±5%. Full scale recorder deflection for the temperature measurement was about 1000 °C, which was roughly the upper temperature limit of the Vycor tube furnace employed.

RESULTS AND DISCUSSION

The TG curves of Co(py)₂Cl₂ are illustrated in Figure 6. The curve for this compound has previously been described

Table I. Weight-Loss Calculations for the Reaction:
$$Co(py)_2Cl_2(s) \rightarrow CoCl_2(s) + 2py(g)$$

Sample No.	Sample wt, mg	Total % wt loss	Difference from theoretical, 54.92%, in %
1	7.26	55.0	0.1
2	6.21	54.6	−0.3
3	5.13	56.5	1.6
4	4.89	54.6	−0.3
5	3.17	55.2	0.3
6	2.72	55.1	0.2

by Allan *et al.* (*15*) and Ocone *et al.* (*16*). The curves obtained here agree well with the thermal dissociation reactions, as given by the equations:

$$Co(py)_2Cl_2(s) \rightarrow Co(py)Cl_2(s) + py(g) \qquad (1)$$

$$Co(py)Cl_2(s) \rightarrow Co(py)_{1/2}Cl_2(s) + \tfrac{1}{2}py(g) \qquad (2)$$

$$Co(py)_{1/2}Cl_2(s) \rightarrow CoCl_2(s) + \tfrac{1}{2}py(g) \qquad (3)$$

Reproducibility of curve shape is demonstrated by the six samples, differing only in sample weight. Calculation of experimental reproducibility of loss of the two molecules of pyridine per mole of complex gives a maximum relative difference from theoretical of 1.6%. Experimental per cent weight losses of the six samples are presented in Table I.

Three successive TG curves of various nickel(II) salt hydrates are illustrated in Figure 7. In the case of $Ni(CHO_2)_2 \cdot 4H_2O$ and $Ni(C_2H_3O_2)_2 \cdot 4H_2O$, the first weight-losses in each curve were due to the hydrate water evolution. The resulting

anhydrous nickel(II) salts then dissociated to yield a residue of NiO. In the case of $NiSO_4 \cdot (NH_4)_2SO_4 \cdot 6H_2O$, the first weight-loss was due to the evolution of water followed closely by the sublimation of $(NH_4)_2SO_4$. The residue obtained was $NiSO_4$.

Sequential TG curves of $Mg(Ox)_2 \cdot 3H_2O$ were obtained to illustrate the utility of automatic atmosphere change and heating rate increase. In Figure 8, the first weight-loss in both oxygen and nitrogen was due to the hydrate water evolution. The anhydrous $Mg(Ox)_2$ then slowly dissociated in nitrogen but rapidly went to the oxide in the oxygen atmosphere.

The obvious advantage of the automated thermobalance system over existing instruments is the ability to determine the weight-loss curves of eight successive samples. Operation of the instrument is completely automatic and once the cycle is begun, the instrument does not require the attention of the operator until the eighth sample curve is completed. The instrument should find use for the routine TG examination of a large number of samples, each to be studied under identical thermal conditions. Since the instrument is completely automated, data reduction or control by a small digital computer could easily be accomplished.

ACKNOWLEDGMENT

Many of the mechanical innovations were designed and constructed by Mr. Ralph Martin. Dr. Colin Williams, of the Cahn Instrument Corp., is acknowledged for the loan of the recording balance.

RECEIVED for review June 1, 1970. Accepted November 13, 1970. The financial assistance of the U. S. Air Force, Air Force Office of Scientific Research, through Grant No. 69-1620, is gratefully acknowledged. The Sun Oil Co. is acknowledged by W.S.B. for a scholarship.

(15) J. R. Allan, D. H. Brown, R. H. Nuttall, and D. W. A. Sharp, *J. Inorg. Nucl. Chem.*, **26**, 1895 (1964).
(16) L. R. Ocone, J. R. Soulen, and B. P. Block, *ibid.*, **15**, 76 (1960).

AUTHOR CITATION INDEX

SUBJECT INDEX

Allophane, heating curve of, 15
Alumina precipitates, ignition tempera-
    ture of, 225
Aluminum silicates, hydrated, 13
Analytical chemistry
   DTA application to, 11
   uses of TG in, 228
Analytical precipitates, effect of heat on,
    197, 199, 218
d'Arsonval galvanometer, 43
Assay methods, 1

$BaBr_2 \cdot 2H_2O$, DTA of, 121
$BaCl_2 \cdot 2H_2O$, DTA of, 121
Balance
   automatic, 248
   Campbell and Gordon, 252
   commercially available recording, 259
   electromagnetic, 255
   Eyraud, 253
   recording, 248
   strain-gage, 251
$BaTiO_3$, DTA of, 180
Benzenediazonium chloride, kinetics of,
    127
Benzoic acid, DTA of, 90, 164
Blowpipe, 1
Boersma cup-type apparatus, 10
Butane, DTA of, 146

$CaCO_3$
   dissociation of, 99
   TG of, 207
$CaCO_3 \cdot MgCO_3$, DTA of, 135

$CaC_2O_4 \cdot H_2O$
   DTA of, 153, 316
   kinetics of, 237
   TG of, 267, 268, 287
Cahn electrobalance, 199, 271
Calcium silicate, DTA of, 181
Ca-montmorillonite, DTA of, 136
Carburized iron, 20
$CaSO_4 \cdot 2H_2O$, TG of, 206
Change of state, determination of by DTA,
    139
Clay minerals
   mixtures, determination of, 82
   quantitative determination of, 81
Clays
   action of heat on, 13
   quantitative analysis of, 84
Clays and minerals
   DTA applied to, 9
   identification of, 76
Coal, pyrolysis of, 9
$CoCl_2 \cdot 6H_2O$, DTA of, 121
Coffee, DTA of, 134
Cooling curves, 25, 26
Cotton fiber, DTA of, 137
CrO_3, TG of, 207
Cupellation, 197
$CuSO_4 \cdot 5H_2O$
   DTA of, 187
   triple point of, 120

Dacron, DTA of, 131
Derivatograph, 200
Dextrose, DTA of, 141

About the Editors

WESLEY W. WENDLANDT is Professor of Chemistry at the University of Houston, where he teaches courses in thermal analysis and first-year chemistry for nonscience major students. He received his B.S. degree from Wisconsin State University in 1950 and his Ph.D. degree from the University of Iowa in 1954. Professor Wendlandt is Editor-in-Chief of *Thermochimica Acta* and received the Mettler Award for his work in thermal analysis in 1970. He has authored or co-authored 20 books and 265 technical papers.

L. WAYNE COLLINS is Senior Research Chemist for Monsanto Research Corporation's Mound Laboratory in Miamisburg, Ohio. He received his B.S. and M.S. degrees from Stephen F. Austin State University in 1968 and 1970, respectively, and his Ph.D. degree from the University of Houston in 1973. Dr. Collins served as a National Academy of Sciences–National Research Council Postdoctoral Associate at the Johnson Space Center from 1973 to 1975. He has authored or co-authored over 20 technical papers.